ENCYCLOPAEDIA OF
HELMINTHES

ENCYCLOPAEDIA
OF
HELMINTHES

Vol. 3

Physiology of Helminthes

By

Arvind N. Shukla

Rajiv Tyagi

ANMOL PUBLICATIONS PVT. LTD.

NEW DELHI - 110 002 (INDIA)

ANMOL PUBLICATIONS PVT. LTD.
4374/4B, Ansari Road, Daryaganj
New Delhi - 110 002
www.anmolbooks.com

Encyclopaedia of Helminthes

First Edition, 2002

ISBN 81-261-1043-0 (Set)

PRINTED IN INDIA

Published by J.L. Kumar for Anmol Publications Pvt. Ltd., New Delhi - 110 002 and Printed at Mehra Offset Press, Delhi.

Contents

Physiology of Helminthes

Preface

Helminthes constitute a large assemblage of *worms* of comparatively simple and varied organization. Helminthes is derived from the Greek word *helmins* means *worms*. The term worms is not correct because it is applied to elongated invertebrates without appendages and with bilateral symmetary. In 18th Century all the worm-like animals were included in an old phylum- *Vermes* (*Linnaneus*, 1735). But this was a very heterogenous group which included a highly diverse assemblage of forms. In the modern classification this group is divided into several independent phyla such as Platyhelminthes, Aschelminthes, Annelida etc.

The term helminthes is restricted to a few phyla of invertebrate animals, all of which are superficially worm like but they differ markedly in their morphology, life-history and bionomics. The helminthes are restricted to two phyla of Animal Kingdom. These are:

1. Platyhelminthes or flat-worms.
2. Aschelminthes or round-worms.

The present work is designed to present all vital information on Helminthes. The entire information is gathered in three volumes namely:

- Platy Helminthes
- Aschelminthes
- Physiology of Helminthes

Profusely illustrated in the light of latest researches, this work will prove a dependable reference work for students, teachers and researchers in the discipline.

—**Authors**

Physiology of Helminthes

Chapter—1

Physiology of Nematodes

Much of the work on nematode physiology has centred around those species readily available or maintained in laboratory animals such as *Ascaris lumbricoides, Nippostrongylus brasiliensis, Trichinella spiralis, Litomosoides carinii, Heterakis gallinae* and *Trichuris vulpis*. The literature in this field is so extensive that it is not possible to give more than a general account here.

Various aspects of nematode physiology have been reviewed by Lee (1965), Crofton (1966), Saz & Bueding (1966), von Brand (1966), Ginger (1969), Fairbairn (1966), Rogers (1969), Rogers & Sommerville (1969), Rothstein & Nicholas (1969), Saz (1969, 1971), Bird & Wallace (1969) and Croll (1970).

Chemical composition

The chemical composition of representative species of nematodes is given in Table 1.1; the protein content in particular has been little studied.

Haemolymph. The body cavity of nematodes is filled with a fluid termed *haemolymph* which is under pressure due to the tonicity of the body wall. Its composition and properties are known largely from studies on *Ascaris* and *Parascaris*. In *Ascaris*, the haemolymph has a pH of 6.2, 6.4 and osmotic pressure of 0.192 M NaCl and contains—in addition to proteins, carbohydrates and fats—small quantities of haemoglobin, various enzymes and several organic acids. Of the carbohydrates present, rather surprisingly, glucose is present only in trace quantities but a disaccharide, trehalose is present (0.77 per cent) as is glycogen (0.4 per cent). In *Ascaris,* the organic acids consist chiefly of 2-methylvaleric, 2-methylbutyric, acetic, succinic with smaller quantities of n-Valeric and propionic.

Carbohydrates. As in trematodes and cestodes, the main polysaccharide in nematodes is a glycogen closely resembling mammalian glycogen. In adults, the greatest part of the stored glycogen occurs in the body-wall tissues especially the muscles. As indicated above, the disaccharide, trehalose, in addition to glucose, also occurs in nematodes, representing 4.0 per cent of the solids in *A. lumbricoides* haemolymph. The eggs frequently contain relatively large polysaccharide stores (7.8 per cent of the dry weight in *A. lumbricoides)* as do the larvae of most species. In *Nippostrongylus brasiliensis,* alkali-stable carbohydrate (glycogen and trehalose) forms 3.4 per cent of the dry weight and acts as reserves which are utilized under starvation conditions. In the larvae of *Nippostrongylus* and the microfilariae of *Acanthocheilonema,* however, glycogen is virtually absent.

Proteins. The proteins of nematodes are of exceptional interest but have been studied by modern analytical methods only in a few instances. The total protein in *Ascaris* makes up eight to nine per cent of the wet weight and the amino-acid composition generally resembles that of its host. The haemolymph of *Ascaris* contains 4.9 per cent proteins of which 2.8 per cent are albumins and 2.1 per cent globulins. The fibrous proteins of the cuticle have been most thoroughly studied but complete agreement has not been reached on their composition. X-ray and chemical analysis of the cuticle have indicated the presence of keratin, collagen and tanned protein although there are differences between species.

Haemoglobin is widespread amongst parasitic nematodes, only three species having been reported to lack haemoglobin; it is localized differently in different species. There is evidence that it plays a respiratory role in some, but not all species. The haemoglobin in the body wall may also serve to transfer oxygen from the cuticle to the body fluid.

Specialized protein granules are also found in the oocytes of nematodes and these play some part in the formation of the egg membranes although their precise function is not known. A further specialized protein, termed *ascaridine,* is synthesized in the testes and present in sperm; it appears to make an important contribution to the synthesis of ribonucleic acid in the egg of *Parascaris.*

Lipids. The detailed composition of the lipids has been reviewed by Fairbairn (1969). The lipid content of some common nematode species is given in Table 1.1. In *Ascaris* the highest concentrations occur in the reproductive system with relatively small quantities in the cuticle, muscle and haemolymph. A notable feature is the high levels of unsaponifiable lipids which consist largely of sterols and *ascarosides,* the latter being a mixture of closely related glycosides. Volatile and non-volatile acids occur as triglycerides whilst the phospholipid fraction consists of lecithins, cephalins and sphingomyelins with the lecithin predominant. Beames (1964) has separately studied the phospholipids of the muscles, cuticle and reproductive system of *A. lumbricoides.*

Nutrition

Adults

The food of nematodes has already been discussed in previous Chapter. Very little detailed information, however, is available on the precise nature of the food requirements in chemical terms. The nature of such demands shows considerable variation during the life cycle. For example, an adult worm producing enormous quantities of eggs in the elementary canal of a homoiothermic host will have greater nutritional demands than its free living larva merely undergoing gradual growth and differentiation at air temperatures.

Table 1.1 Chemical composition of nematodes.

	Glycogen		Lipid		Protein	
	% *f.w.*	% *d.w.*	% *f.w.*	% *d.w.*	% *f.w.*	% *d.w.*
Ancyclostoma caninum	1.6	—	—	—	—	—
Ascaridia galli♂	4.2*	17.9*	1.2	—	—	—
Ascaridia galli♀	4.8*	19.8	1.3	—	—	—
Ascaris lumbricoides♂	3.3-8.7	14-24	1.3	—	—	48-57
Ascaris lumbricoides♀	—	—	1.6	8.9	—	—
Dipetalonema gracilis	0.2	—	—	—	—	—
Dirofilaria immitis	1.9	10	—	—	—	—
Dirofilaria uniformis	1.7	—	—	—	—	—
Eustrongylides ignotus (larvae)	6.9	28	1.1	4.4	—	—
Heterakis gallinae	2.7	11.8	—	—	—	—
Litomosoides carinii	0.8	5	—	—	—	—
Parascaris equorum	2.1-3.8	10-23	—	—	—	—
Porrocaecum decipiens (larvae)	—	55	—	3.5	—	—
Strongylus vulgaris	3.5	—	—	—	—	—
Trichuris vulpis	2.3	—	—	—	—	—
Nippostrongylus brasiliensis larva	—	5-6†	—	15-20†	—	73-76†
Nippostrongylus brasiliensis adult	—	—	—	11.9	—	—
Trichinella spiralis	—	—	—	5.5	—	—

Carbohydrates and proteins form the main nutritional requirements of adult worms, with fats playing a relatively unimportant role. A high carbohydrate diet is thus beneficial to the growth of most nematodes. Hormones, iron and certain unidentified growth factors may also be essential in most species. As is to be expected, a deficiency in the host diet may be reflected in the inhibition of the growth and reproduction of the worm. The extent and speed of this inhibition will depend, almost entirely, on the food reserves of the worm. Host starvation, for example, is reflected within 24 hours by a drop in the glycogen reserves of *Ascaridia galli,* followed by a fall in egg production. Continued host starvation may result in the passing of the worm in two to four days. With the depletion of its main energy source, carbohydrate, a worm is no longer able to energize its muscles and so, maintain its position in the intestine.

Deficiency in one or several nutritional components may provide unexpected results. Thus a deficiency of vitamin A in the host diet leads to decreased resistance to infection of rats of *Nippostrongylus brasiliensis* and dogs to *Ancylostoma caninum* and *higher* laboratory infections. Considerations such as these mean that it is seldom possible to state with certainty whether or not an observed effect is *directly* due to the nutritional deficiency under consideration. Such conclusions can only be reached on the results of carefully controlled experiments *in vitro,* where host reactions do not operate.

Unfortunately, media for successful *in vitro* culture contain so many components that this approach has, to date, provided little data on the precise nutritional requirements.

With a few exceptions, the food material of adult nematodes appears to be solid or semi-solid, and since the majority are intestinal parasites, this consists of semi-digested food and debris.

A number of digestive enzymes or enzymes related to feeding processes, have been detected in nematodes. These include amylase, invertase, pectinase, glycosidases, chitinase, cellulase, proteases, esterases and lipases. Extra-corporal digestion occurs as a preliminary to ingestion in many species. Detailed information on the secretion or activity of the digestive enzymes is scarce. In *A. lumbricoides,* holocrine secretion apparently occurs in the anterior intestine; in this process, part of the cell detach from the gut and disintegrate in the lumen. Merocrine secretion also occurs in this species. Adult nematodes thus appear to be well equipped to deal with ingested food materials and contain batteries of enzymes.

In the strongylids, blood and tissue form the main diet and many species, particularly hookworms, draw a plug of mucosa into their buccal cavity and suck blood and tissues. In some forms, blood is almost continuously being drawn into the intestine and passed out through the anus. Both the plasma and corpuscles undergo at least partial digestion. The heartworm, *Dirofilaria immitis,* likewise feeds exclusively on blood. Lungworms feed on the inflammatory exudate evoked by their own presence, as do some tissue forms.

In species with a capillary pharynx such as *Trichuris,* the lumen is not of sufficient diameter to permit the ingestion of solid food material and this nematode presumably takes in more or less liquid food.

Larvae

The nematode egg contains sufficient endogenous reserves to enable a fully-formed active larva to develop without the absorption of further nutriment. Thereafter, food is required to supply energy and materials for synthesis of new tissue. Nematodes are peculiar in one respect, since, with a few exceptions, there is *no* somatic cell multiplication after hatching in spite of very considerable increase in size. This phenomenon of cell constancy is known as *eutely*. The major growth requirements of nematodes prior to maturation—are thus for cytoplasmic rather than nuclear synthesis, and nucleic acids occur in nematodes in small quantities relative to the total tissue weight. In ovaries, the RNA/DNA ratio is about 4:1.

Free-living stages of nematodes (e.g. hookworms) must feed largely on bacteria and such stages are readily cultured in faecal-charcoal cultures in which abundant bacteria are normally available.

Larvae which penetrate skin or tissue have been shown, in some cases, to produce a 'spreading factor' with an activity similar to that of hyaluronidase, in addition to lipases and esterases, and a collagenase-like enzyme.

An unusual nutritional adaptation occurs in the so-called 'entomophilic' nematodes - a group which parasitizes insects. In some species, e.g. *Bradynema* sp. in the fly *Megaselia halterata* the worm develops a mouthless, gutless body, the cuticle is absent and the body wall possesses microvilli through which, presumably, nutrient is directly absorbed. The outside covering thus more nearly approaches that of a cestode than a nematode.

Respiration

General account

As the life cycles of the majority of nematodes involve complex environmental changes, it is to be expected that marked differences between the respiratory metabolism of the different stages will exist. A nematode egg, for example, may embryonate and hatch in an aerobic environment, pass through its later larval stages similarly in an aerobic environment (tissues of lungs) and finally mature and produce eggs in the anaerobic environment presented by the gut lumen. However, adult nematodes may be divided into *(a)* those which are obligate aerobes and *(b)* those which lack an oxygen requirement *for energy metabolism* in at least one stage. Most members of the latter group have a functional requirement for carbon dioxide and accumulate succinate (or related compounds) as end-products of carbohydrate metabolism. It should be stressed, however, that the fact that a particular species lives in an aerobic environment or utilizes oxygen *in vitro* does not necessarily mean that it utilizes oxygen *in vivo*. It has already been pointed out that the blood fluke *Schistosoma mansoni*, although living in an essentially aerobic environment, has a predominantly anaerobic type of metabolism.

Nematode species as obligate aerobes

Included in this group are *Litomosoides carnii*, *Nippostrongylus brasiliensis*, and *Trichinella spiralis*. *L. carnii* loses its mobility in nitrogen within 60-90 minutes and clearly appears to have an oxygen requirement. Nevertheless, like many other 'aerobic' parasites, this species does not completely oxidize all its carbohydrate and a Krebs cycle does not appear to operate, at least not completely. However, the respiration is inhibited by cyanine, suggesting an electron transport system involving a heavy metal. Respiration is also inhibited by cyanine dyes which are known to inhibit enzyme systems concerned with the oxidative mechanism. In contrast, the human filarial worm *Wuchereria bancrofti* is unáffected by cyanine dyes suggesting that this species is not dependent on an oxidative system.

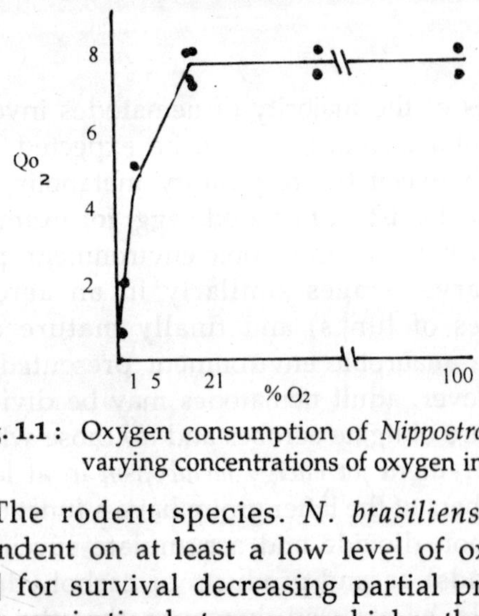

Fig. 1.1 : Oxygen consumption of *Nippostrongylus brasiliensis* at varying concentrations of oxygen in the gas phase.

The rodent species. *N. brasiliensis* appears to be dependent on at least a low level of oxygen [pO_2>8 mm (1%)] for survival decreasing partial pressure resulted in lower respiration but pressures higher than air had no effect. A curious features of this organism is the fact that oxygen is of little importance in the carbohydrate metabolism which is almost entirely fermentative. The oxygen thus appears to be required for other metabolic activities (e.g. biosynthesis).

Nematode species lacking an oxygen requirement at some stage.

Ascaris lumbricoides : adult

Although *Ascaris* takes up oxygen if offered to it, it survives equally well under aerobic or anaerobic conditions, indicating that energy is obtained via anaerobic pathways. Cytochrome oxidase is lacking but a flavin system of terminal oxidation appears to operate. The complex pathways probably involved in this are discussed further below.

Ascaris lumbricoides : egg and larval stages

Ascaris eggs and larval stages appear to be obligate aerobes, at least below 32°C. Early eggs lack cytochrome oxidase but probably possess a flavin-mediated terminal oxidase; cytochrome oxidase appear after the 8-cell stage of development. Only one malate dehydrogenase is found in a 1-day egg but infective 21-day eggs have three dehydrogenases. Lipids comprise 50 per cent of the protoplasmic solids of *Ascaris* eggs and the energy for embryonic development appears to be obtained largely from oxidation of fat.

Trichuris vulpis

Although this species—the adult whipworm of dogs—lives in close proximity of the mucosa, it derives its energy entirely from anaerobic reactions (Saz, 1969). Survival is longer in nitrogen/carbon dioxide than in either air or oxygen/nitrogen/carbon dioxide. Carbon dioxide is apparently fixed and is necessary for optimal survival. A cytochrome system which is functional under aerobic conditions has been detected. There is no evidence, however, that this system is operational in the (anaerobic?) environment of the adult worm.

Carbohydrate metabolism

General account

The carbohydrate metabolism of nematodes has been extensively studied and only a general review will be given here. Since nematodes have a well-developed gut complex, carbohydrate materials can be used as food and the end-products of digestion taken up. Thus, they are not dependent on the provision of a carbohydrate in the form of monosaccharides as in cestodes. In fact, few studies on the uptake of monosaccharides have been made.

All nematodes studied appear to be able to take up glucose, although the subsequent metabolic pathways differ between species. Thus the aerobic species *Litomosoides carinii* can utilize glucose, fructose or mannose. Under aerobic conditions, a net polysaccharide synthesis takes place. In contrasts, under anaerobic conditions, there is a net loss of polysaccharide; i.e. there is a pronounced Pasteur Effect. In the absence of exogenous carbohydrate, polysaccharide reserves (glycogen) are utilized. In the aerobic *N. brasiliensis*, utilization of endogenous carbohydrate is twice as rapid anaerobically as aerobically. In contrast, larval *Trichinella spiralis* consumes the same amount of glycogen under both aerobic and anaerobic conditions. The fowl nematode *Heterakis gallinae* behaves similarly. Beames (1969) has shown that uptake of the sugar 3-0-methyl glucose across the isolated gut of *Ascaris* is dependent on an energy source and the presence of carbon dioxide; uptake was inhibited when a gas phase other than 95 per cent nitrogen/ 5 per cent carbon dioxide was employed.

Glycolysis in nematodes

Parasitic nematodes obtain part of their energy from a process of glycolysis which in general follows the Embden-Meyerhof pathway of vertebrate tissues. However, the pathway normally terminates in pyruvate, lactic acid rarely being produced. Exceptions to this, however, are *Litomosoides, Dracunculus, Dirofilaria, Ascaridia, Trichinella* and *Trichuris* in which some lactic acid is produced. The presence of only low levels of pyruvate kinase and lactic dehydrogenase appear to be responsible for the lack of lactate in other species. In some species, e.g. larval *Haemonchus contortus*, pyruvate kinase is entirely lacking. Hence the system, operating in 'normal' glycolytic systems (i.e. those terminating in lactic acid) whereby the reduced NAD (NADH) produced in glycolysis is reoxidize, is prevented from operating.

The Krebs or citric acid cycle in nematodes

Some nematodes have been shown to contain all the enzymes of the Krebs cycle, others very few and some (e.g. *Ascaris*) virtually none at all. The occurrence of enzymes, as emphasized elsewhere does not necessarily indicate a functional Krebs cycle unless it can be demonstrated that they operate in the right sequence. The high carbon dioxide production from some species, suggests, however, that a high level of aerobic respiration is operating via this cycle. According to Ward & Schofield (1967*b*), however, in *Haemonchus contortus,* in which a full Krebs cycle appears to operate, some of the mitochondrial enzymes are not found in the mitochondria but in the cell 'sap'.

Since nematodes, with some exceptions, do not produce lactate, NADH cannot be oxidized by lactic dehydrogenase and must therefore be achieved by other means. Nematodes, like some cestodes utilized the enzyme succinic dehydrogenase as a hydrogen acceptor. Under these (anaerobic) conditions, this enzyme thus acts as a 'fumarate reductase' transferring 2H from fumarate to succinate which is then excreted. The fumarate required for this system is probably provided by CO_2-fixation involving carboxylation of phosphoenolpyruvate (PEP) or pyruvate leading, by a process demonstrated in several species (e.g. *Haemonchus contortus,* to the formation of malate or oxaloacetate. As in cestodes there is thus a reversal of the Krebs cycle, and this system is believed to operate in *Ascaris*. The enzyme phosphoenolpyruvate carboxykinase appears to be the most active CO_2-fixing enzyme.

End-products of carbohydrate metabolism

As already emphasized, lactate is not generally the main excretory product of nematodes and a wise range of acids is produced by some species. The conditions under which the material is collected and maintained, however, may have an important bearing on the nature of the excreted material and there seems no doubt that many results using *Ascaris* under mere 'survival' conditions are open to question. Thus Harpur & Leigh-Browne (1971) showed that tiglic acid is not present in the haemolymph of *Ascaris* freshly removed (i.e. within half an hour) from the pig host whereas if removed four hours after autopsy, this acid is present. Seven hours after autopsy, small quantities of n-caproic and two unidentified acids have been detected. The main acid in *Ascaris* haemolymph is 2-methylvaleric but 2-methylbutyric, succinic, acetic and n-valeric are also produced in quantity.

According to Saz (1971, 1972) succinate and also acetate and other acids are formed, as indicated above by the reversal of reactions which occur in aerobic tissues. In addition to *Ascaris,* succinate is also produced by *Heterakis gallinae, Trichuris vulpis,* and larval *Trichinella spiralis.*

Since pyruvate kinase is apparently absent in this species, it can be asked 'How then is acetate formed by this species' Saz (1972) has speculated that a possible mechanism is as follows: *Ascaris* fixes carbon dioxide into PEP to form oxaloacetate, the reaction catalyzed by the enzyme PEP carboxykinase. This is rapidly reduced, by glycolytically formed NADH, to malate by what appears to be the most active of all *Ascaris* enzymes, malate dehydrogenase. This reaction (like the lactate dehydrogenase reaction in mammalian tissues) serves to regenerate NAD so that glycolysis can continue. The malate which is thus formed in the cytoplasm now crosses over into the mitochondrion where it becomes a mitochondrial substrate. Within the mitochondrion, the oxidative decarboxylation of malate to pyruvate and carbon dioxide occurs, thus giving rise to pyruvate in the absence of pyruvate kinase. Pyruvate may then serve as a precursor of acetate within the mitochondrion. The site of fatty acid synthesis appears to be unknown.

Table 1.2: Organic acids present in haemolymph* of *Ascaris lumbricoides* up to 8 hours after the death of the pig host.

	Time after death of host (hr) *[and in vitro time (hr)]*			
Acid radicals	*0.5[0.0]*	*4.0[7.5]*	*8.0[7.5]*	*P†*
Acetic	13.6	17.0	8.4	0.05
Propionic	2.4	0.1	1.6	<0.05
2-Methylbutyric	28.2	32.0	36.5	0.05
n-Valeric	8.8	8.9	3.9	<0.05
2-Methylvaleric	36.5	26.9	20.6	0.05
Tiglic	ND‡	3.0	1.8	0.05
n-Caproic (?)	ND	ND	0.2	
Unidentified I	ND	ND	+§	
Unidentified II	ND	ND	+	
Succinic	13.7	8.9	5.4	<0.01
				0.05
Total (mEquiv)	116.9	105.7	83.8	0.05

- • Each figure is the mean concentration (mM) of acids in samples from 10 worms.
- † At the given probability a mean with an unbroken underline is not considered different from a mean similarly marked by the same line.
- ‡ ND = Not detected.
- § + = Trace detected.

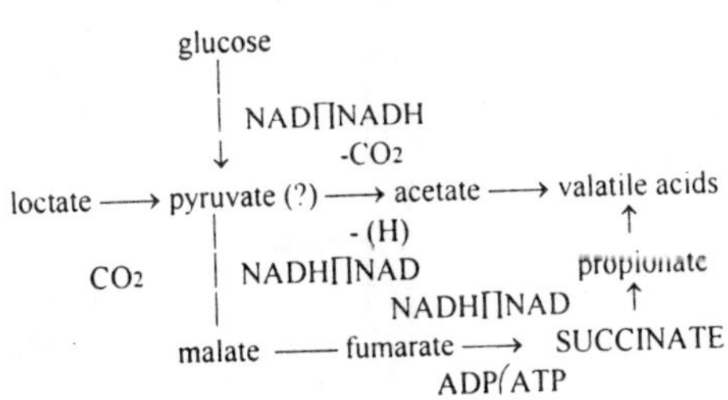

Fig. 1.2 : Possible pathway for the formation of succinate and volatile acids in muscle of *Ascaris lumbricoides*.

Electron transport system

Much of the work on electron transport in nematodes has centered on *Ascaris*, presumably because this species is readily available in quantity. As stressed earlier the way in which the material is stored before use may have an influence on the results of experiments, so that the data from different workers are often not in agreement. Thus cytochrome oxidase is generally said to be lacking in *Ascaris* and yet according to Smith (1969*a*) failure to detect cytochrome oxidase in significant amounts in essentially a failure in technique; in particular, processes such as deep freezing or sonification may greatly affect oxidase activity. The methods of attempting to prepare pure preparations of mitochondria also differ and may produce different results after different protocols.

As already pointed out, the reoxidation of the NADH produced during glycolysis occurs by a reversed Krebs cycle reaction involving the reduction of fumarate to succinate and succinic dehydrogenase acting as a fumarate reductase, as discussed above. This reaction thus operates as an electron transport system under anaerobic conditions and closely follows the pattern believed to occur in some cestodes.

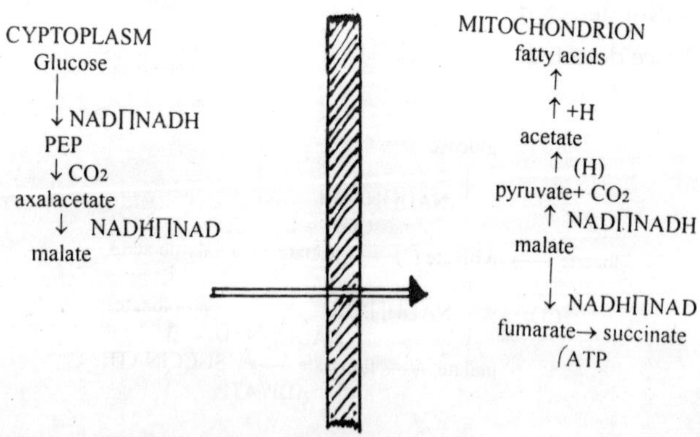

Fig. 1.3 :. Possible pathway of carbohydrate dissimilation in muscle of *Ascaris lumbricoides*.

Under aerobic conditions the situation is less clear, although in some species the levels of cytochrome a_3 (cytochrome oxidase) and *c* components may be so low as to be almost undetectable in *Ascaris*. Functional *cytochromes* a_3 *o*, *c* and b_{556} have been detected by Cheah (1972) and this worker concludes that *Ascaris* is aerobic as evidenced by the existence of a_3.

There is also thought to be a second branching pathway in *Ascaris* and other species in which a cytochrome *b* component reacts with oxygen directly to give hydrogen peroxide. Such an alternative system has also been demonstrated in cestodes.

In addition to *Ascaris*, some data is available for other species e.g. *Ancyclostoma caninum, Necator americanus, Metastrongylus apri, Haemonchus contortus, Nippostrongylus brasiliensis.*

It is likely that different systems operate in different species since various components of these alternative transport systems have been identified. Thus, it appears that, depending on the availability of oxygen, alternative electron transport systems can be followed. The basic problem is that the precise pO_2 in the natural site is poorly known and it may be that as local fluctuations occur a switch from one pathway to another takes place.

It is important to note that the metabolic pathways may change substantially during the life cycle, especially where major environmental changes occur. Thus, in the eggs and larvae there is evidence of a Krebs cycle and a cytochrome - linked terminal pathway which would be in keeping with the well-known fact that oxygen is required for embryonation and early larval development in *Ascaris;* after the third-stage larva, however, a switch from aerobic to anaerobic metabolism apparently occurs.

Table 1.3 Comparison of the concentrations* of respiratory pigments in *Moniezia, Ascaris* and ox-neck muscle mitochondria.

	Concentration (nmol/mg protein)		
Respiratory Components	*Moniezia*	*Ascaris*	*Ox-neck*
Cyt b_{556} (-196°C.)		0.073	
Cyt b	0.120	-	0.247
Cyt c_1		-	
Cyt c		0.089	
Cyt cc_1	0.139		0.582
Cyt aa_3	0.064	0.003	0.478
Cyt o	0.331	0.012	—

* The concentration of respiratory pigments was estimated from difference spectra at 20°C except those of *Ascaris*-muscle mitochondria. The cytochromes in *Ascaris* where calculated from difference spectra recorded at - 196°C with α-glycerophosphate as substrate.

Although the 'free' amino acids in a number of species have been examined, little information on the 'dietary' and 'essential' amino acids is available, nor has the amino acid metabolism been studied in any detail. Transaminase activities have, however, been identified in *A. lumbricoides*. The microfilariae of *Dirofilaria immitis* are able to incorporate exogenous amino acids into their protein; the rate of incorporation of dicarboxylic acids (e.g. glutamic acid) being lower than that of neutral amino acids such as leucine or phenylalanine.

The lipid metabolism has been relatively little studied. Fairbairn (1966) has received available data. The neutral lipids have been analysed for a few worms e.g. *Ascaris*. In larvae leading an aerobic life, such as those of *Nippostrongylus, Haemonchus, Ancylostoma, Necator,* stored lipid is used as an energy reserve and frequently glycogen is absent Moreover, larvae which use up their lipid reserves are apparently not as infective as those containing substantial amounts of lipid. No evidence appears to be available regarding the utilization of tissue lipids in nematodes.

Table 1.4 Comparative rates of uptake and incorporation into protein in various ^{14}C-labelled amino acids by *Dirofilaria immitis in vitro.*

Uniformly ^{14}C-labelled amino acid	*cpm recovered per mg dry tissue per hr*		*Total incorp. into protein (μmol/mg dry tissue/hr)*
	TCA-soluble	*Protein*	
Glutamic acid (a)	460	50	$1.4X10^{-7}$
Glycine (b)	3,020	320	$1.5X10^{-6}$
Leucine (e)	12,360	3500	$8.0X10^{-6}$
Lysine (c)	11,620	3470	7.5X10-6
Phenylalanine (d)	24,400	7240	$1.0X10^{-5}$

(a) 1.0μCi per $5.0X10^{-3}$μmol; (b) 1.0μCi per $8.6X10^{-3}$μmol;
(c) 1.0μCi per $4.0X10^{-3}$μmol; (d) 1.0μCi per $2.6X10^{-3}$μmol.

In *Ascaris,* the lipid catabolism is limited, as evidenced by the fact that five days' starvation produces no appreciable fall. Nearly two-thirds of the total lipids occur in the female reproductive system and the uterus contains some 35 per cent lipids. Barrett (1968) examined the lipids of larval and adult *Nippostrongylus* and differences only amounted to higher levels of triglycerides in the adult than in the larvae, and higher levels of free fatty acids in the larvae. Wilson (1965) reported that the lipid in the infective larvae disappeared at the rate of 0.9 per cent of their total dry weight per day.

During the embryonation of *Ascaris* eggs lipids are utilized, 17 per cent of the total lipids disappearing up to the vermiform stage. Only the triglycerides appear to be involved and esterified volatile and non-volatile acids disappear at nearly equal rates. The disappearance of these acids cannot be accounted for by their complete oxidation and it has been speculated that a considerable quantity of the lipid becomes converted to glycogen and trehalose.

Chapter—2

Physiology of Nutrition

The Ecologist, Charles Elton, once remarked that parasites Live off their capital whereas predators live off their interest—thus suggesting that parasites maintain themselves at the expense of their hosts. In reality, there are very few examples where this conclusion is borne out by experimental observation. Parasites, in general, do not eat themselves out of a home.

The nutritional physiology of parasites is a highly complex subject. Parasites inhabit a wide variety of tissues in both invertebrate and vertebrate animals, including the alimentary canal, the blood, the nervous system, various body cavities and organs, such as the liver and the eyes. Many parasites have exceedingly complicated life cycles, involving up to three different hosts, with two or more free-living stages concerned with transmission, i.e. dispersion and infection of the next host. The transmission stage are often non-feeding phases of the life cycle (e.g. miracidia and cercariae) and there are also non-feeding parasitic larvae that are quiescent, resting stages (e.g. metacercariae). With this wide diversity of form and habitat, it becomes increasingly difficult to make general statements concerning the nutrition of parasites.

Most of the information on the nutritional physiology of parasites derives from studies on the helminth parasites that inhabit either the gut or the blood of vertebrates, and from the blood-dwelling protozoan parasites of vertebrates. These parasites inhabit a microenvironment rich in dissolved organic molecules. Rather less is known about the nutrition of parasites that live within the tissues of their host.

An important concept in the nutritional physiology of parasites, and one that we shall explore in this chapter, is the *host-parasite interface* or *interfacial space*. This is the region of contact between the parasite and its host, and is the site at which the molecular transfer of nutrients occurs. In nutritional terms the interface involves both the external surfaces and, if present, the alimentary canal of the parasite: in the majority of the parasitic nematodes the alimentary canal alone forms the nutritional interface.

In vitro culture of parasites

The culture of a parasite *in vitro* requires the parasite to be grown and maintained outside the host, in conditions that emulate its normal surroundings and support its continued development. The culture of parasites serves two main purposes: to provide parasite material for study in the absence of the host; and as a means of investigating the nutritional requirements of a parasite under completely controlled conditions. In fact, only a small number of parasites have been grown in chemically-defined conditions, containing known quantities of each nutrient. More usually, an undefined addition-such as whole serum, liver extract of yeast extract—is required for successful maintenance of the culture. There is an extensive literature on the *in vitro* culture of protozoan parasites, particularly the haemoflagellates, but rather less is known about the culture of helminth parasites.

Much of the data that we shall discuss in this chapter has been obtained from parasites that are maintained in culture, normally for short periods to time. It must be stressed that a cautious interpretation of these data is essential, since we cannot assume that events measured *in vitro* will accurately reflect the processes that take place *in vivo*. The culture of parasites *in vitro* is a powerful tool for parasitologists but, to be of any lasting service, it requires wisdom in its use.

The alimentary canal of helminth parasites

The Monogenea, Digenea and Nematoda posses as alimentary canal, which varies widely from group to group in both form and function. We shall briefly consider the morphology of the gut in these parasites, and at the same time discuss the importance of this system to the nutrition of parasites. There is no gut Protozoa, or Acanthocephala.

Monogenea

The gut in the Monogenea conforms to the basic pattern of a mouth, pharynx, oesophagus and either a simple or a highly-branched blind-ending caecal system or intestine. The mouth is anterior, either at the extreme margin (as in *Microcotyle* and *Polystoma*), or slightly behind the anterior margin on the ventral surface—the latter representing the more usual arrangement. The mouth, which may be simple or elaborated into labia, has associated with it an oral sucker. The buccal cavity leads into the pharynx, a powerful muscular structure composed of three distinct groups of muscle fibres. Glandular structures may be associated with the pharynx; *Polystoma integerrimum* produces a proteinaceous secretion from its pharyngeal glands, but this species may be unusual among the Polyopisthocotylea in the possession of pharyngeal glands, since other members of the group lack such structures. Within the Monopisthocotylea, some species posses pharyngeal glands while in others the pharynx is wholly muscular. The glandular pharynx of *Entobdella hippoglossi,* for example, reflects the extent to which the pharynx itself is used as an organ of food acquisition by being protruded through the mouth and appended to the surface of the host. In *E. hippoglossi* and *E. soleae,* the secretions from the pharyngeal glands are proteolytic.

The pharynx leads to a short oesophagus, which is some groups is furnished with openings from glands whose function is assumed to be digestive, while in other groups (Microcotylidea and Hexabothriidae) the oesophagus is non-glandular. Little seems to be known about the function of oesophageal glands in the Monogenea.

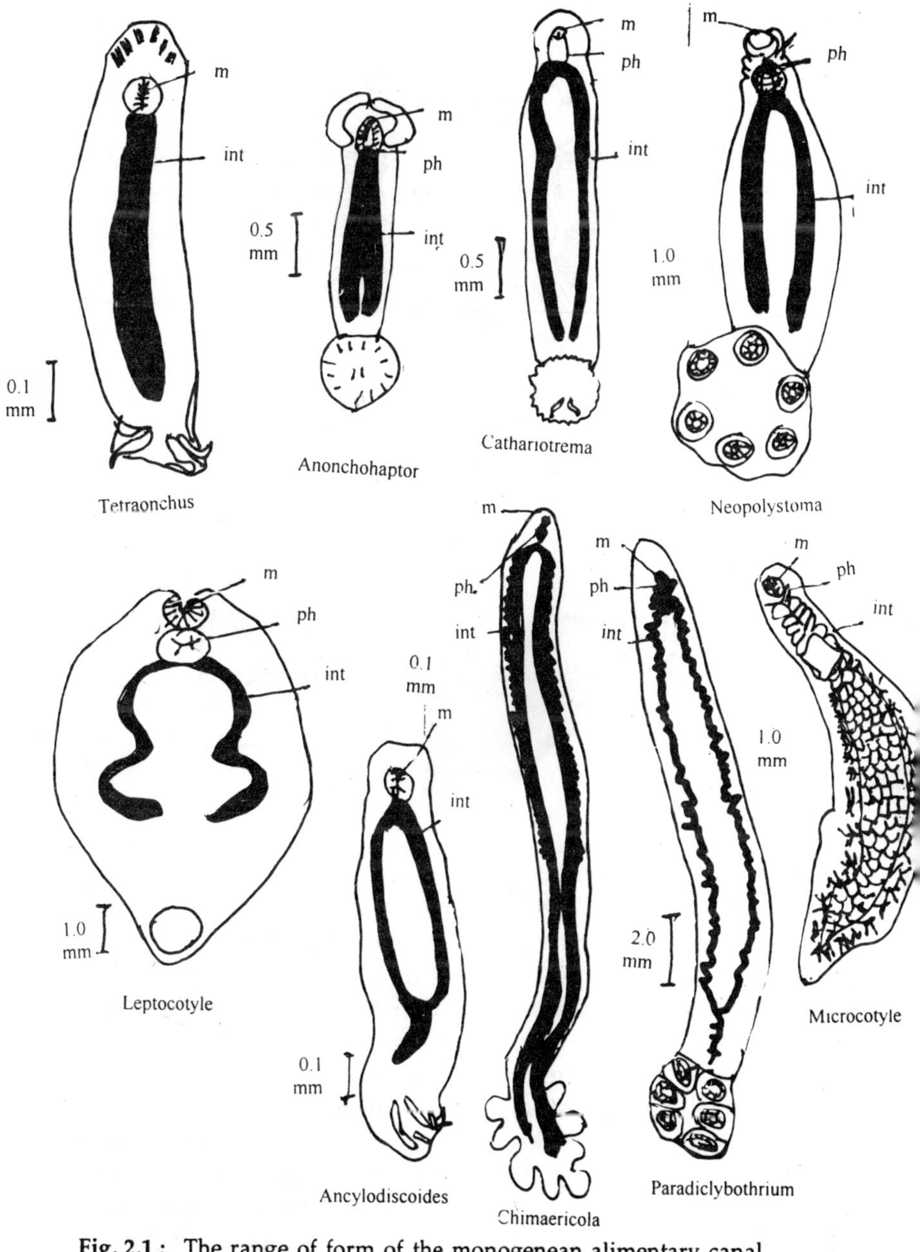

Fig. 2.1 : The range of form of the monogenean alimentary canal. m—mouth; ph—pharynx; int-intestinal caeca.

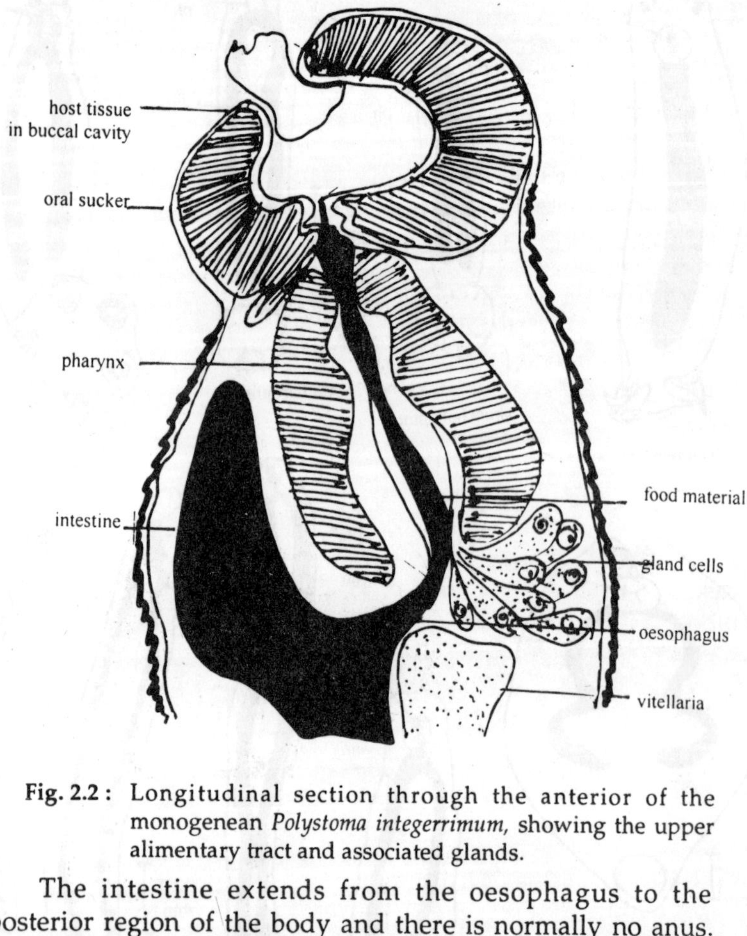

Fig. 2.2 : Longitudinal section through the anterior of the monogenean *Polystoma integerrimum,* showing the upper alimentary tract and associated glands.

The intestine extends from the oesophagus to the posterior region of the body and there is normally no anus. The intestine itself varies greatly in form. At its simplest it is a single tube, as in the Tetraonchidae. More usually, however, it is a bifurcates structure. In either case, a substantial degree of secondary branching may occur. According to Bychowsky, the simple form of intestine is characteristic of the smaller monogeneans, while the larger worms tend to possess the more elaborate pattern of intestine, culminating in the extensive anastomoses typical of the large marine monogeneans. Unfortunately, the relationship between the diet and the gross morphology of the monogenean intestine has attracted little attention. Somewhat more information is available, however, on the cellular structure of the gut and its relationship of worm nutrition.

The two orders of the monogenea, the Monopisthocotylea and the Polyopisthocotylea, differ distinctly in their diet. The Monopisthocotylea feed on the epidermal tissues and mucoid secretions of the host, while the Polyopisthocotylea appear to be exclusively blood feeders. This difference is reflected in the cellular construction of the intestinal epithelium, the gastrodermis. (In invertebrates the epithelial lining of the intestine is called the gastrodermis and is analogous to the mucosa of vertebrates). The tissue and mucus feeders possess a continuous gastrodermis, whereas the sanguinivorous monogeneans are normally characterized by a *discontinuous* or *deciduous* gastrodermis in which the entire epithelium is replaced with every meal ingested—though this is not the case with *Diclidophora*. This condition is also found in some blood feeding Digenea.

There is also a divergence in the digestive processes of the two orders. The Monopisthocotylea digest their food material extracellularly within the lumen of the intestine (or, more rarely, actually outside the body, as in the case of *Entobdella*), while the digestion in the Polyopisthocotylea is both extracellular and intracellular. The digestive processes in the polyopisthocotylean *Diclidophora merlangi*, a common gill parasite of whiting in British waters, have been examined in some detail by Halton and his co-workers. This fluke feeds on host blood, and digestion of the blood meal follows a biphasic pattern. The extracellular phase commences in the prepharynx, assisted, presumably, by the secretions from at least one of the three glands located in the anterior gut. The second phase of digestion takes place intracellularly within the haematin cells of the caeca. The caecal gastrodermis is not discontinuous but consists of alternating haematin cells and connecting cells, the latter providing

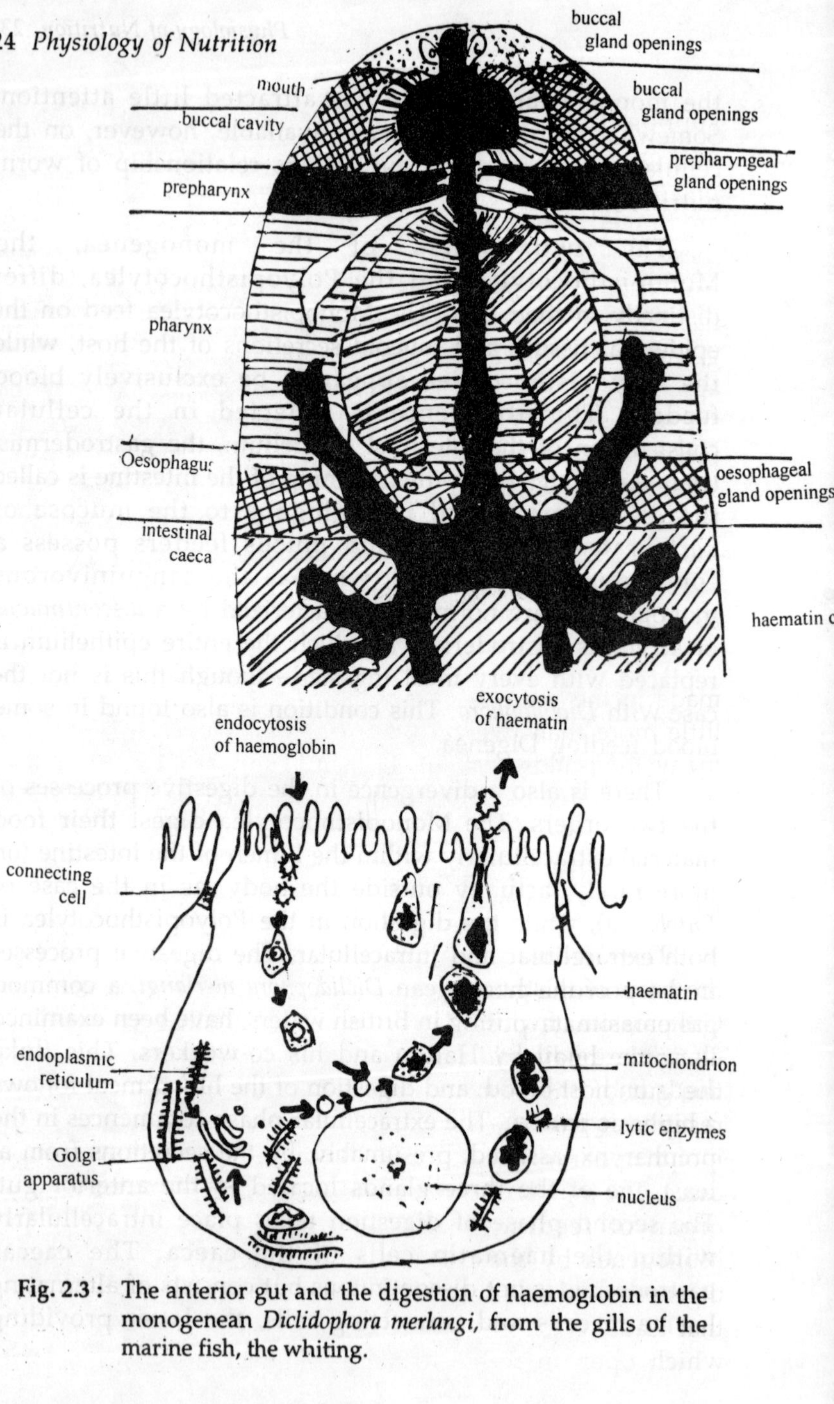

Fig. 2.3 : The anterior gut and the digestion of haemoglobin in the monogenean *Diclidophora merlangi,* from the gills of the marine fish, the whiting.

support for the former. Absorption of haemoglobin is accomplished by endocytosis (pinocytosis) across the microvilli of the haematin cells, during which process vesicles are formed. There is a direct numerical relationship between these vesicles and the presence of host blood in the adjacent lumen of the caeca. Once within the haematin cell, haemoglobin is digested by enzymes originating from the granular endoplasmic reticulum, and packaged in the Golgi apparatus. The waste haematin residues are removed from the digestive cells by continuous exocytosis (reverse pinocytosis) rather than the alternative method involving sloughing off of the entire cell layer. A great deal remains to be learned about the digestive processes in *Diclidophora;* morphological studies show that the foregut has great complexity but the physiological significance of this remains to be elucidated.

One word of caution is necessary. The generalisations made in the preceding paragraphs are based on studies of little more than twenty species of monogeneans out of the many thousands that are known. The overall applicability of these statements therefore must await confirmation or refulation by additional investigations in what is a much neglected area of research.

Digenea

The Digenea are separated into two major groups on the basis of the gross structure of the alimentary canal. The Gasterostomata possess a ventral mouth situated part way along the body and have a simple sac-like intestine while the gut of the second group, the Prosostomata, closely resembles that of many monogeneans with an anterior mouth, pharynx, oesophagus and paired (bifid) digestive caeca. The mouth of the prosostome Digenea forms the major organ of feeding—a role partially fulfilled in many Monogenea by the pharynx. The pharynx in prosostome digeneans, when present, it a powerful muscular structure that leads to the oesophagus. Some digeneans have glands which open in to the oesophagus but their function is

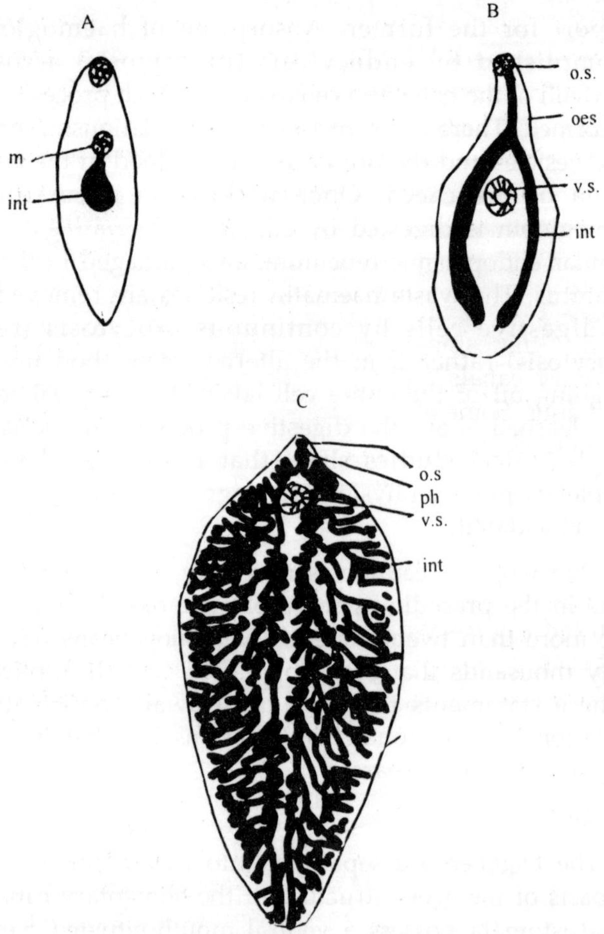

Fig. 2.4 : The range of form of the digenean alimentary canal. int-intestinal caeca; m-mouth; oes-oesophagus; o.s. -oral sucker; ph-pharynx; v.s.—ventral sucker.

unknown. The basic pattern of paired intestinal caeca (or diverticula) may become modified either by varying degrees of branching or by fusion. There appears to be no obvious relationship between the form of the intestine and the diet of the parasite: e.g. *Haematoloechus medioplexus* and *Haplometra cylindracea* both feed on the blood of their amphibian host and possess a simple bifid intestine, while *Fasciola hepatica,* which is probably in part sanguinivorous, has an extensively branched intestinal network; the caeca in the schistosomes are anteriorly branched and fuse to form a single structure posterior to the ventral sucker; the Sanguinicolidae, which are blood parasites of poikilothermic vertebrates, are characterized by the possession of an H-shaped intestine.

The caeca in most Digenea are blind-ending, but in some species they open to the outside by way of an anus. In certain echinostomes, for example, the caeca open into the excretory canals and the excretory pore therefore doubles as an anus. Some of the Digenea of fishes (e.g. *Acanthostomum* spp. and *Schistorchis carneus*) have a true anal opening, or even paired and pores. The functional significance of an anus in the Digenea is obscure, since waste products from feeding and digestion are normally voided via the mouth.

It is rather surprising to find that little attention has been paid to the dietary requirements of digeneans and the best studied species, not unnaturally, are those of economic importance, such as *Fasciola hepatica* and *Schistosoma mansoni*. As with the Monogenea, digeneans ingest either the epithelial tissue of the host or host blood. Sanguinivorous forms include *Azygia* spp., *Haematoloechus medioplexus, Haplometra cylindracea* and *Schistosoma* spp. Tissue feeders include *Opisthoglyphe ranae, Gorgoderina vitelliloba, Pleurogenes* spp. and *Plagiorchis* spp. Some species, such as *Diplodiscus subclavatus* and *Fasciola hepatica,* seem equally able to feed on both blood and other tissues.

The process of feeding in many of the Digenea is assisted by the secretions emanating from glands associated with the ventral sucker, e.g. *Haplometra cylindracea* secretes a non-specific esterase thought to hydrolyse host tissues during feeding; *Opisthoglyphe ranae* produces a mucoid secretion, that may serve to protect the worm from the host digestive enzymes.

The intestinal gastrodermis of digeneans has a range of morphology similar to that seen in the Monogenea. The epithelium is either continuous or discontinuous and, in all species that have been examined, the surface area of the intestine is greatly enlarged by the presence of microvilli. Digestion of food takes place primarily within the lumen of the gut, but this process may be completed intracellularly within the gastrodermal cells following absorption of partially digested material.

Feeding and digestion in Fasciola and Schistosoma. Since our knowledge of feeding and digestion in the Digenea depends to a large extent upon studies of these processes in *Fasciola hepatica* and *Schistosoma mansoni,* it is worthwhile considering the nutrition of these two species in some detail.

Schistosomes are exclusively sanguinivorous, and it has long been known that host red blood cells are ingested and haemolysed rapidly within the lumen of the intestinal caeca. The male partner of the permanently paired schistosomes ingests some 30000 red cells per hour *in vivio,* while the female worm has an appetite approximately ten times greater. This has been demonstrated by the use of radio-chromium labelled red blood cells introduced into infected mice. The schistosome gut is characteristically blackened by the presence of the pigment haematin (produced by haemoglobinolysis) and this waste material is periodically removed by regurgitation; *S. mansoni* completely empties its intestine every three to four hours, the female worm producing four times more haematin than her male partner. Experimental studies on *S. mansoni* in mice indicate that the worm can incorporate amino acids originating from mouse red cell haemoglobin into its own protein, and it has been proposed that peptides, produced by the incomplete lysis of haemoglobin, may be absorbed across the gastrodermis of the worm's gut and incorporated, unaltered, into yolk protein. However, there remains some doubt concerning the precise role of haemoglobin in the nutrition of schistosomes, since only a single proteolytic enzyme has been isolated from the gut of a schistosome and partially purified-this enzyme has a high affinity for native globin and functions optimally at pH 3.9. The products of globin hydrolysis are not amino acids, as might be expected, by small peptides. The female schistosome has a large requirement for proteins and amino acids in her diet as she may produce and liberate up to 300 eggs per day and can do so for the several years of her life span; this rate of egg production costs the female worm ten per cent of her body weight per day. Evidence obtained from the *in vitro* culture of *S. mansoni* supports the view that red cells, or more particularly globin, are necessary for the adequate growth and survival of the worm, i.e. worms will not grow in culture in the absence of red blood cells; the addition of purified globin to the culture increases worm survival. Interest in the nutrition of schistosomes is more than simply academic, since a through knowledge of the worm's feeding requirements may eventually allow therapeutic intervention of egg production in infected human beings. This would be of profound importance, as the disease schistosomiasis is primarily associated with the numbers of schistosome eggs that are liberated and that subsequently become trapped within the human body tissues.

Blood feeding in *Fasciola* is less well documented than in *Schistosoma* and is in fact disputed by several authorities. It is probably reasonable to assume that *Fasciola* can and does ingest both epithelial tissues of the bile duct and red blood cells. It has been suggested that the liver fluke browses upon the hyperplastic epithelium produced in the bile duct as a reaction to the presence of the parasite. *In vitro* studies carried out on the feeding of *Fasciola* show that the worm is capable of digesting erythrocytes (red blood cells) and excreting iron as a waste product. Digesting is both extracellular and intracellular, in contrast to the schistosomes where red cells are digested solely in the intestinal lumen. A curious feeding cycle occurs in *Fasciola* during which the gastrodermis is seen to undergo radical morphological alteration. The gastrodermis is highly irregular in appearance,

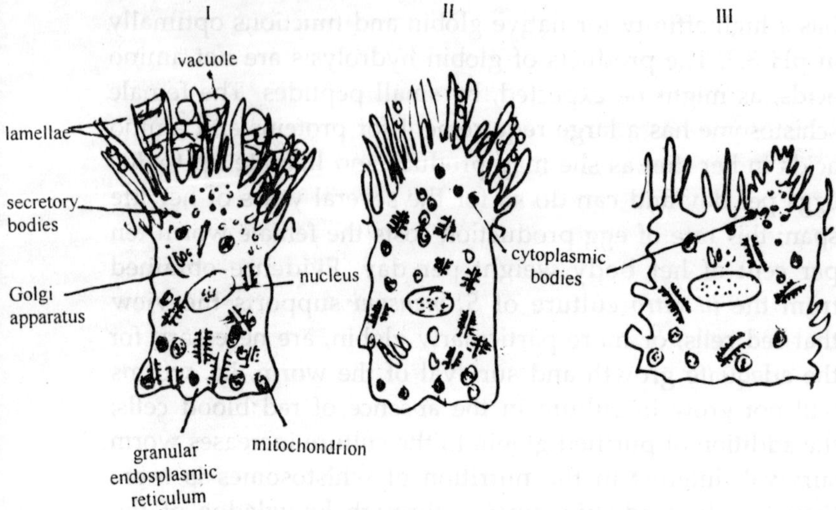

Fig. 2.5 : Variation in the morphology of gastrodermal cells lining the intestine of the liver fluke, *Fasciola hepatica*. Group I cells: most abundant type, occurring in clusters of 4 to 7 cells; columnar cells 25-40 μm tall; apices with numerous lamellae separated by vacuoles; granular endoplasmic reticulum abundant; secretory granules increase in number from early to late development. Group II Cells: columnar cells 20-35 μm tall; apices lack secretory bodies but with numerous cytoplasmic bodies containing coiled membrane material; numerous lamellae separated by vacuoles; in late development of this cell type there is a reduction in the number of lamellae.

Group III cells: cuboidal cells, 10-20 μm x 10-20 μm; short lamellae; numerous mitochondria; apical secretory bodies present.

Group I cells are predominantly secretory in function and Group II cells are predominantly absorptive in their function. The sequence of cell types from I to III represents a cycle of functional activity in a single gastrodermal cell.

the cells ranging in shape from columnar to squamous—the starved worm is characterized by a columnar gastrodermis in which digestive enzymes (non-specific esterases) are located in the basal region of the cells. These enzymes gradually accumulate in the apical portions of the gastrodermal cells and, on ingestion of a blood meal, the enzyme-rich cells rupture at their apices and release blebs of enzymes into the lumen of the intestine. These "packets" of enzymes are rapidly mixed with the gut contents and digestion proceeds. The actual presence of food in the gut appears to be the stimulus for the release of digestive enzymes. Partially digested haemoglobin is fully hydrolysed in the gastrodermal cells, and the major waste material, iron, is transported through the parenchyma and removed via the excretory system. Schistosomes solve the problem of unwanted iron from the digestion of haemoglobin rather differently, by regular regurgitation of gut contents; this is usually considered to be a more advanced adaptation to the sanguinivorous habit. In schistosomes the gastrodermis is non-deciduous, whereas in *Fasciola* there is a high rate of turnover of gastrodermal tissue along with the need to excrete iron from the body tissues. There is some evidence that *Fasciola* has proteolytic enzymes which hydrolyse globin to free amino acids, the enzymes functioning optimally at a low pH. Clearly we need to extend out knowledge of the gut physiology of these two digeneans in particular and other representatives of the group in general, but the practical problems, though not insuperable, do offer a serious challenge to the researcher.

Nematoda

The alimentary canal of both free-living and parasitic nematodes conforms to a single basic pattern, comprising three distinct regions: the stomodaeum, which includes the mouth, labia, buccal cavity and pharynx; the intestine, which is a straight tube and is the site of digestion and absorption; and the proctodaeum, which includes the rectum and anus. The stomodaeum and proctodaeum are lined by cuticle which is therefore shed with each successive moult. The anterior region of the nematode gut shows the greatest degree of morphological variation and, in general, this reflects the feeding habits of the work. The oral region of parasitic nematodes is commonly less elaborate than that found in the free-living forms but is complex in groups such as the hookworms. Nematodes feed by the suction produced by the powerful pharyngeal musculature and by "teeth", in those forms that possess them, such as the hookworms (*Ancylostoma*) parasitic in the mammalian gut. The intestine is lined with a microvillous gastrodermis that is both secretory and absorptive.

The animal parasitic nematodes can be divided into four groups according to their feeding habits: feeders on the contents of their host's gut (e.g. *Ascaris*); epithelium feeders (e.g. *Ancylostoma*); tissue feeders involving tissue penetration (e.g. *Trichuris*); and fluid tissue feeders (e.g. *Wuchereria*).

There is virtually no evidence for the transcuticular absorption of nutrients in nematodes and the gut must be regarded as the functional interface. Exceptions to this rule are the entomophilic (insect-parasitic nematodes, such as *Mermis nigrescens,* parasitic in the haemocoel of locusts, and *Bradynema,* a parasite of dipterans; in the latter species the mouth, anus and gut degenerate and the cuticle develops a microvillous surface, through which nutrients can pass, while the cuticle of the former is permeable to nutrients despite the presence of the alimentary canal.

The role of the external surfaces of parasites in their nutrition

The external surfaces of all endoparasites form a potential nutritional interfaces between the parasite and its host. In the case of the Cestoda and the Acanthocephala, the complete absence of an alimentary canal renders the body surface of the parasite the major site of nutrient uptake, while the surfaces of all other parasites, with the virtual exception of the Nematoda, may play a significant nutritional role.

The evidence for the nutritional role of parasite surfaces comes from both morphological and physiological studies,

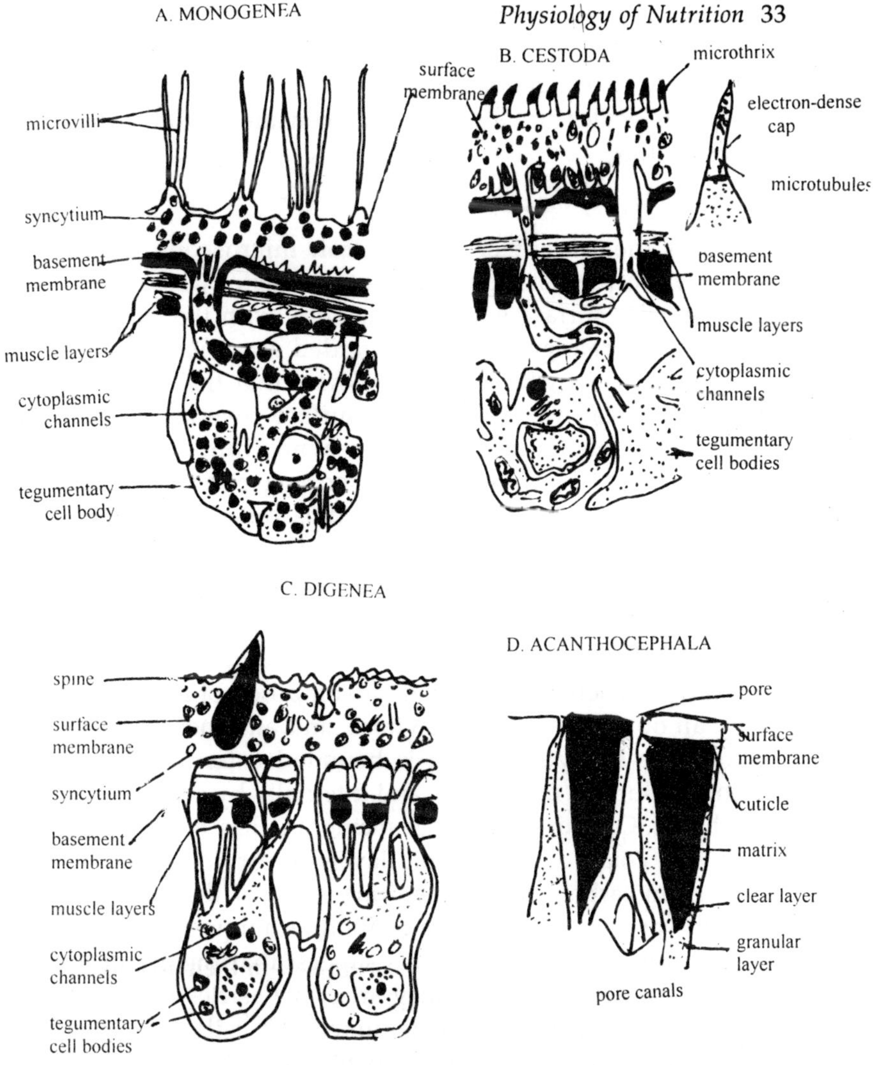

Fig. 2.6 : The structure of the surfaces of helminth parasites (excluding nematodes) based on ultrastructural studies using the transmission electron microscope. Platyhelminths are covered by a metabolically active layer and not acuticle as was formerly thought to be the case, the acanthocepohalans and nematodes do possess a cuticular covering, which, in the former group, is richly supplied with pores. The Protozoa are covered simply by a plasma membrane.

particularly investigations of membrane transport of small organic molecules and ions. We shall consider both the these aspects in turn.

The morphology of the parasite surface

Until the advent of the electron microscope, it was generally accepted that platyhelminth parasites were covered by a cuticular surface which, by analogy with the cuticle of other invertèbrate groups, was regarded as an impermeable and protective structure. We now know that this is not the case, and instead we find that the surface covering of these worms is a metabolically—active syncytium whose cell bodies are sunk deep into the underlying parenchymatous tissue of the worm. The surface is thus called a tegument, though the terms integument or epidermis appear to be interchangeable. The similarities between the tegumentary structure in the Monogenea, Digenea and Cestoda encourage us to consider them together.

The platyhelminth tegument is composed of two distinct layer: an outer anucleate syncytium, and an inner nucleated region. The syncytial region is bounded externally by the plasma or surface membrane, and internally by the basement membrane. Outside the surface membrane lies the *glycocalyx* or "fuzzy coat', a polyanionic layer of varying thickness, and rich in carbohydrate—the dimensions of the glycocalyx seem to vary according to the fixation techniques employed. Studies using radioactive tracers suggest that the glycocalyx and the surface membrane may arise in Vesicles formed by the Golgi apparatus of the tegumentary cells. The gross surface topography is highly variable in the helminths, not only from group to group but in different regions on the body of an individual worm, e.g. in *Fasciola hepatica,* some regions of the surface show deep invaginations while other areas are quite smooth. The surface of the Digenea is typified by the presence of spines which are embedded into the syncytial region of the tegument and are covered, externally, by the plasma membrane. The density of these spines frequently differs over the surface of the worm and they tend to be least numerous posteriorly. The cestode surface is unusual in the possession of numerous microvillous projections from the surface called *microtriches;* these structures are found in all groups of tapeworms that have been examined, from the most primitive Proteocephalidea to the Cyclophyllidea. In general, the density of the microtriches is lower in the scolex and neck region of the worm than in the more mature regions of the body. The cestode microtriches resemble the brush border found on many cell surfaces (e.g. the mammalian mucosal cell), but they differ in the possession of electron—dense tips, the function of which remains obscure. While the microtriches serve to increase the surface area of the tapeworm (by a factor of 3-6 times) they are not as effective in this regard as are the microvilli on the mammalian mucosal cell and it may be that they have alternative functions, such as attachment and maintenance of position in the host gut. Microtubules run longitudinally within the shaft of the microtriches and these probably have a function in the translocation of nutrients. Rather surprisingly, the surface of monogeneans is also microvillous despite their external position on the host and this may be best explained on phylogenetic grounds rather than in functional terms, since it is unlikely that external parasites would make use of their surfaces for nutrient acquisition to any significant extent.

In all the platyhelminths the syncytial tegumentary layer, frequently referred to as the distal or pronuclear cytoplasm, is richly supplied with mitochondria which have characteristically few cristae, endoplasmic reticulum, various vacuoles and vesicles that contain replacement surface membrane and glycocalyx. These vesicles have been studied in detail in *Schistosoma mansoni,* a digenean with a complex surface structure whereby the membranes of the surface proliferate during the development from larva to adult in the mammalian host. The plasma membrane of the schistosome cercaria, the invasive larva, is trilaminate but it becomes heptalaminate and deeply folded within three hours of penetration of the host. There is also thought to be a continual process of surface membrane turnover in these worms.

The tegumentary cells of platyhelminths are located within the parenchymal tissue below the circular and longitudinal muscle layers. The cytoplasm of these cells is connected to the syncytial layer by an extensive network of tubules. The cellular region of the tegument contains the nuclei, Golgi apparatus, mitochondria and endoplasmic reticulum. It is probable that all platyhelminths have a relatively high turnover rate of surface membrane, with new membranous material assimilated within the nucleated portion of the tegument.

The overall structure of the platyhelminth tegument, as revealed by electron microscopy, is clearly consistent with a nutritional and metabolic function rather than simple projection. As we shall discuss later there is an accumulating body of evidence to show that, in the Cestoda and Digenea and to a lesser extent the Monogenea, the tegument is an active region of nutrient uptake.

The body surface of adult Acanthocephala is rather different from that of the parasitic Platyhelminthes, comprising five distinct layers. Externally lies a thin mucopolysaccharide-containing epicuticle, internal to which is a cuticle, a striped layer, a fibrous layer and an innermost radial layer, residing on the basement membrane. The surface region, or body wall, is syncytial and metabolically active, as indicated by the presence of nuclei, mitochondria, ribosomes and folded plasma membranes, and it is rich in organic material such as glycogen and lipid. An interesting feature of the acanthocephalan surface is the presence of numerous pores that penetrate through the cuticular layer and extend throughout the body wall. Each pore is lined with plasma membrane and is enclosed within a granular layer, itself contained within what appears to be a clear zone. The canal derived from the pores divide considerably as they extend internally serving to provide an enormously increased surface area. It is not know how these pores and their canal system are involved in nutrient acquisition in the Acanthocephala, but their role must be important considering the absence of a gut and the impermeable nature of the cuticular surface.

The great majority of nematodes are covered by a thick, often multiple layered, cuticle which is impermeable to solutes. The cuticle of *Ascaris* comprises three major regions: an outer cortex, which may contain keratin and collagen; a central matrix, which is made up of a number of proteinaceous layers; and an innermost fibrous layer bounded by a basement membrane. The matrix layer of *Nippostrongylus brasiliensis* is fluid-filled and is rich in both enzymes and haemoglobin. There is no evidence to suggest that the nematode cuticle functions as a site for nutrient absorption, except in the entomophilic species.

For most nematodes, this activity must reside exclusively with the gut and its associated structures.

Transtegumentary absorption of nutrients

In discussing the functional morphology of the alimentary canal of parasites (where it is present) it will have become evident that current knowledge is far from complete concerning the digestive and absorptive activities of the gut. We can probably assume that the gut in the Monogenea and Digenea is involvedly in the digestion and subsequent absorption of host tissue and red blood cells, but studies on the occurrence and activity of alimentary enzymes suggest that the gut may be somewhat limited in its potential, e.g. *Schistosoma mansoni* possesses, as far as we can tell, only a single intestinal proteolytic enzyme, hydrolysing globin to peptide fragments. One can reasonably question, therefore, the precise origin of free amino acids for somatic protein synthesis. We cannot provide an answer to this question at present, but we are in a position to suggest that the transtegumentary absorption of low molecular weight nutrients, known to occur, may be of great significance to schistosomes, and indeed all internal parasites, with the notable exception of most nematodes—obviously it must represent the major route in the Cestoda and Acanthocephala where there is no gut.

Direct evidence for the absorptive capabilities of the external surfaces of parasites has been obtained by study in the membrane transport of radioactively labelled solutes *in vitro*. Surprisingly, as a result of such studies, more is known about tegumentary absorption than that occurring through the intestine of parasites. This is probably due to the excellence of the adult tapeworm as an experimental system, exploited with considerable success by Read and his colleagues during the last fifteen years. There have been many similar transport studies on the parasitic Protozoa, particularly the trypanosomes.

Mechanisms of solute entry

The plasma membrane of all cells serves not only to maintain the integrity of the cell but also to regulate the passage or flux of dissolved or insoluble molecules. The chemical structure of the plasma membrane has been the focus of considerable attention for many years and a number of models have been proposed, none of which has gained complete general acceptance. The major membrane components are lipid (60%) and protein (40%), and there is good evidence that the lipid exists in the form of a bilayer with the hydrophobic ends of the molecules orientated to the inside of the membrane. The relationship of the protein component to the lipid bilayer is in some doubt, but proteins probably extend through the lipid layer.

As we have discussed already all parasites, with the exception of nematodes, are bounded by an external plasma membrane and there is no evidence suggesting that this is unique in its structure in parasitic animals. Therefore we may assume that the surface membranes of parasites provide selective barriers to solutes and they may be studied using techniques and approaches developed for other transporting systems.

Solutes move across membranes by a number of different processes: simple diffusion, facilitated or mediated diffusion, active transport and pinocytosis (endocytosis or exocytosis). These modes of transport are readily distinguishable by their biochemical and kinetic characteristics which can be determined under *in vitro* conditions in the laboratory.

Simple *diffusion,* sometimes termed passive transport, obeys Fick's law which states that the rate of diffusion of a solute is directly related to the concentration difference on either side of the membrane. A solute will always migrate from a region of high concentration to a region where the concentration is lower. When the concentration of the solute is the same on both sides of the membrane the net flux of the solute in zero. Fick's Law can be expressed as follows:

$$\frac{dn}{dt} = DA\frac{dc}{dx}$$

where *dn/dt* is the number of molecules of the solute (*dn*) moving across the membrane with a surface area *A* in the time *dt* and where a concentration difference of *dc* exists over the distance *dx.D* is the coefficient of diffusion. Movement of a solute by simple diffusion requires no energy expenditure on the part of the cell, is unaffected by other solutes or by metabolic poisons and shows no predilection for a particular stereoisomer.

Facilitated or *mediated diffusion* differs from simple diffusion in the involvement of a binding process at some point during transport and hence it exhibits saturation kinetics. Frequently the term "carrier" is used to indicate the unknown structure that binds the solute and transports it across the membrane, releasing it at the appropriate place on the other side. However, since there is little convincing evidence for the existence of these carriers *per se,* it is probably preferable to use alternative terminology.

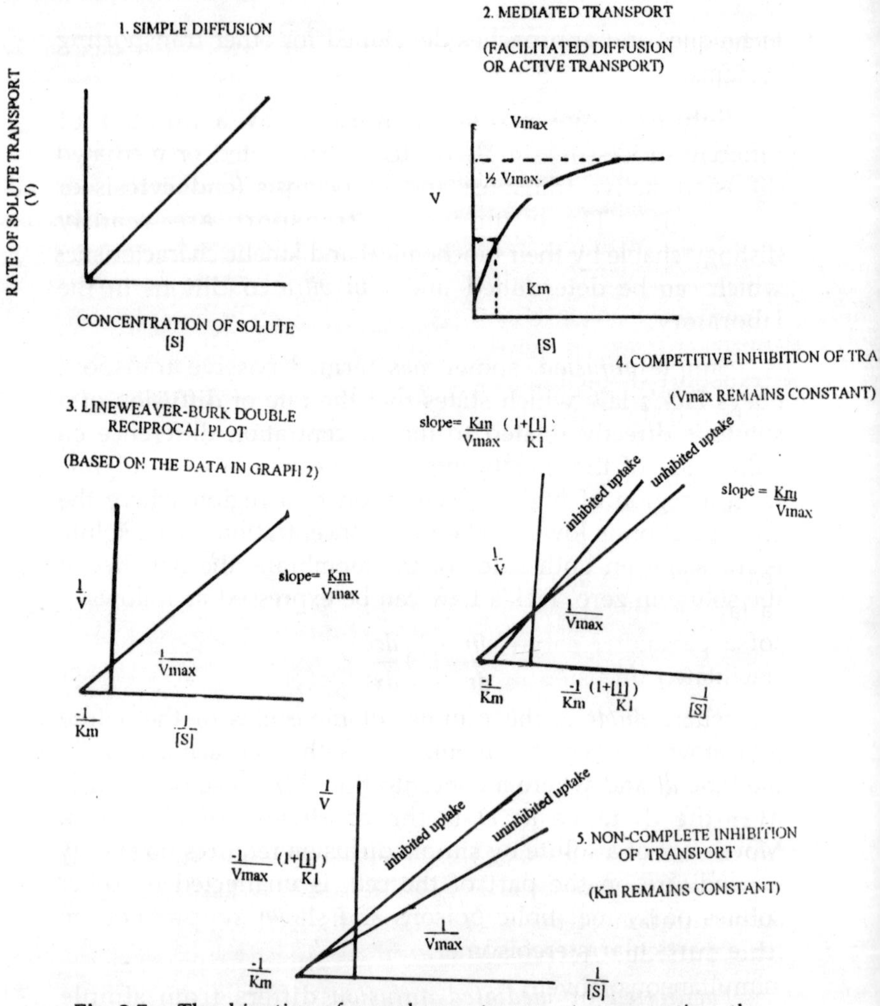

Fig. 2.7 : The kinetics of membrane transport. In practice, the uninhibited uptake of low molecular weight moeities by intact parasites *in vitro* often fits a curve which is a combination of 1. and 2. above. Mediated transport is more important at low substrate concentrations, but at higher concentrations more substrate enters by simple diffusion. K_I-inhibitor constant (can be determined experimentally as in 5. or 6.); [I]-concentration of inhibitor; K_m-transport or Michaelis constant; V_{max}-maximum rate of transport (K_m=½ V_{max}).

Parasitologists prefer to use the term *site* or *locus* in place of carrier, whereas other workers have employed terms such as *permeases*. In this book we shall use the terms site or system, thereby leaving open the nature of the transport process. Facilitated diffusion, like simple diffusion, involves solute movement down an electrochemical gradient, so that no accumulation of solute on one side of the membrane is possible. Accordingly, no energy is used to transport molecules in facilitated diffusion. The graphical relationship between the concentration of the solute and its rate of transport in faciliated diffusion is non-linear, and as the concentration of solute increases so the rate of uptake approaches an asymptote. Solute transport by facilitated diffusion can be inhibited by a variety of chemically similar molecules which compete or interfere with the binding of the solute at the transport site. The degree of specificity of each transport site may vary widely, as we shall see, being a facet of the array of molecular species that inhibit uptake of a particular solute. Stereospecificity is characteristic of facilitated diffusion. An advantage of this type of transport process is that the rate of solute movement may be much greater than could occur by simple diffusion.

Active transport is kinetically similar to facilitated diffusion, but in the active process solutes are transported *up* an electrochemical gradient and concentration of accumulation of a solute inside a cell when the initial concentration of the solute inside the cell exceeds that on the outside. Active transport frequently involves the simultaneous movement of sodium ions, and solute transport can often be interrupted by altering sodium transport, e.g. by adding ouabain. Active transport also shows stereospecificity and various degrees of inhibition as in facilitated diffusion, but differs in being sensitive to metabolic poisons.

Pinocytosis, often called endocytosis or exocytosis depending upon the direction of the movement, is a macromolecular transport process whereby large molecules (e.g. proteins) are transported in membrane bound vesicles. Pinocytosis may show saturation kinetics since it involves the initial binding of the macromolecule to the membrane prior to vesicle formation. Metabolic inhibitors may interfere with transport by this process.

An important consideration, germane to discussion of transport of molecules into parasites, is the use of intact organisms as opposed to the wide use of isolated tissues, e.g. Erhlich ascites tumour cells, bacterial preparations or isolated intestinal loops or rings, for the many transport studies, in non-parasites. As a consequence the whole-animal approach used by parasite physiologists lends itself to problems of interpretation not encountered in the isolated cell systems, particularly with regard to the internal concentrations of naturally occurring molecules, and no account can be taken of concentration differences within the various body tissue of the intact parasite. Compartmentalization of molecules is a problem which has not been overcome to any satisfactory degree. Ideally, transport of solutes across the isolated helminth tegument should be examined, but the diffuse structure of the surface in metazoan parasites precludes this—this criticism, however, does not apply to the protozoan parasites. Nevertheless, a great deal of important information on transtegumentary transport in helminth parasites has been obtained by the use of carefully controlled experimental conditions.

Michaelis-Menten Kinetics: In 1913 Michaelis and Menten formulated an equation to account for the rate of an enzymic reaction. The equation, known universally as the Michaelis-Menten equation, can be expressed as follows:

$$V = \frac{V_{\max}[S]}{[S]+K_m} \quad \text{or} \quad V = \frac{V_{\max}}{\frac{K_m}{[S]}+1}$$

where V is the rate of the reaction, V_{max} is the maximum velocity and K_m the Michaelis constant or concentration of substrate S required to yield half maximum velocity. This equation applies to a saturable enzymic reaction and assumes the reversible formation of a complex between the substrate and the enzyme:

$$E + S \rightleftharpoons ES \rightleftharpoons E + P$$

where E is the enzyme, S the substrate, ES the transition complex and P the product of the reaction. It has been subsequently noted that this equation will apply equally well to the saturable process of membrane transport, such as occurs in facilitated diffusion and active transport:

$$C + S \rightleftharpoons CS \rightleftharpoons C + S$$

where C is the hypothetical membrane carrier, S the substrate transported, and CS the presumed complex formed between the carrier and the substrate, in which the substrate is presumed to be transported across the membrane. There are two important differences concerning the Michaelis-Menten equation when applied to membrane transport rather than to an enzymic reaction: first, there is probably no chemical change of the substrate during its transport, and secondly, unlike enzymes, membrane carriers await identification and isolation and must therefore be regarded as hypothetical entities whose presence is only implied by the occurrence of saturable transporting systems.

While the application of the Michaelis-Menten equation to transport systems provides no insight into the actual nature of the transport process itself, it does provide a very convenient means of comparing the kinetic parameters of different systems, or similar systems in different organisms.

Transport of molecules into parasites

Carbohydrate transport The uptake of carbohydrates has been investigated in only a small number of parasites including three species of haemoflagellate Protozoa (*Trypanosoma lewisi, T. equiperdum* and *T. gambiense*), two species of Digenea (*Fasciola hepatica* and *Schistosoma mansoni*), four species of Cestoda (*Hymenolepis diminuta, H. microstoma, Taenia crassiceps* and *Calliobothrium verticillatum*), two Acanthocephala (*Polymorphus minutus* and *Moniliformis dubius*) and across the intestinal wall of two species of Nematoda (*Ascaris suum* and *Trichuris vulpis*).

The cestode *Hymenolepis diminuta* has provided an ideal model system for transport studies since it can be maintained in the laboratory with ease, is not pathogenic to man and, as a system, it yields highly reproducible data. Not surprisingly, considerable information has been amassed on the transport of nutrient molecules into *H. diminuta,* while much less is known about other species of helminth parasite. Even so, most studies on *H. diminuta* have utilized only ten-day-old worms; these worms are immature, since egg production does to commence until the sixteenth day in the rat gut. Little is known about transport in mature worms or in the younger stages, and there have been just two studies on transport in the cysticercoid larvae obtained from the intermediate host (a flour beetle).

In vitro, glucose is actively transported across the tegument of *H. diminuta.* Absorption of glucose is competitively inhibit by some sugars (D-allose, ∝-methyl glucose, 6-deoxyglucose, 3-O-methylglucose and 1-deoxyglucose) but not by others (fructose, L-fucose, 1-5 anhydro-D-mannitol) (table 2.1). The uptake system is stereospecific, temperature dependent, with a Q_{10} (for the temperature range 15-40°C) of 2.4, and is inhibited by metabolic poisons (p-mercuribenzoate, iodacetate and 2, 4-dinitrophenol). The worm can accumulate glucose-if incubated in 5mM glucose for 60 minutes, an internal concentration of 25 mM is attained. It has been proposed that galactose enters *H. diminuta* via the glucose transport system. The cysticercoid larvae of *H. diminuta* possess a comparable transport system for hexoses which is not surprising in view of the structural similarities between the adult surface and the larval cyst wall. The bite duct tapeworm of mice, *H. microstoma,* also actively transports glucose.

In vivo studies on carbohydrate transport by *H. diminuta* have also employed isolated intestinal loops removed from infected rats. Three concurrent entry processes are reported to occur using this technique: these are active transport, simple diffusion, and a third process known as *solvent drag.* The relative contributions of each vary with alternations of pH. It is not possible to compare directly the transport of carbohydrates *in vitro* and *in vivo,* but is should be stressed that, while the use of an *in vitro* approach is of undoubted value for examining isolated systems, it may tell us little about what takes place *in vivo,* where great chemical complexity prevails.

Table 2.1 Selected kinetic data on carbohydrate transport in cestodes

Tapeworm species	*Molecule transported*	K_t *(mM)*	V_{max} ($\mu mol\ g^{-1}h^{-1}$)	*Inhibitors*
Hymenolepis diminuta	Glucose	0.7-1.6	400-789	phlorizin galactose α-methylglucoside 2-dexyglucose 3-O-methylglucose 6-deoxygluclse allose
	galactose	5.0		phlorizin
	glycerol	0.2-0.4	84-91	1,2-propanediol α-glycerophosphate
H. microstoma	Glucose	2.0	1080	galactose phlorizin Phloretin α-methylglucoside 1-deoxyglucose
Taenia crassiceps (larvae)	Glucose	0.3	20	α-methylglucoside β-methylglucoside galactose 6-deoxygalactose phlorizin
	galactose	0.8	70	—
Calliobothrium verticillatum	glucose	0.6	610	α-methylglucoside galactose phlorizin

Hymenolepis diminuta absorbs glycerol by both facilitated diffusion and simple diffusion. The mediated system is inhibited by 1, 2-propanediol, α-glycerophosphate, iodoacetate and arsenite; uptake of glycerol is also affected by the presence or absence of Na^+, by temperature variation and by the pH of the medium. Glucose uptake in all cestodes examined is dependent upon the concentration of Na^+.

Carbohydrate transport has been investigated in only two species of the Digenea, *Fasciola hepatica* and *Schistosoma mansoni*. Glucose, fructose, mannose, glucosamine and ribose enter *F. hepatica* by mediated transport processes, probably facilitated diffusion, and xylose enters by simple diffusion. Glucose, galactose, 2-deoxyglucose, glucosamine, mannose, ribose and 3-O-methylglucose (a non-metabolizable analogue of glucose) enter *S. mansoni* by simple diffusion and a mediated saturable system. Cestodes and digeneans may differ in their permeability to fructose, *H. diminuta* being impermeable and the Digenea studied permeable, to this hexose.

Moniliformis dubius (an acanthocephalan) has saturable transport systems for several sugars (glucose, 2-deoxyglucose, mannose, N-acetylglucosamine, 3-O methylglucose, fructose and galactose). Glucose and 2-deoxyglucose share the same site for entry. Glucose entry is not dependent upon the presence of Na^+ and is not inhibited by phlorizin, in contrast to glucose transport in other helminths. *Polymorphus minutus* also absorbs glucose by a system that exhibits saturation kinetics and may possibly be active transport.

Carbohydrate transport has been examined in a small number of parasitic Protozoa and particular attention has been paid to haemoflagellates of the genus *Trypanosoma*. The mammalian bloodstream forms, *T. gambiense* and *T. lewisi*, transport a variety of carbohydrates by saturable systems. *T. gambiense* transports glucose, mannose, glycerol, 2-deoxyglucose and fructose but not the glucose analogue 3-O-methylglucose. Kinetic data suggest that there are two carbohydrate transport systems in *T. gambiense;* the 'glucose site" and the "fructose site". The latter exclusively translocating fructose and glucosamine. Fructose transport in inhibited by a number of carbohydrates (glucose, glycerol, glucosamine, mannose and N-acetylglucosamine) which themselves are not transported through this site. This phenomenon is called *Non-productive binding,* where molecules bind to, but are not translocated by, a transport system. In *T. gambiensè* the glucose site is markedly specific for the hexose carbon atoms in the 1,3, 4 and 6 positions (demonstrated by measuring inhibitory efficiency as a function of molecular structure) and the "fructose site' requires carbon atom number 5 of the sugar to remain unaltered. *Trypanosoma lewisi* also has more than one carbohydrate transport system; in this organism, fructose and 3-O-methylglucose inhibit glucose transport, in contrast with *T. gambiense.* There is only a single carbohydrate transport system in *T. equiperdum.* There is some evidence that the rate of transport of carbohydrates. This has been demonstrated in *Hymenolepis diminuta* and *T. gambiense, amino acid transport.* Unlike carbohydrates, amino acids are to usually involved in the energy metabolism of parasites. Their importance lies in their role in protein synthesis, which is commonly high in helminth parasites due to the large numbers of eggs produced. Amino acid transport has been studied extensively in the cestode *H. diminuta* by Read and his co-workers, while the Digenea, haemoflagellate Protozoa, and Acanthocephala have received rather less attention.

Amongst the Protozoa, studies on lysine transport in *Trypanosoma cruzi,* the organism responsible for Chagas disease, reveal some departures in site specificity from those typical of other organisms. Notably, in the culture form of *T. cruzi* (the stage analogous to the insect stage of the parasite in nature), uptake of lysine (a basic amino acid) occurs through three kinetically distinguishable sites. Moreover, other basic amino acids do not share these routes of entry with lysine. Inhibition of lysine uptake is effected by neutral amino acids as follows: at site 1, phenylalanine, tyrosine; at site 2, proline, alanine, methionine and cysteine and at site 3, glycine, valine, leucine and isoleucine. The partially competitive nature of these inhibitions suggests some degree of overlap between amino acid entry through the three lysine sites. In the mammalian gut, by contrast, there are no such interactions between neutral and basic amino acids. Arginine transport by the culture form of *T. cruzi* is also atypical in its degree of specificity; neutral amino acids are competitive inhibitors of arginine transport, while the other basic amino acids are without effect. *Trypanosoma gambiense* transports a number of amino acids by saturable processes and there also appears to be a substantial diffusion component, predominant at higher amino acid concentrations.

Any study of solute transport in the Digenea is complicated by the presence of the gut. It is possible, however, to ligature the pharynx and thereby examine the uptake of amino acids and carbohydrates that occurs through the tegument, though in practice ligaturing has been found to be unnecessary since, in short-term studies, there is no difference in the rate of uptake with either the gut open or ligatured. *Schistosoma mansoni* uses a combination of mediated uptake and simple diffusion to absorb alanine, arginine, glutamate, glycine, methionine and tryptophan; proline enters exclusively by a saturable mediated system. There are indications that amino acid transport in schistosomes is unusual in that neutral amino acid transport is inhibited by basic amino acids. Additionally, there are quantitative changes in permeability to amino acids in *S. mansoni* during development: the free swimming cercariae are virtually impermeable to methionine, recently penetrated schistosomula are somewhat more permeable, but by the time to worm has been domiciled in its for three weeks the adult transport system has become developed.

By contrast with *S. mansoni,* amino acid transport in *Fasciola hepatica* takes place by simple diffusion only, an unusual situation deserving of further attention.

In *Hymenolepis diminuta* there are at least six kinetically distinguishable amino acid transport systems, the specificities of which vary widely. The four neutral amino acid systems show relatively little overlap with basic or acidic amino acids. On the other hand, some neutral amino acids interact with the acidic amino acid site, while the basic amino acid system has a very narrow specificity. All amino acid entry into *H. diminuta* occurs via mediated transport, although simple diffusion may also feature in the uptake of some, such as histidine and proline. These, along with α-aminoisobutyric acid and cycloleucine, are absorbed by active transport and are therefore accumulated against a concentration gradient. No such information is available for the other amino acids.

The complexity of amino acid transport of *H. diminuta in vivo* has been described by Read, who derived a formulation, based upon the Michaelis-Menten equation, to account for the transport of a given amino acid in the presence of a realistically large number of other amino acids (analogous to the situation pertaining in the mammalian gut):

$$V_1 = \frac{V_{max}}{\frac{K_t}{[S]} + 1 + \frac{K_t[S_1]}{K_{t_1}[S]} + \frac{K_t[S_2]}{K_{t_2}[S]} + \ldots + \frac{K_t[S_n]}{K_{t_n}[S]}}$$

where V_1 is the velocity of the uninhibited uptake of amino acid S, K_t is the Michaelis or transport constant, V_{max} is the maximum velocity of the uninhibited uptake and $[S_1]$, $[S_2]$ ... $[S_n]$ are the concentrations of different inhibitory amino acids. This formulation has been tested by examining alteration in the uptake of methionine (V_1) by *H. diminuta* incubated in various mixtures of amino acids. The predicted value of V_1 is met by the observed value only when the K_t value is replaced with the K_i) (inhibitor constant with respect to methionine) for each amino acid in the mixture. Good correlation is then obtained. The inhibitor constants (K_t) are numerical indications of the affinity of the inhibitor molecule for the transport site; they are determined by a variety of experimental procedures.

There are a number of important differences between amino acid transport systems of cestodes and those of mammals. For instance, mammalian systems maintain an almost exclusive stereospecific preference for the Lenantiomorph, whereas some tapeworms show an equal affinity for both D and L amino acids (*H. diminuta, H. citelli* and larval *Taenia crassiceps*). Another divergence is the apparent lack, in cestodes, of an ion (Na^+/Cl^-) requirement coupled to amino acid transport-most mammalian systems require ion coupling for efficient amino acid transport.

There is a paucity of data on amino acid uptake in the Acanthocephala. *Moniliformis dubius* and *Macracanthorhynchus hirudinaceus* both absorb amino acids by mediated systems and by simple diffusion. Leucine and methionine are actively transported and the site specificity of the methionine system (as judged by inhibitor studies) is narrow, being limited to neutral amino acids alone. As with the cestodes, there appears to be no sodium ion requirement for amino acid transport.

The majority of nematodes, as we have already noted, do not absorb nutrients across their external surfaces. There have been but a few studies on uptake of amino acids by the nematode intestine.

Lipid transport This is a much neglected area of research in parasite nutrition and biochemistry. In general, parasites do not synthesize their long chain fatty acids *de novo* and therefore rely on the host to supply them with their basic needs. *H. diminuta* cannot synthesize long chain fatty acids but can lengthen the chain of absorbed fatty acids by one or two methylene units, by adding acetate units; hence fatty acid absorption becomes all-important. There are two distinct systems for fatty acid transport in *Hymenolepis diminuta*—one specific for short chain fatty acids (C_2-C_8) and the other specific for longer chain acids (C_{14}—C_{24}). These two systems have been designated, somewhat arbitrarily, the "acetate site' and the "palmitate site" (table 2.2). Acetate is absorbed in *H. diminuta* at a rate four times greater than palmitate. Bile salts are required for long chain fatty acid transport but not for short chain transport. The role of bile acids may be indirect, since they are thought not to be absorbed by *H. diminuta* but instead become absorbed to the tegument. Both of the fatty acid transport systems are specific to the fatty acids of appropriate chain length, and uptake is unaffected by any other class of compound. Cholesterol is absorbed by *H. diminuta* from solutions of mixed micelles with mono-olein and sodium taurocholate. The worm cannot synthesize cholesterol, a block residing at the step of conversion of farnesol (farnesyl pyrophosphate) to squalene. The formation of micellar solutions by bile acids is clearly of importance to the efficient absorption of a number of compounds by *H. diminuta,* including free fatty acids, monoglycerides and cholesterol. On the other hand, glucose transport can be inhibited by sodium taurocholate. This is an area of some interest since it may in part explain the postulated role of bile salts in determining host specificity of certain parasites, though a great deal more data need to be collected before any assertions can be made.

Table 2.2. Inhibitors and inhibitor constants (K_i) of fatty acid precursor transport in the tapeworm *Hymenolepis diminuta*.

Substrate	*Inhibitor*	*K_i(mM)*
Short chain fatty acids		
acetate	formate	6.9
	propionate	1.3 (adult); 5.3(cysticercoid)
	butyrate	3.6 ("); 16 (")
	valerate	5.0
	octanoate	5.1
butyrate	acetate	5.4
Long chain fatty acids		
palmitate	pentadecanoate	4.9
	heptadecanoate	2.3
	stearate	2.3
	arachidate	3.5
	oleate	4.5
	linoleate	3.8
	linolenate	5.6
Nucleic acid precursors		
uracil	hypoxanthine	1.2
	thymine	-4.2*
hypoxanthine	adenine	6.0
adenine	hypoxanthine	1.0
uridine	adenosine	0.1
	AMP	0.1
	ATP	0.3

* Stimulation of uptake rather than inhibition

Fatty acid transport in other groups of parasites has been virtually over looked. *Fasciola hepatica* obtains short chain fatty acids (acetate and butyrate) by simple diffusion, but here is also an indication that acetate may enter by a mediated transport system shared by propionate, butyrate and valerate.

Purine and pyrimidine transport. Purine and pyrimidine uptake has been examined only in *H. diminuta*. The transport

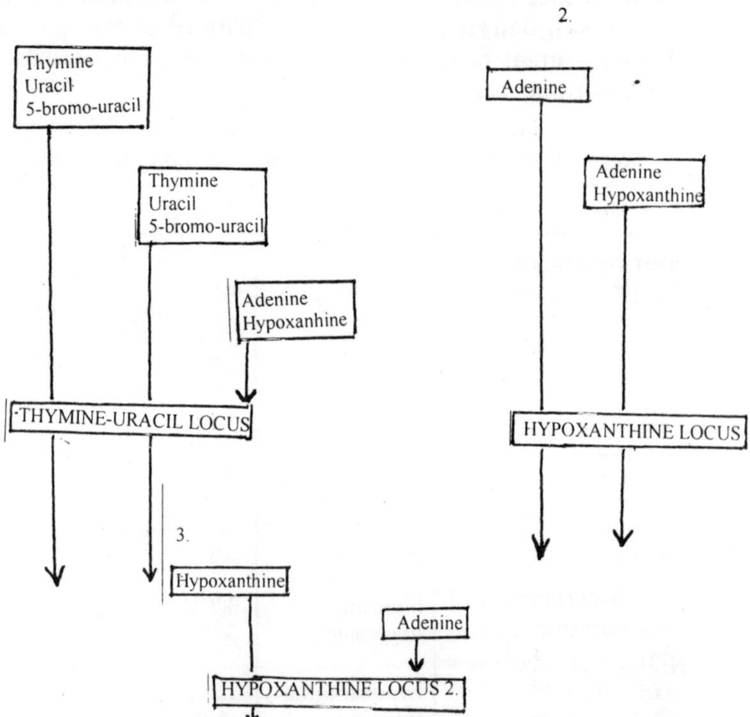

Fig. 2.8 : Proposed model for the transport of purines and pyrimidines in the cestode *Hymenolepis diminuta.* Where the arrows pass through the transport locus actual entry of the substrate occurs; arrows that do not pass through the locus indicate the occurrence of non-productive binding with no transport. Some solutes appear to bind to a single locus at two different sites.

of these molecules is highly complex, compared with the other systems that we have discussed, in that there are at least three separate systems operating and two of these involve multiple binding sites. Additionally, inhibitory or stimulatory compounds can bind non-productively to binding site, but are not themselves transported. Pappas, Uglem and Read proposed a model to account for the data obtained. The thymine-uracil system transports thymine and uracil via an allosteric interaction and is thought to comprise a minimum of two binding sites. Non-productive binding of adenine and hypoxanthine occurs at this site. There are two additional sites concerned with the transport of hypoxanthine, designated locus 1 and locus 2. Hypoxanthine, guanine and adenine are transported through locus 1 and adenine binds at two separate sites on the locus. Hypoxanthine is also transported via locus 2, which binds adenine non-productively. It is not clear why such a complex transporting system is found in *H. diminuta* and studies on purine and pyrimidine transport in other parasites would be of value in assessing general patterns.

Vitamin transport. Very little experimental data on the vitamin requirements of parasites have been collected, let alone data on the modes of vitamin absorption. The B_{12} requirement of the broad tapeworm of man (*Diphyllobothrium latum*) is well known, but the mechanism of entry has not been examined. Thiamine and riboflavin enter *H. diminuta* by mediated transported processes with narrow specificity, while pyridoxine and nicotinamide enter by simple diffusion.

Macromolecular transport (pinocytosis). The uptake of macromolecules, such as proteins, across the surfaces of parasites has been the subject of only limited investigation. We can generalized by recording that many of the parasitic Protozoa obtain macromolecular nutrients by endocytosis, whereas helminth parasites do not seem to favour this process. There is, for example, no evidence of the endocytotic acquisition of the macromolecules ferritin and horseradish peroxidase by *Hymenolepis diminuta,* though both of these molecules are adsorbed into the surface of the worm. Neither of the digeneans *Fasciola hepatica* and *Haematoloechus*

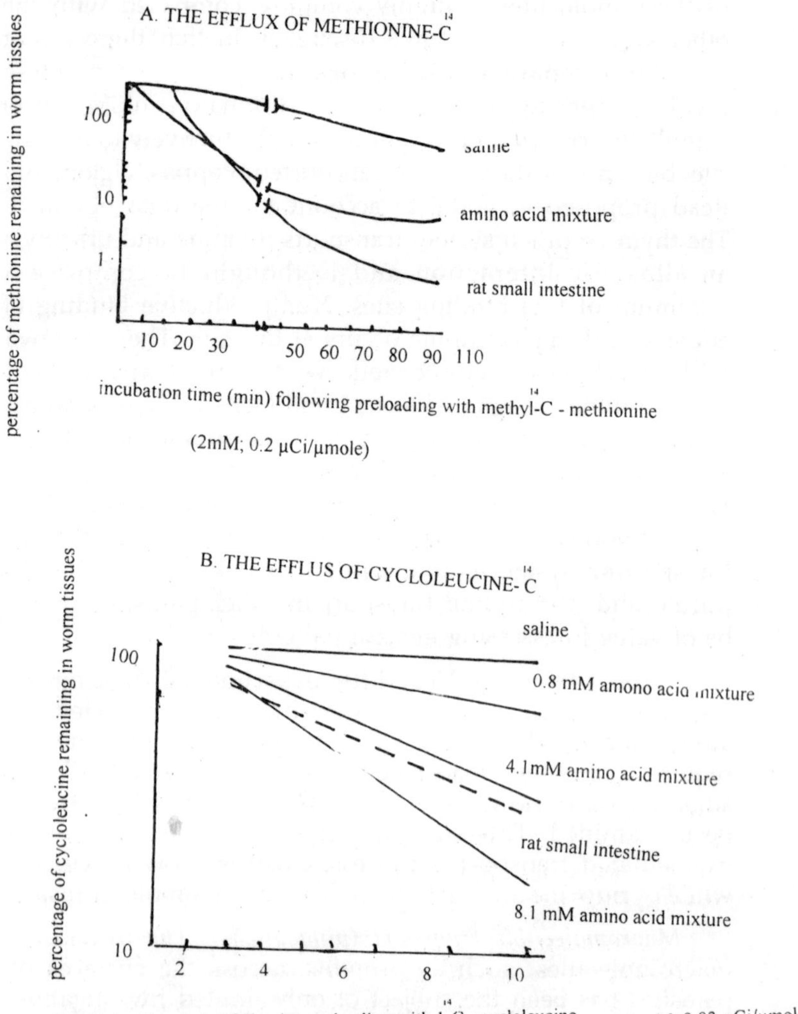

Fig. 2.9 : The kinetics of the *in vitro* and *in vivo* efflux of amino acids from the cestode *Hymenolepis diminuta*. In both of these experiments, tapeworms were incubated in radiolabelled amino acid and the subsequent efflux of the preloaded acid was determined either *in vitro* (in saline or mixtures of amino acids in the ratios found in the small intestine of the rat) or *in vivo* following surgical transplantation of the worms into the rat gut.

medioplexus absorb colloidal material across the tegument, but the latter species acquires ferritin by endocytosis across the caecal gastrodermis. *Schistosoma mansoni*, on the other hand, absorbs horseradish peroxidase by endocytosis across the tegument.

Endocytosis has been examined in a number of species of parasitic Protozoa. The malaria parasite (*Plasmodium* sp.) feeds on host haemoglobin during the intraerythrocytic phase of the life cycle. The haemoglobin is absorbed intact through the cytostome by a process often referred to as *phagotrophy* (as opposed to osmotrophy, which is the absorption of low molecular weight molecules by simple diffusion or mediated transport). Phagotrophy, or endocytosis, in *Plasmodium* involves the formation of vesicles in which the haemoglobin is both absorbed and digested; these vesicles (or vacuoles) contain digestive enzymes and, ultimately, become filled with haem, the end-product of haemoglobin digestion. The formation of endocytotic vesicles takes place exclusively at the cytostome in *Plasmodium*. Other sporozoan parasites (gregarines and *Babesia rhodaini*) also have a cytostome. Among the parasitic amoebae, *Entamoeba histolytica* employs a surface lysosome to obtain the macromolecular contents and, presumably, dissolved nutrients, of the host cell. A cytostome is present in one group of haemoflagellates, the Kinetoplastida, but not in the Salivaria. Intact protein molecules are ingested in vesicles formed at the base of the flagellar invagination in the kinetoplastid trypanosomes and the cytostome itself may be involved . In *Crithidia fasciculata*, ferritin is absorbed in cytostomal vesicles. In the salivarian trypanosomes, on the other hand, macromolelcules are absorbed only within vesicles formed in the flagellar invagination. Endocytosis may occur at localized regions of the body, such as the cytostome, when present, or over the entire surface of the parasite.

Bidirectional fluxes of nutrients

Movement of many nutrients, e.g. amino acids, can take place both into and out of parasites. The efflux of labelled methionine from *H. diminuta* is more rapid when worms are placed in a solution containing a mixture of amino acids resembling the proportions of amino acids found in the rat

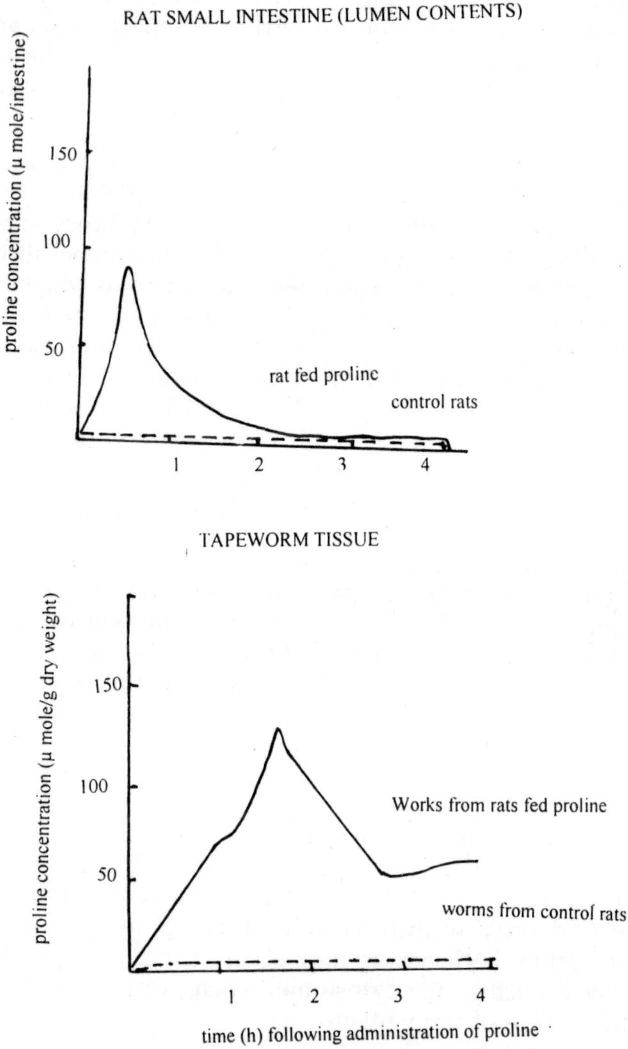

Fig. 2.10 :0 The differential handling of an experimentally-induced imbalance of a single imino acid (proline) by the rat and its tapeworms (*Hymenolepis diminuta*). Rats, infected with 30 worms each, were fed a single dose of 250 mg proline; the levels of proline in the rat gut and in the tapeworm tissue were determined by amino acid analysis at intervals up to 4h post-feeding.

gut, than when incubated in a simple saline solution; the rate of efflux into the mixture is comparable to the rate of loss of labelled methionine in worms that are surgically implanted into rats. This suggests the occurrences of counterflow mechanisms and demonstrates the mediated nature of the efflux process. Additional evidence for the dynamic balance that exists between the tapeworm and its host comes from the distribution of the radioactively labelled non-metabolizable amino acids, cycloleucine and µ-aminoisobutyric acid, between rat and parasite following their injection. Both these amino acids become distributed in a steady state between host and parasite, whereas amino acid imbalance *in vivo,* induced by feeding a single amino acid, is rapidly restored in the host but not in the worm (figure 2.10). This may indicate that tapeworms are deficient in regulatory mechanisms and therefore, to some extent, parasitize the homoiostatic mechanisms of their hosts. Despite this notion, there is evidence that amino acid imbalances may have little or no effect upon worm protein synthesis and overall growth. The notion of parasites existing in dynamic equilibrium with their hosts is an exciting one that has been sadly ignored.

Table 2.3. Kinetics of membrane transport in the digenean, *Schistosoma mansoni.*

Substrate	K_m*(mM)*	$V_{max}$$(\mu mol^{-1}\ h^{-1})$
Carbohydrates		
glucose	0.8	147
Galactose	0.2	10
2-deoxyglucose	1.7	270
Amino acids		
methionine	0.3-1.9	120-192
glycine	0.6-1.1	165-225
Proline	1.7-2.0	750
arginine	0.1-0.4	45-60
glutamic acid	0.3-0.5	75
tryptophan	0.5-1.0	90-180
Nucleic acid precursors		
hypoxanthine	0.9	45
adenosine	0.5	52
uridine	0.4	22
adenine	0.05	10

Surface enzymes in parasites

Enzymes bound to the surfaces of parasites may be of parasite origin (intrinsic) or derived from their hosts (extrinsic).

Intrinsic surface enzymes

Enzymes of parasite origin may play a role in digestion, as well as in penetration and migration through host tissues. We shall discuss the latter enzymes in a later chapter. Cestodes were long though to rely exclusively upon the digestive enzymes of their hosts to provide them with nutrient molecules in a state available for absorption, but recent work has shown that tapeworms may possess their own digestive enzymes bound to their surface. Phosphohydrolase activity is present in the tegument of *H. diminuta,* and its action permits the absorption of otherwise impermeable phosphate esters following hydrolysis, e.g. fructose 1, 6-diphosphate is hydrolysed at the surface to yield inorganic phosphate that can then be absorbed; fructose itself is unable to enter *H. diminuta*. The spatial proximity of the glucose transport system in *H. diminuta* to membrane-bound phosphohydrolases could confer a kinetic advantage to the absorption of carbohydrates. The same may be said of the ribonuclease activity associated with the tegument of *H. diminuta* and the hydrolysis of RNA to yield nucleosides that can then be acquired by the parasite.

Extrinsic surface enzymes

Ugolev coined the terms *membrane* or *contact digestion* to account for the enhancement of starch digestion by pancreatic ∝-amylase in the presence of isolated segments of vertebrate intestine. It was proposed that the enzyme became adsorbed on to the glycccalyx of the mucosal cell surface, in which condition amylolytic activity was increased. Contact digestion also occurs with tapeworms, since the activity of pancreatic ∝-amylase is also enhanced *in vitro* by the presence of living tapeworm tissue (*H. diminuta, H. microstoma* and *Moniezia expansa*). Enhancement does not occur if the tapeworms are killed with ethanol. None of these worms have intrinsic amylolytic activity and it is therefore assumed that , *in vivo,* enzymes of host origin may become bound to the glycocalyx of the worms, providing the spatial advantage of proximity of the products of enzyme activity to the worm surface. The acanthocephalan *Moniliformis dubius* can bind host amylase, but not trypsin, chymotrypsin or lipase.

Inhibition of host enzymes by parasites

Antienzymes, with an inhibitory effect upon their host's enzymes, are reported to be common amongst intestinal parasites. Antienzymes from ascarids are inhibitors of trypsin or pepsin. *Ascaris suum* is thought to possess an antitrypsin and an antichymotrypsin located in its cuticle, intestine and other tissues. *Hymenolepis diminuta* possesses a number of antienzymic properties, e.g. trypsin, β-chymotrypsin and ∝-chymotrypsin are irreversibly inactivated *in vitro* and pancreatic lipase undergoes a reversible inhibition. No antienzyme as such has been isolated from *H. diminuta* and it has been suggested that these enzyme inactivations take place within the glycocalyx of the worm. Their precise mechanisms, however, is unknown. While there are obvious advantages to a parasite that inhabits the upper intestine of vertebrates in having the ability to inhibit host digestive enzymes, the high rate of turnover of the parasite surface may render antienzymic properties redundant.

Chapter—3

Metabolism of Carbohydrates

Polysaccharide. An almost universal feature of endoparasitic organisms is their dependence upon anaerobic carbohydrate metabolism to obtain energy, regardless of the amount of environmental oxygen available. Accordingly, a great many parasites store polysaccharides which can be oxidised (though rarely to completion) to yield ATP. The most common polysaccharide reserve is glycogen, occurring in protozoans and helminths alike. Glycogen is composed exclusively of glucose molecules joined by α -1 4, and α -1, 6 linkages, the ratio of 1, 4:1, 6 bounds ranging from 12:1 to 15:1.

The glycogen content of parasites varies quantitatively according to species. Only 1% of the dry body weight of fish gill monogeneans is glycogen, whereas higher amounts are present in the tissues of internal parasites. Digeneans, for example, may store between 2% and 30% of their dry weight as glycogen, and cestodes tend to contain even greater quantities (20-60% of dry weight). Nematodes and acanthocephalans also contain relatively large amounts of stored glycogen, varying between 10% and 60% of their dry weight. Even within a single species, sexual dimorphism of stored glycogen content may occur, e.g. male schistosomes stores much more glycogen (14-30% of dry weight) than do female worms (3-5% of dry weight). The metabolic significance of this difference is unclear. Sexual difference in the quantities of stored glycogen is also a feature of several other parasites: *Macracanthorhynchus hirudinaceus, Moniliformis dubius* (Acanthocephala); *Ascaris lumbricoides, Ascaridia galli, Parascaris equorum* and *Angiostrongylus cantonensis* (Nematoda).

Utilization of glycogen as an energy store can be demonstrated readily by experimental deprivation of either carbohydrate or total food of the infected host, whereupon rapid depletion of stored glycogen within the parasite occurs. The chicken tapeworm, *Raillietina cesticillus* reduces its glycogen store from approximately 25% of its dry weight to 1.5% in just 24 hours if the host chicken is denied access to food. *Hymenolepis diminuta* shows a similarly dramatic glycogen loss if the host rat is starved of carbohydrate. Glycogenesis, the reverse process, takes place rapidly if experimentally starved worms are then given access, either *in vitro* or *in vivo,* to a utilizable glycogenic substrate (monosaccharide, pyruvate, citrate, succinate, glutamate, isoleucine, valine or glyoxylate, depending upon the parasite species).

Helminth parasites store glycogen in their parenchymal cells, but significant quantities also occur in the musculature. Generally speaking, however, determinations of the glycogen content of parasites are made on extracts of the whole body and no attempt is made to distinguish between the various tissues. A feature of glycogen determination, presumably pertaining to all animal tissues, is that the molecular weights of extracted glycogen vary quite widely according to the extraction procedures used, e.g. aqueous extraction of glycogen from *Fasciola hepatica* yields a broad spectrum of different molecular weight moieties, whereas most of the higher molecular weight glycogen is lost using potassium hydroxide as an extractant. Their terms lyo- and desmoglycogen, referred to in some studies, are now considered erroneous, the different forms representing artifacts of the extraction procedures.

Some of the parasitic Protozoa store glycogen, but a alternative polysaccharide reserve, amylopectin- a highly branched polymer occurs in many species.

Simple sugars. The presence of simple sugars has been detected in many parasites, but these tend to occur at rather low concentrations. Glucose, fructose, galactose, mannose, xylose and ribose are monosaccharides that occur most frequently. The disaccharide trehalose, most abundant in the eggs of *Ascaris lumbricoides,* is a common constituent of protozoan and helminth tissue. Nematodes may also contain glycosides, whose sugar component is a unique hexose, ascarylose (3, 6-dideoxy-L-arabino hexose). These glycosides, termed *ascarosides*, occur in the egg shells of a number of nematodes, including *Ascaris lumbricoides, Parascaris equorum, Ascaridia galli* and *Toxocara cati.*

Glycolysis

Glycolysis is a fundamental pathway of carbohydrate catabolism common to almost all animal tissues. It is sometimes referred to as the Embden-Meyerhof-Parnas pathway after those workers who, in the early decades of this century, described the breakdown of glucose to pyruvate under anaerobic conditions. Glycolysis is used by free-living animals as a relatively rapid means of obtaining energy, usually for muscular contraction, from stored glycogen. During this process an "oxygen debt" is accumulated which much be "repaid" under aerobic conditions, in order to oxidise the reduced products of glycolysis, such as lactate, whose accumulation becomes toxic. During repayment of the oxygen debt, part of the lactate is reoxidised to provide energy for glycogen synthesis from the remainder. Glycolysis, therefore, is a short-term pathway in free living animals, e.g. in teleost fish white muscle, which is anaerobic and can only be used for short bursts of rapid swimming. Parasites, by contrast, may rely on anaerobic metabolism to a much greater extent than most free living animals. Many internal parasites, particularly those inhabiting various regions of the vertebrate and invertebrate alimentary canal,

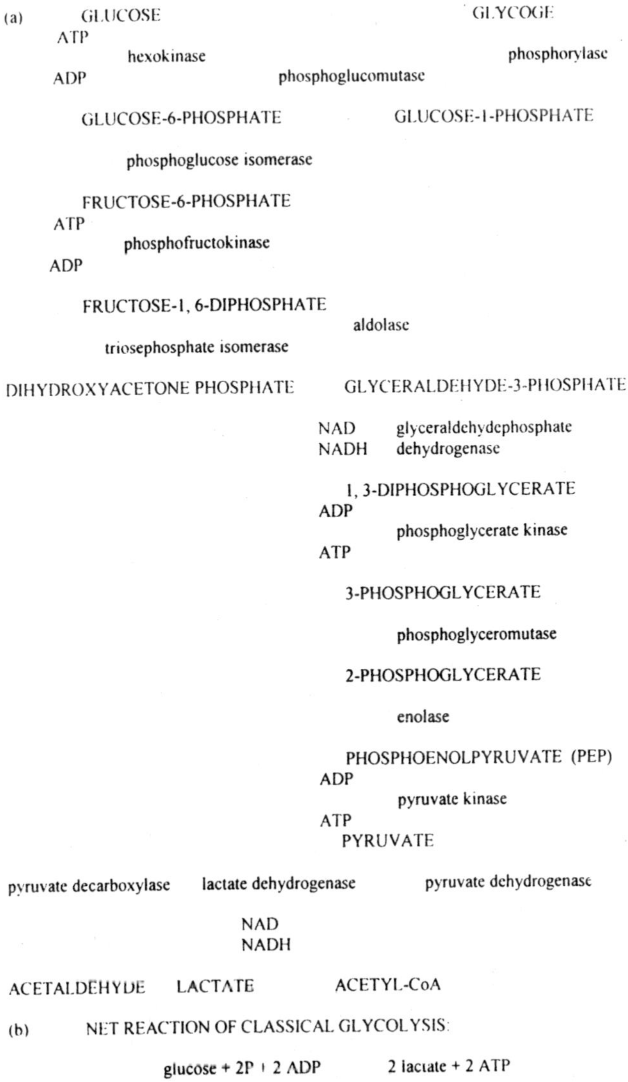

Fig. 3.1 : (*a*) the glycolytic pathway of carbohydrate catabolism. (b) net reaction of classical glycolysis.

live under conditions of greatly reduced oxygen tension. As a consequence we find that these parasites often rely on glycolysis as a major route of carbohydrate catabolism, rather than as an emergency source of energy. The net result of this is that these parasites only partially oxidise substrates such as hexose sugars, and excrete reduced energy-rich products, which, perhaps, may then be used by their hosts.

The glycolytic pathway has been examined in a wide variety of parasites, and, in most cases, a typical array of enzymes and intermediate metabolites is found i.e. parasite glycolysis appears to function as it does in other, non - parasitic, animals. However, in some parasites various enzymes of the glycolytic pathway have not been detected, but whether this alone is sufficient evidence to indicate that glycolysis is an unimportant source of energy in these parasites is a matter for conjecture.

End - products of carbohydrate catabolism

The typical end-product of glycolysis is lactic acid. The reduction of pyruvate to lactate serves to reoxidise the reduced cofactor NADH produced by glyceraldehyde-3-phosphate dehydrogenase. Despite this general statement, parasite catabolism is typified by the production of a wide variety of end-products-many more than are produced by vertebrates or micro-organisms.

Parasites that are homolactic fermenters, such as *Schistosoma mansoni, Brugia pahangi* and *Dipetalonema viteae,* excrete lactic acid almost exclusively, under both anaerobic and aerobic conditions. Many other parasite excrete volatile fatty acids-trypanosomes excrete acetate, and some ciliates excrete acetate, propionate and butyrate aerobically. Among the helminths examined *Fasciola hepatica, Moniliformis dubius, Ascaris lumbricoides* and *Trichinella spiralis* excrete a variety of volatile fatty acids, including acetate, propionate, butyrate and, to a lesser extent, valerate, isovalerate, isobutyrate and caproate. The proportions of fatty acids produced varies with the prevailing conditions, e.g. *Fasciola* excretes less propionate

Table 3.1. The end products of carbohydrate catabolism in selected parasites.

Species	*Conditions*	*End-products*
Protozoa		
Crithidia fasciculata	aerobic	pyruvate, succinate, ethanol, CO_2
	anaerobic	lactate, succinate, ethano, CO_2
Trypanosoma brucei	aerobic	pyruvate, acetate, succinate, glycerol, CO_2
Trypanosoma lewisi	aerobic	lactate, pyruvate, acetate, succinate, glycerol, ethanol.
	anaerobic	lactate, pyruvate, acetate, succinate, ethanol.
Trypanosoma gambiense	aerobic	acetate, pyruvate, CO_2.
	anaerobic	lactate, pyruvate, acetate, succinate, CO_2
Plasmodium berghei	aerobic	lactate, pyruvate, acetate, succinate, CO_2.
Digenea		
Fasciola hepatica	aerobic	lactate, acetate, propionate, α-methylbutyrate, α-methylvalerate.
	anaerobic	lactate, acetate, propionate, α-methylbutyrate, α-methylvalerate, succinate.
Cestoda		
Hymenolepis diminuta	aerobic	lactate, acetate, succinate.
	anaerobic	lactate.
Acanthocephala		
Moniliformis dubius	aerobic	lactate, formate, acetate, propionate, α-methylbutyrate, succinate, ethanol.
Nematoda		
Ascaris lumbricoides	aerobic	lactate, formate, α-methylbutyrate α-methylvalerate, C6 acids, succinate.
	anaerobic	lactate, formate, acetate, propionate, α-methylbutyrate, succinate, acetone.

and proportionally more acetate with increasing oxygen tension. Ethanol is an end-product of carbohydrate catabolism in some parasites (larval scolices of *Echinococcus granulosus*, larval and adult *Taenia taeniaeformis* and

Moniliformis dubius.) This is reminiscent of alcoholic fermentation found in the yeasts.

Glycolytic enzymes of parasites

There is insufficient space to consider in detail the characteristics of parasite glycolytic enzymes and readers are directed to the comprehensive account by Von Brand (1973) for further reading. Suffice it here to indicate areas of particular interest concerning parasite glycolysis.

Hexokinases are a group of enzymes concerned with phosphorylating hexoses and thereby allowing their entry to the glycolytic pathway. *Schistosoma mansoni* and *Echinococcus granulosus* contain a number of different specific hexokinases, while trypanosomes, posses a single hexokinase with a broad spectrum of phosphorylating activity. In general, cestodes metabolize a limited number of sugars, usually only glucose and galactose. Interestingly, few parasites metabolize sugars other than monosaccharides, implying the absence of disaccharidases from parasite tissue. Noteable exceptions to this generalization are *Moniliformis bubius* and *Ascaris lumbricoides*. It should be recognized that the inability of a particular parasite to utilize a given sugar may not necessarily reflect the absence of the appropriate hexokinase, e.g. while many parasitic flagellates and some helminths metabolize fructose, the cestode *Hymenolepis diminuta* cannot, due to its impermeability to that hexose rather than any enzyme deficiency.

Partial purification of parasite glycolytic enzymes reveals the occurrence of several forms (isoenzymes) of a single enzyme. Lactate dehydrogenase (converting pyruvate to lactate) may occur in from one to five electrophoretically different forms depending upon parasite species. Although it is often difficult to ascribe a function for several physico-chemically different enzymes that catalyze the same chemical conversion, it may be that these isoenzymes variously functions at different stage of the parasite life-cycle, or only in certain tissues.

Carbon dioxide fixation

Many parasites, particularly gut dwellers, are exposed to environmental conditions with a characteristically high CO_2 tension. It is no surprise therefore, to find that CO_2 can be fixed via certain well described metabolic routes, serving both a biosynthetic and energetic function. The presence of radiocarbon derived from $C^{14}O_2$, in both the end-products of carbohydrate catabolism and in stored polysaccharide, indicates the overall direction of CO_2 fixation. Current research has brought to light some of the specific pathways used for CO_2 fixation by parasites. These involve either "malic enzyme" (ME) or phosphoenolpyruvate carboxykinase (PEPCK).

"Malic enzyme" occurs in a variety of forms, one of which (L-malate: NAD oxidoreductase (decarboxylating) E.C. 1.1.1.39) is found exclusively in *Ascaris lumbricoides* and catalyses the carboxylation of pyruvate to form malate. This enzyme is NAD -dependent and the reaction it catalyzed is reversible . Another form of the enzyme (L-malate: NADP oxidoreductase (decarboxylating) E.C. 1.1.1.40), which is found in many animal tissues, is dependent upon the cofactor NADP. Phosphoenolpyruvate carboxykinase, now usually referred to as PEP carboxylase, carboxylates pyruvate to form oxaloacetate and is dependent upon either guanosine triphosphate (GTP) or inosine triphosphate (ITP). The distribution of ME and PEPCK in parasites is shown in table 3.2, which effectively illustrates the paucity of data currently available. In fact, the details of the mechanisms of CO_2 fixation are derived from studies on a very small number of parasites indeed.

Radiocarbon from $C^{14}O_2$ can be detected in both glycogen and end-products of catabolism, such as pyruvate and succinate, in isolated *Ascaris* muscle. The reaction scheme places "malic enzyme" within the mitochondria, and PEP carboxykinase in the cytosol of the cell. Malate is formed from oxaloacetate (catalyzed by NADH-dependent malate dehydrogenase) and diffuses across the mitochondrial membrane, where a dismulating reaction is thought to occur, cleaving malate to pyruvate and fumarate, by the respective actions of "malic enzyme" and fumarate hydratase. The NADH generated by 'malic enzyme' is reoxidized during the reduction of fumarate to succinate at which step one mole of ATP is produced per mole of fumarate reduced. Typically, parasites capable of CO_2 fixation excrete succinate and pyruvate, but each of these may be further metabolized, succinate to volatile fatty acids and pyruvate to lactate, alanine or volatile fatty acids.

Table 3.2. Occurrence of "malic enzyme" (ME) and PEP carboxykinase (PEPCK) in some helminth parasites.

Parasite species	*ME*	*PEPCK*
Digenea		
Dicrocoelium dendriticum	?	+
Fasciola hepatica	+	+
Schistosoma mansoni	?	+
Cestoda		
Echinococcus granulosus	+	+
Hymenolepis diminuta	+	+
Moniezia expansa	+	+
Acanthocephala		
Moniliformis dubius	+(?)	+
Nematoda		
Ascaris lumbricoides	+	+
Dictyocaulus viviparus	-	+
Haemonchus contortus	+	+
Nippostrongylus brasiliensis	-	+
Syphacea muris	-	+
Trichinella spiralis	+	+

\+ = present

– = not detected

(?) = not determined or uncertain presence

The cestode *Hymenolepis diminuta* excretes succinate and there is good evidence that, like *Ascaris*, it is also fixes CO_2 via "malic enzyme' and PEP carboxykinase, to yield succinate and synthesize limited amounts of glycogen. The reactions of CO_2 fixation in *Hymenolepis* are essentially identical to those of *Ascaris,* with the exception of the occurrence of a transhydrogenase, responsible for reducing NAD in the presence of NADPH in the former species. This is essential for redox balance, since the "malic enzyme" of *Hymenolepis* is NADP- dependent while that of *Ascaris* is NAD-dependent.

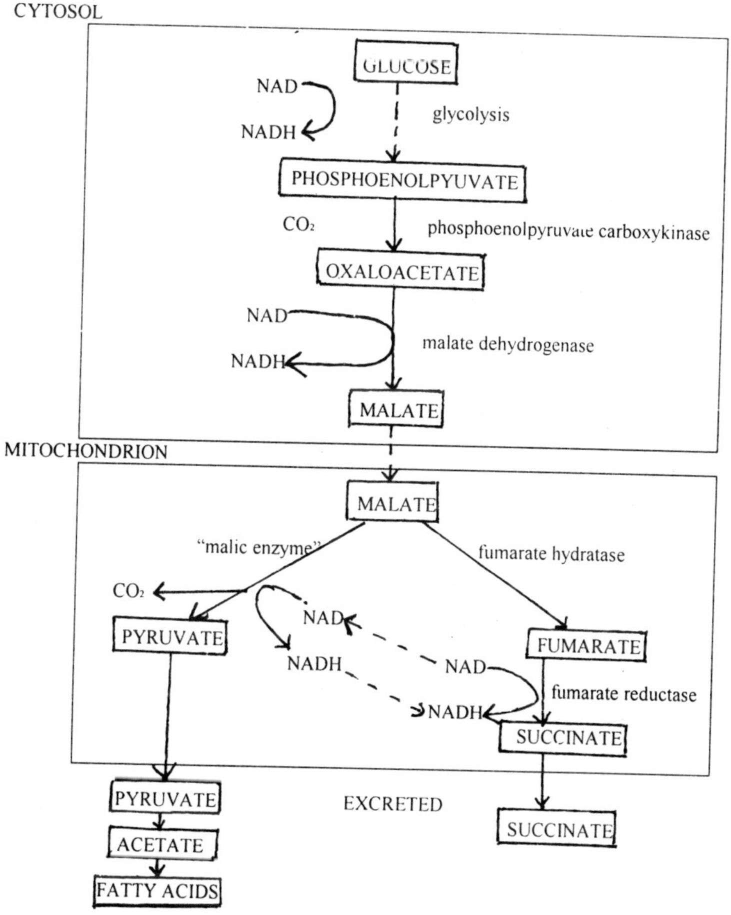

Fig. 3.2 : The roles of "malic enzyme' (ME) and phosphoenolpyruvate carboxykinase (PEPCK) in CO_2 fixation by the nematode *Ascaris lumbricoides*.

The advantage of these mixed fermentations that occur is many parasites may be considered as threefold. First, the decarboxylations of succinate to propionate, and pyruvate to acetate will yield potential metabolic energy, since both propionyl-CoA and acetyl-CoA are high energy compounds. Secondly, energy budgets are such that homolactic fermentations, such as occur in *Schistosoma,* yield only 2 moles of ATP per mole of glucose oxidised; this yield increases by one mole of ATP if CO_2 is fixed, as in *Ascaris* and *Hymenolepis*. Also 2 moles of ATP will be obtained if pyruvate and succinate are decarboxylated to the appropriate end-product. Mixed fermentations generate and utilize reducing equivalents (NAD↔NADS↔, NADP↔NADPH) allowing for some degree of flexibility in balancing reactions coupled to redox potentials. The third possible advantage of mixed fermentations is related to volatile fatty acid excretion; these end-products are likely to be less damaging to the host tissues due to their low pKa values.

Table 3.3. Ratios of the activities of pyruvatekinase (PK) to phosphoenolpyruvate carboxykinase (PEPCK) in selected helminths.

Parasite	*PK : PEPCK*	*Major end-product*
Digenea		
Dicrocoelium dendriticum	0.05	lactate, propionate, succinate, acetate
Schistosoma mansoni (adult) male	5.0	Lactate
female	9.7	lactate
Cestoda		
Hymenolepis diminuta (adult)	0.18	succinate, lactate
Nematoda		
Ascaris lumbricoides (adult)	0.04	succinate, volatile fatty acids
Nippostrongylus brasiliensis (adult)	1.86	succinate, lactate
Dirofilaria immitis (adult)	10.6	lactate
Acanthocephala		
Moniliformis dubius (adult)	0.34	lactate, succinate
(larvae)		lactate, succinate

Regulation of carbohydrate catabolism

The ratio of the activities of two key enzymes in carbohydrate catabolism, pyruvate kinase (PK) and PEP carboxykinase, is responsible, in part, for the nature of the metabolic end-products excreted. High PK activity will lead to the formation of lactate from PEP and high PEPCK activity will lead to the production of succinate or its metabolites. Here we can see that PEP forms a major branch point in carbohydrate metabolism. As expected, homolactic fermenters, like schistosomes, have a high ratio of these enzymes, while succinate producers have a much lower value. One current hypothesis proposes that control of the direction of metabolism from PEP to either succinate or lactate, in gut parasites, is mediated through the partial pressure of CO_2 in the host's intestinal tissue.

The tricarboxylic acid cycle.

Complete oxidation of carbohydrates, and other metabolites, such as amino acids and fatty acids, is carried out aerobically via the Tricarboxylic Acid (TCA) Cycle, also referred to as the Krebs or Citric Acid Cycle. The cycle functions both catabolically to yield ATP, and biosynthetically as a fundamental source of carbohydrates, proteins and lipids.

The classical methodology for examination of the TAC cycle (isolation and purification of the component enzymes, isolation and identification of intermediate products, radiocarbon distribution in intermediates of the cycle and in end-products, and the use of specific inhibitors and stimulatory molecules) has been applied to many parasites tissues. Nevertheless, the data available on the TCA cycle in parasites are incomplete and restricted to a small number of species. We can, however, tentatively make some general statements. Some parasites, living under conditions of relatively high oxygen tension, may possess only some of the cycle functional cycle, while others appear to possess only some of the cycle enzymes and intermediates. A small

(a)

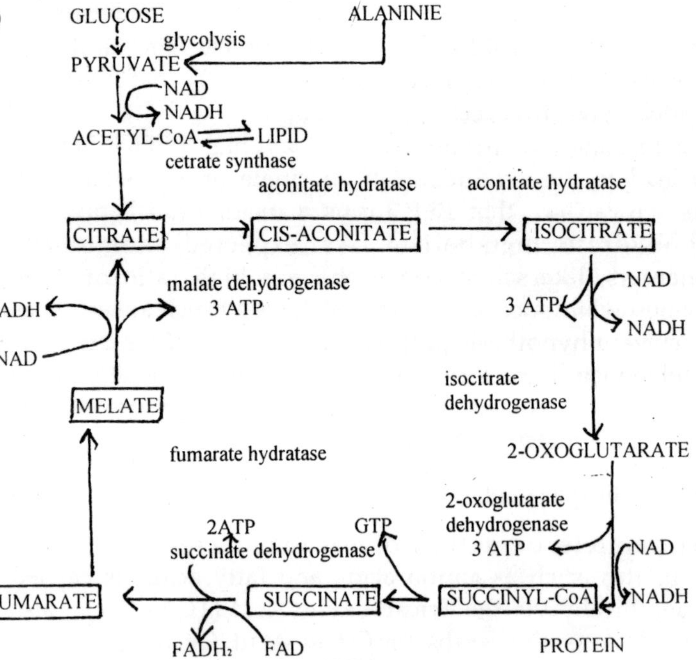

(b) NET REACTION OF THE CYCLE:

Acetyl-CoA+3NAD+FAD+GDP+P+2H2O-2CO2+CoASH+3NADH+FADH2+GTP

11 moles of ATP are produced by the TCA cycle when it leads to electron transport and oxidative phosphorylation.

Fige. 3.3 : (a) The tricarboxylic acid cycle (citric acid or Krebs cycle). (b) Net reaction of the cycle. 11 moles of ATP are produced by the tricarboxylic acid cycle when it leads to electron transport and oxidative phosphorylation.

number of parasites, on the other hand, do not have a demonstrably functional TCA cycle at all. Because our knowledge of the TCA cycle in a sufficient number of different parasites is lacking, it is not possible to make meaningful generalizations on the presence or absence of the cycle relative to parasite habitat. Even the implication of a non-functional cycle may be due to the failure to demonstrate a particular enzyme rather than to its actual absence.

Organisms with a fully operational TCA cycle include avian malarial parasites, some species of trypanosomes (particularly forms grown in culture) and *Trichomonas gallinae,* among the Protozoa, and larval schistosomes, along with many nematodes, in the helminths. The absence of the cycle has been established unequivocally in *Hymenolepis diminuta, Moniezia expansa, Fasciola hepatica* and *Ascaris lumbricoides* (adults but not eggs). In these species there can be little doubt that the cycle makes no significant contribution to the energy budget of the parasite, though it has been suggested that a vestigial pathway could have a biosynthetic function. Even though the complete cycle is absent in these parasites, some species (*Hymenolepis* and *Ascaris*) utilize part of the cycle in the reverse direction, to reduce malate to fumarate and thence to succinate; this would appear to be a conservative adaptation to the special conditions of partial anaerobiosis experienced, particularly, by gut-dwelling parasites. It is worth nothing that similar biochemical adaptations may occur in free-living animals whose environment is characterized by periods of oxygen depletion, e.g. intertidal molluscs.

The role of oxygen in parasite energy metabolism

Our discussion of energy metabolism, os far, leads us to the inevitable conclusion that many parasites obtain their energy from anaerobic pathways; such parasites tend to inhabit regions of low oxygen tension, such as the lumen of the vertebrate alimentary canal. Yet almost all parasite species do consume oxygen. Furthermore, metabolic end-products frequently vary according to whether aerobic or anaerobic conditions prevail. This, then, raises the important question of the role of oxygen in organisms whose metabolism is primarily anaerobic. Some oxygen will undoubtedly be used biosynthetically, e.g. in egg-shell synthesis, nevertheless most workers are agreed that oxygen also plays a significant role in the energy metabolism of many parasites.

The key questions would appear to be:

1. What are the mechanisms of oxygen uptake?
2. Does oxygen uptake lead to the production of ATP?

3. If this latter process (i.e. oxidative phosphorylation) does occur, what overall contribution does it make ot the energy budget of the parasite?

Some of the better studied large helminths are unusual in that they are oxygen conformers, the rate of oxygen uptake depending upon its partial pressure. Hydrogen peroxide is often formed from oxidative metabolism, and cyanide, normally an inhibitor of oxidative respiration, is either without effect of it may even stimulate oxygen uptake. These observation point to the occurrence of flavoprotein oxidases rather than a conventional cytochrome system, but the latter, as we shall discuss shortly, does function in these worms.

We must not loose sight of the fact that much of the data obtained for parasite energy metabolism derives from experimental conditions containing high oxygen tensions, such as those of air (160mm Hg)—the parasite in its natural habitat may be confronted with much lower oxygen tensions than this.

Considerable controversy surrounds the role of oxygen in parasite energy metabolism. One view is that many of the gut-dwelling nematodes, and possibly other helminths, are in fact aerobic in their natural habitat, despite the apparent importance of glycolysis discussed earlier. The argument continues by nothing that the anaerobic metabolism determined *in vitro* may well be erroneous, due, in part, to the fact that worms removed from their hosts may rapidly become moribund, and data are therefore obtained from unhealthy rather than "normal" tissue. Unfortunately, satisfactory procedures for the examination of parasite metabolism *in vivo* have yet to be developed. Nevertheless, while Smith's (1969) cogent line of argument urges caution in the interpretation of metabolic data obtained from *in vitro* studies and should not be ignored, much of the current evidence strongly suggests that helminth parasites rely to varying extents on anaerobic metabolism, though many species suitable environmental conditions. Very few parasites are actually intolerant of oxygen; some rumen ciliates, flagellates from the gut of termites, several trichomonad species and adult *Ascaris* appear to find high oxygen tensions inimical have the capability to switch to normal aerobic metabolism under.

Pasteur and Crabtree effects

In most animal tissues that degrade glucose to pyruvate via glycolysis, the rate of fermentation is more rapid anaerobically than aerobically. This inhibition of glycolysis by oxygen is termed the *Pasteur Effect.* Its converse is the *Crabtree Effect,* whereby oxygen consumption is inhibited by glucose.

The fact that glycolysis in many helminth parasites in the presence of high oxygen tensions (i.e. limited or no Pasteur Effect) lends support of the view that anaerobiosis is indeed fundamental to their energy metabolism. A small number of species do exhibit a Pasteur Effect including larval *Schistosoma mansoni,* some larval nematodes such as *Ascaris lumbricoides, Eustrongyloides ignotus, Litomosoides carinii* and *Nippostrongylus brasiliensis.* The Crabtree Effect is regarded as a likely occurrence in those parasites with a demonstrably active glycolytic pathway.

Electron transport and terminal oxidations

Oxidative phosphorylation (aerobic ATP synthesis) takes place within mitochondria by means of a sequence of electron acceptors and donors (electron transport). The sequence commences at NAD, and ends with oxygen, which acts as the final electron acceptor. Between these two lie a chain flavoproteins, including flavin mononucleotide (FMN) and flavin adenine-dinucleotide (FAD), and the cytochromes. The latter, in mammals, are designated b, c_1 c, a and a_3 according to their absorption characteristics rather than their sequence in the cytochrome chain. Interestingly, the pioneering studies of Keilin (1925) on animal respiratory systems and the discovery of the cytochromes included observations of the nematode *Ascaris*. Quinone coenzymes, coenzyme Q (CoQ) or ubiquinone (UQ), are also characteristic components of respiratory systems. A series of enzymes, flavoprotein and cytochrome oxidases, catalyze the transfer of electrons along the respiratory chain.

The ever-growing literature on electron transfer and terminal oxidation in parasites in complicated, and space available does not permit a detailed account here.

The cytochrome system can operate at oxygen tensions as low as 5 mm Hg and, it is, therefore, possible for an active system to function in many parasites, even vertebrate gut inhabitants. In the latter, as we have already discussed,

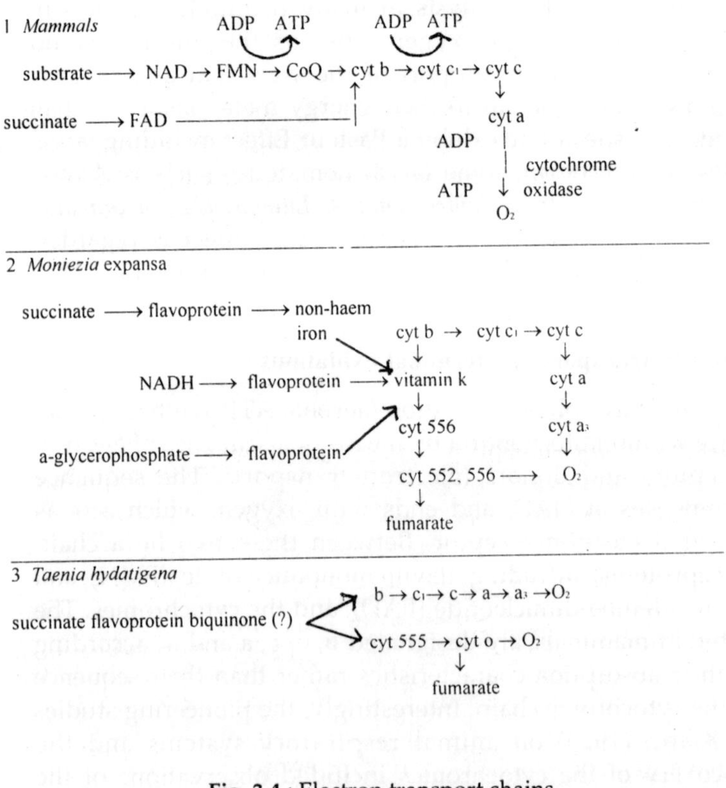

Fig. 3.4 : Electron transport chains.

CO_2 fixation and the reduction of fumarate to succinate by fumarate reductase, along with the excretion of a variety of reduced end-products, is an important feature of metabolism. Nevertheless, helminths such as *Fasciola hepatica, Hymenolepis diminuta, Moniliformis dubius* and *Ascaris lumbricoides* possess a functional cytochrome chain. It differs from the classical mammalian cytochrome system by being branched and possessing multiple terminal oxidases. One branch resembles the mammalian system by terminating in cytochrome a_3 and cytochrome oxidase, while other branches may terminated in cytochrome o (*Fasciola* and *Moniezia*) or, in addition, cytochrome a_1 (*Ascaris*)- the latter may be due to bacterial contamination, however. These parasite cytochrome systems are capable of oxidative phosphorylation apparently only through the mammalian-type branch; none occurs through the branch terminating in cytochrome o. The mechanisms responsible for controlling the direction of the flow through such branched cytochrome systems remains to be elucidated. Cytochrome o may be the source of hydrogen peroxide in some helminths and number of peroxidases have been identified. Hydrogen peroxide may, additionally, be produced by other cytochrome oxidases or by flavoprotein oxidases.

The P:O ratio, a measure of oxidative phosphorylation efficiency, is three for the classical mammalian cytochrome chain, but it is reduced if one of the alternative parasite pathways is used. P:O ratio of one will be obtained if NADH is oxidized via cytochrome o. Branched cytochrome chains are not unique to helminth parasites and they are also found in many micro-organisms as well as parasitic Protozoa.

The succinoxidase system of *Ascaris,* involving succinic dehydrogenase and a terminal respiratory chain, has been examined in some detail. This nematode contains the classical mammalian-type cytochrome chain but produces hydrogen peroxide from the oxidation of succinate. Furthermore, cytochrome inhibitors such as azide, cyanide and Antimycin A, have little or no effect upon respiration. If the cytochrome o system of *Ascaris* plays a role in the reduction of fumarate to succinate, while at the same time the mammalian-type chain can operate normally under certain conditions, then respiration in this helminth is typical of that of a facultative, rather than obligate, anaerobe. This situation may well pertain to other helminths in which there are both normal cytochrome chains and alternative branched pathways.

Cytochromes and terminal oxidations have been examined in a limited number of protozoan parasites, with trypanosomes, trichomonads, and malarial parasites attracting the greatest attention. The cytochrome systems of some trypanosome species are sensitive to azide and cyanide (e.g. *Trypanosoma cruzi*), whereas partial or complete insensitivity to these inhibitors of the classical cytochrome system is demonstrable in other species (*T. vivax, T. congolense, T. rhodesiense*). One of the more interesting features of trypanosome electron transport systems resides in the operation of different systems at various stages of the parasite life-cycle. The applies particularly to the African trypanosomes (*T. brucei* and *T. rhodesiense*) which have morphologically and biochemically distinct forms living in

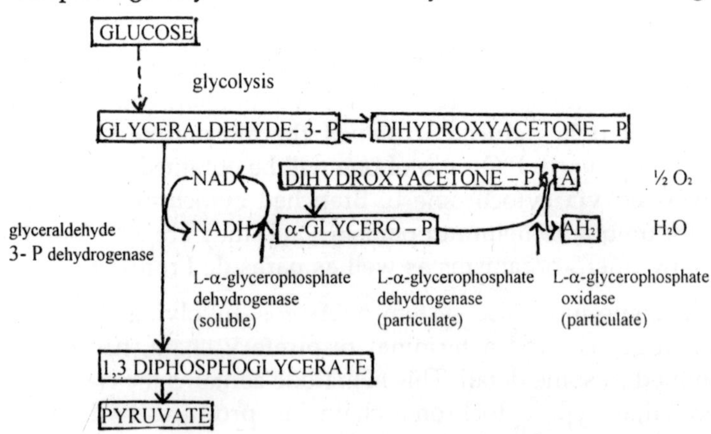

Fig. 3.5 : Proposed L-a-glycerophosphate oxidase pathway of trypanosomes (from Hill, 1976).
* = phosphate, A = unknown component

the vertebrate blood-stream and the insect midgut-the culture form of the parasites morphologically resembles the insect midgut—the culture form of the parasites morphologically resembles the insect midgut form. The bloodstream trypanosomes possess an a-glycerophosphate oxidase system, oxidising L-a-glycerophosphate to dihydroxyacetone phosphate, using oxygen as the terminal acceptor (figure 3.5). The prominent feature of respiration in culture forms is the partial sensitivity to cyanide. The L-a-glycerophosphate oxidase system may account for as much as 95% of respiration in bloodstream trypanosomes, whereas additional oxidases are produced on transformation of bloodstream forms in culture. These culture-forms oxidases have a higher affinity for oxygen than the a-glycerophosphate oxidase system, a feature possibly related to the lower oxygen tension of the insect midget compared with the mammalian bloodstream.

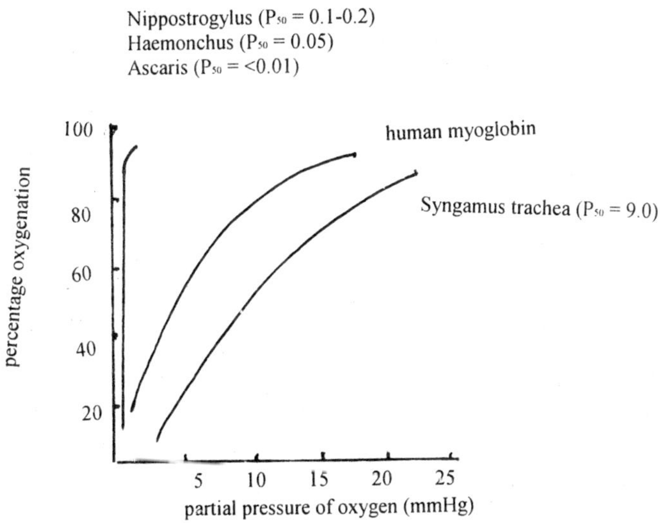

Fig. 3.6 : Oxygen affinities of some nematode haemoglobins. The values for P_{50} are the pressures of oxygen required to give 50% saturation of the haemoglobin. The figures are for *Ascaris* bodywall Hb and *Haemonchus contortus*, *Nippostrongylus brasiliensis* and *Syngamus trachea* perienteric (pseudocoelomic) fluid Hb.

Respiratory pigments in parasites

Haemoglobin occurs in a small number of parasites, notably digeneans and nematodes. Interesting recent studies on the haemoglobins in animal-parasitic nematodes have revealed that the presence of these respiratory pigments probably allows the parasite to inhibit regions of low oxygen tension and yet still obtain oxygen for use in certain tissues rather than for the whole animal. *Ascaris* contains two distinct types of haemoglobin; one form of the pigment is found in the perienteric fluid and serves as a source of haematin, while a second form is found in the body wall and is probably truly respiratory in function. Characteristically, the respiratory pigments of many nematode parasites (*Ascaris, Haemonchus* and *Nippostrongylus*) possess a high affinity for oxygen (figure 3.6). By contrast some species, such as *Syngamus trachea* and perhaps *Camallanus trispinosus,* contain haemoglobin with a low affinity for oxygen, and this is thought to be related to the relatively high oxygen tensions of the environments inhabited by these parasites, e.g. the trachea of birds in the case of *Syngamus*.

We may assume that parasites with respiratory pigments that have a high affinity for oxygen possess an adaptive advantage for life in habitats of low oxygen tension, such as the vertebrate alimentary canal. However little information is available on the release of oxygen from such haemoglobins. Partial deoxygenation of some nematode haemoglobins has been demonstrated *in vitro* in conditions which might limit oxygen consumption, but no data are available for the release of oxygen *in vivo*. Studies on the pharyngeal haemoglobin of a free-living marine nematode, *Enoplus brevis,* indicate that deoxygenation of haemoglobin will take place both at low oxygen tensions and during periods of activity by the worm—in this way the respiratory pigment could guarantee oxygen release at the tissues in a time of need. Clearly more information must be gathered on the respiratory pigments of parasitic animals before any substantial generalizations can be made.

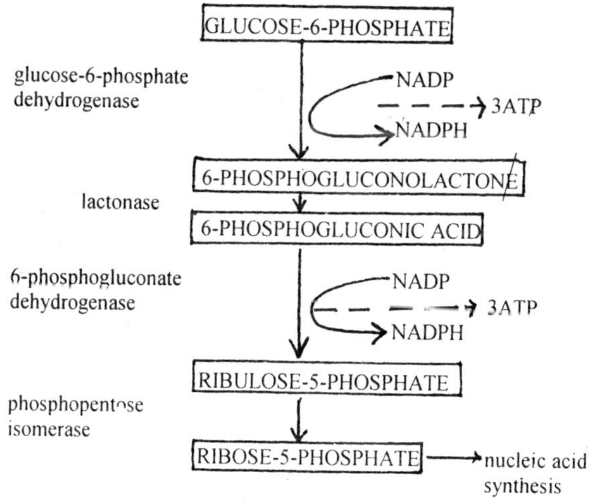

Fig. 3.7 : Outline of the pentose phosphate pathway.

Pentose-phosphate pathway in parasites

This pathway is of relatively minor significance in carbohydrate catabolism. Hexoses are converted via a series of reactions to pentoses, with a net yield of two moles of NADPH, and hence six moles of ATP, per mole of glucose metabolized. The biosynthetic nature of the pathway is probably more important, particularly with regard to nucleic acid synthesis (figure 3.7).

Parasites that possess a functional pentose phosphate pathway include *Trypanosoma cruzi, Plasmodium berghei, Echinococcus granulosus* and *Ascaris lumbricoides*. Unfortunately there has been little attempt to verify the presence of the pathway in other parasites and its significance to the parasitic mode of life remains obscure.

The glyoxylate pathway

Many micro-organisms and higher plants use the gloxylate cycle for carbohydrate synthesis from two-carbon precursors, notably from lipid via acetyl-CoA. The key enzymes of the pathway, isocitrate lyase and malate synthetase, are not found in higher animals. As far as parasites are concerned, this cycle is known to operate in only two species, *Fasciola hepatica* (adult) and *Ascaris lumbricoides* (eggs), though complete evidence for a functional cycle in the former species is lacking.

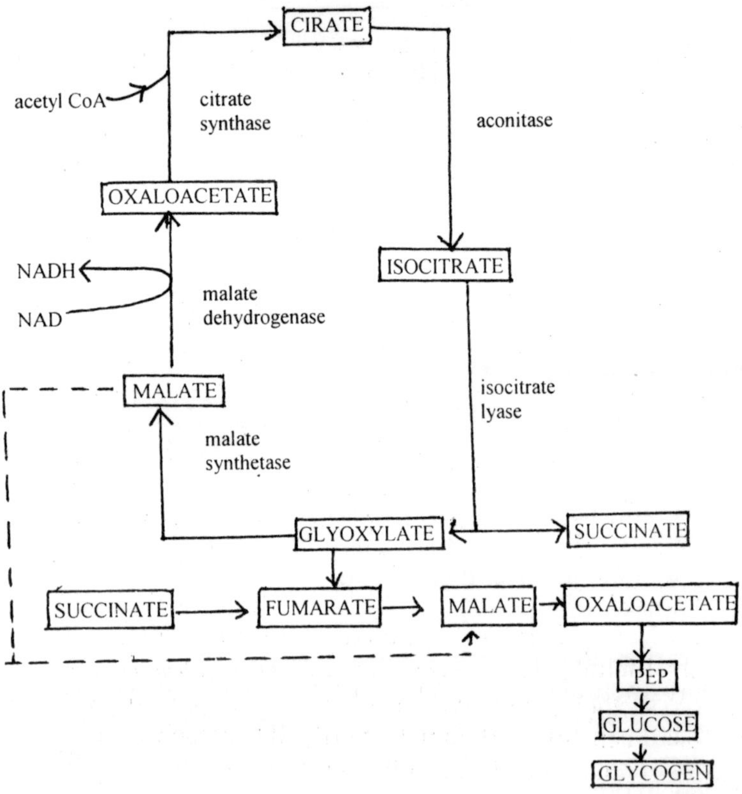

Fig. 3.8 : The glyoxylate cycle.

Chapter—4

A Metabolism of Proteins, Lipids and Nucleic Acids

Although a detailed account of the b' chemistry of proteins, lipids and nucleic acids is outside the scope of this book, it is nevertheless considered essential that the student of parasitology should be aware of the general direction of much of modern research into parasite organisation at the macromolecular level. What must also be made clear is that the biochemistry of these fundamental macromolecules poorly understood when we compare the information available for parasites with that for bacteria and mammals.

The chapter provides a brief insight into the major features of macromolecular biochemistry of parasites. One of the basic interests in parasite biochemistry is the quest for outstanding differences in the essential biochemical pathways of parasites and their hosts. Such differences could then be exploited chemotherapeutically and serve to eradicate or minimize the effects of pathogenic parasites of man and his domestic animals.

Proteins

Protein composition

Proteins are composed of long chains of L-α-amino acids, each linked by a peptide bond. Only twenty of the known amino acids commonly occur in proteins, nevertheless protein structure shows considerable variation, both in amino acid

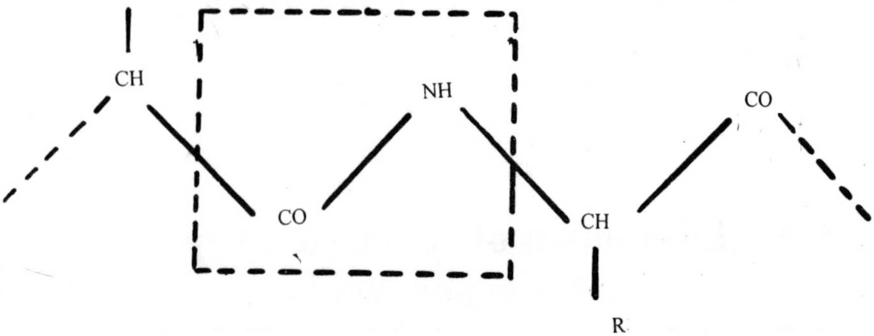

Fig. 4.1 : The peptide bond linking adjacent amino acids in a protein chain. R^1 and R^2 will be different amino acid residues, such a an aliphatic or an aromatic group, depending upon the particular amino acid.

sequence and in gross morphology. Proteins are involved in a number of fundamental functions in all tissues, i.e. in a structural and support role, in the formation of contractile elements, and as enzymes, hormones, antigens and antibodies. Some proteins, such as the glycoproteins and the lipoproteins, contain carbohydrates and lipids respectively, whereas others may have a metallic prosthetic group, e.g. iron in haemoglobin.

The total protein content of parasite tissues varies from around 20% of the dry weight to as much as 80%. Protein content is derived either from the total nitrogen content, multiplied by a factor of 6.25 (accuracy will depend upon the amounts of non-protein nitrogen present), or by direct chemical determination using the Lowrie colourimetric method.

Tissue proteins fall into two broad categories—soluble and insoluble, or particulate. The particulate proteins are associated with the cell membranes or with intracellular membranous structures. The soluble proteins include enzymes, hormones and antigens (parasites do not have antibodies). The bulk of the particulate proteins are involved in a structural and support role and include collagen, keratin and the scleroproteins. Collagen is a common constituent of

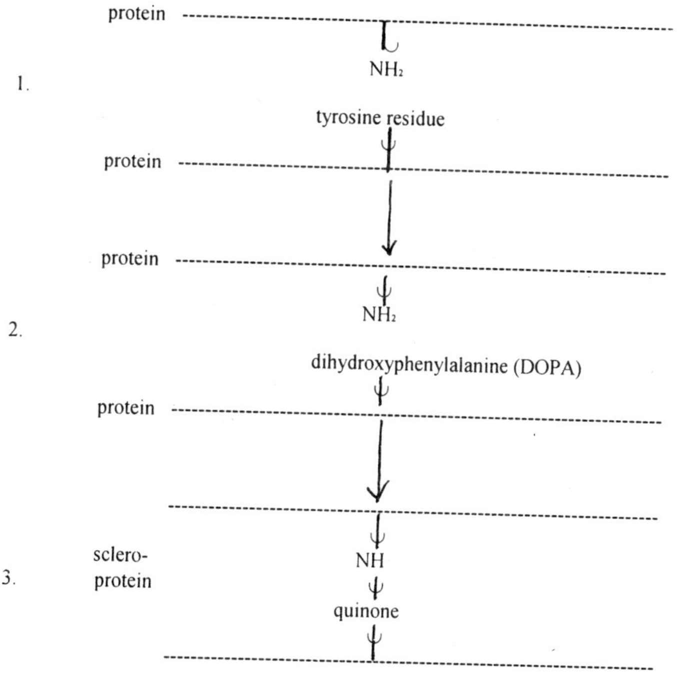

Fig. 4.2 : Simplified scheme of quinone tanning to form protective scleroproteins. Two adjacent protein chains are cross-linked when the O-quinone of one chain reacts with the free amino (-NH_2) residues on the other protein chain.

parasite tissues, occurring particularly in the nematode cuticle and egg shell, where it confers a high degree of mechanical strength and impermeability. Chemical analysis of *Ascaris* cuticular collagen shows it to differ from vertebrate collagens in having only small quantities of hydroxyproline and hydroxylysine; *Ascaris* muscle collagen, on the other hand, more resembles the vertebrate forms in its chemical composition. Keratin itself is not found in parasites but keratin-like proteins do occur in the tissues of some nematodes and acanthocephalans, as well as in the hooks of some cestode scolices. Scleroproteins are constituents of the walls of some parasitic protozoan spores but are more common as the quinonetanned form, sclerotin, found in the egg shells of digeneans and pseudophyllidean cestodes and in the metacercarial cyst walls of some digeneans. Quinone tanning is brought about by the action of a phenolase which converts an o-phenol to the equivalent o-quinone in the presence of oxygen. The o-quinone then cross links with an adjacent protein by reacting with free NH_2 groups to produce stable sclerotin, which is typically brown in colour. The precise mechanisms involved in quinone tanning are not understood, but it is widely used in the helminths as a means of protecting the otherwise fragile eggs.

Amino acid content

The amino acid content of parasite proteins and of the free pool of unbound acids shows no major departure from

Table 4.1 Amino acid biosynthesis in selected protozoans parasites.

Organism	*Source molecule*	*Amino acids synthesized*
Trypanosoma rhodesiense	carbohydrate	alanine, glycine, serine, aspartic acid, glutamic acid.
T. cruzi	serine	alanine; aspartic acid, glutamic acid, glycine, cysteine, threonine.
T. gambiense	glucose	alanine.
	alanine	cysteine, taurine,
	aspartate	serine.
	glitamic acid	cysteine, taurine, threonine.
	arginine	taurine, glutamic acid, proline.
T. brucei	proline	alanine, aspartic acid.
Crithidia fasciculata	methionine, cysteine	threonine.
Leishmania tarentolae	arginine	proline.
L. donovani	proline	arginine.
Plasmodium knowlesi	carbohydrate	alanine, aspartic acid, glutamic acid.
	methionine	cysteine.
	serine	methionine.
Entamoeba histolytica	carbohydrate	alanine, aspartic acid, glutamic acid.

that in non-parasitic animals. The requirement for amino acids by parasites has not been thoroughly examined, as a consequence of which, it is not possible to list the essential amino acids for any parasite species. Probably the greatest body of information no requirement for, and biosynthesis of amino acids concerns the parasitic Protozoa, particularly the trypanosomes and the malarial parasites. Despite this, even for the Protozoa, there is only limited information available at present. For instance, *Leishmania tarentolae* cannot synthesize arginine from proline, the former therefore being an essential requirement. On the other hand, *L. donovani* does carry out this synthesis. *Crithidia oncopelti* is unusual in that it will grow in culture with methionine as its only source of amino acid, but this property is thought to be due to the presence of the "bipolar body" in this parasite, which recent evidence suggests is probably an endosymbiotic bacterium. Other trypanosomatid parasites require much more complex mixtures of amino acids for growth in culture.

It has proved difficult to cultivate metazoan parasites *in vitro* and therefore out knowledge of essential amino acids and biosynthetic pathways remains scanty. The cestode *Hymenolepis diminuta* has been cultured from cysticercoid to adult in a highly complex medium, containing many amino acids. However, subsequent attempts to repeat Berntzen's (1961) culture procedures with this tapeworm have met with no success. Culture media that have been used to grow other species of helminth parasite usually contain crude preparations such as horse serum, yeast or liver extract, or protein hydrolysates. Each of these is provided with a full complement of amino acids and therefore provide on clue as to which amino acids are required and which can be synthesized from precursors.

In many parasites, transamination reactions, involving aminotrans-ferases, are common systems for the interconversion of amino acids, e.g. α-ketoglutarate to glutamic acid and pyruvate to alanine. For both these reactions a wide range of amino acids can act as amine donors. The α-ketoglutarate system operates in *Trypanosoma cruzi, L. donovani, Entodinium shaudini, Schistosoma japonicum, Fasciola hepatica, Hymenolepis citelli, H. diminuta, H nana* and *Ascaris lumbricoides*. Other keto acids can also act as transamination substrates, e.g. oxaloacetate is involved in amino acid biosynthesis in *H. diminuta, F. hepatica, S. japonicum* and *A. lumbricoides*.

1. Genral reaction:

$$\text{amino acid}_1 + \text{keto acid}_1 \rightleftharpoons \text{amino acid}_2 + \text{keto acid}_2$$

2. α-detoglutarate: glutamic acid transamination:

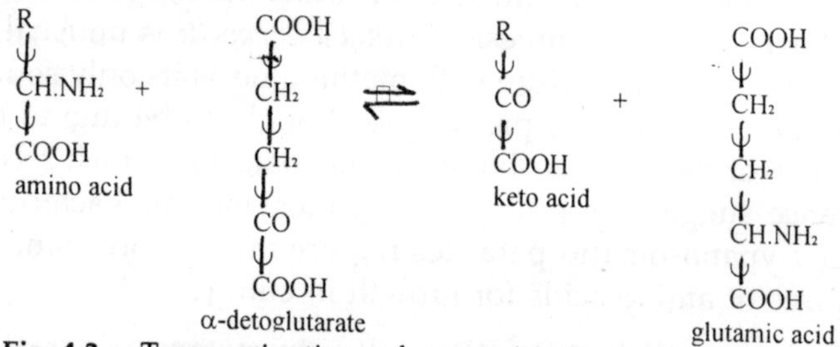

Fig. 4.3 : Transamination schemes. (1) General reaction. (2) α-ketoglutarate: glutamic acid transamination.

Protein synthesis

Studies on the synthesis of proteins by parasites have proceeded on two fronts, one involving the study of incorporation of radioactively-labelled precursors into the protein of intact organisms, the other utilizing cell-free ribosomal preparations. Once again we find that the majority of studies have utilized the parasitic Protozoa, especially the trypanosomes. These haemoflagellates, like mammals, possess both mitochondrial and cytoplasmic protein synthesis. The mitochondrial system, like that of bacteria, is sensitive to the protent inhibitor chloramphenicol, whereas the cytoplasmic system is inhibited by cycloheximide. This state of affairs holds true for the majority of protozoans studies, with the possible exception of mitochondrial protein synthesis in *Crithidia fasciculata*.

Ribosomal preparations of *Crithidia oncopelti*, containing 86S ribosomes, can activate a number of amino acids at pH

6.8 in the presence of Mg^{++}. The cytoplasmic ribosomes of *C. fasciculata,* ranging in size in from 83 to 88S, also depend upon Mg^{++} for amino acid activation. Additional studies on the malarial parasites, *Plasmodium berghei* and *P. knowlesi,* show that there is considerable similarity between their cytoplasmic protein synthesis and that of mammals.

Virtually nothing is known of the details of protein synthesis in helminth parasites. The should, however, prove to be an interesting area of research since many helminths must have a remarkably sigh rate of protein synthesis associated with their prolific level of egg production, i.g. the female schistosome sheds 10% of her body weight every twenty four hours in egg production alone. Many helminths are known to absorb and incorporate a wide variety of substrates into their tissue proteins, but in general such an approach tells us little about the mechanisms involved. There is , however, no reason to believe that the fundamental mechanisms of protein synthesis differ in parasitic and non-parasitic animals.

Protein and amino acid metabolism

Although the majority of internal parasites rely heavily upon carbohydrate metabolism as their basic source of energy, a few examples are known in which protein and amino acid metabolism achieve some degree of importance, particularly under emergency conditions. Incubation *in vitro,* in the absence of carbohydrate, leads to an increased production of ammonia in a number of parasites (*Trypanosoma cruzi, Trichomonas vaginalis,* and *Plasmodium gallinaceum*) suggesting active protein metabolism. Starved *Entodinium caudatum* may also use protein metabolism for energy production and *Leishmania tarentolae* shows an increased oxygen uptake in the presence of some amino acids. This latter organism is unusual in its requirement for the amino acids, proline, as a major energy source, oxidising it first to glutamate and finally to aniline and CO_2. Other protozoan parasites, including *Trypanosoma rhodesiense* and *Plasmodium lophurae,* can also utilize amino acids for energy metabolism and there is a suggestion that some nematode larvae metabolism and there is a suggestion that some nematode larvae metabolize proteins under emergency conditions.

The major end-products of nitrogen metabolism will be considered in chapter 5. Suffice it to say here that the majority of protozoans excrete ammonia while metazoan parasites excrete ammonia and urea to a greater or lesser extent. Many parasites excrete amino acids, but the mechanisms involved and the significance of this remain obscure.

Respiratory proteins

A wide variety of parasites contain haemoglobin. In the platyhelminths, haemoglobin occurs within the tissues. In the nematodes, it may also be found in the pseudocoelomic (perienteric) fluid— a situation reminiscent of the annelids. Parasite haemoglobins, particularly those of the nematodes, are characterized by their high affinity for oxygen, suggesting a true respiratory function. Despite this, it has been observed that the body-fluid haemoglobin of some nematodes is curiously reluctant to release its bound oxygen, and this has given rise to considerable speculation as to the real function of this family of proteins. Alternative functions have been postulated and they include : peroxide elimination via methaemoglobin; the formation of a pool from which other haem compounds might be synthesized; and photosensitivity) as in *Mermis subnigrescens*). Other authors have concluded that these parasite haemoglobins are, in fact, vestigial and without function. Clearly this is an interesting area or research which might benefit from continued attention.

The cytochromes are respiratory pigments which contain a haem prosthetic group. The structure of cytochromes has been examined in only a few parasites, including the trypanosomes and a small number of helminths (*Moniezia* and *Ascaris*). Spectral and amino acid analyses reveal certain differences between parasite and vertebrate cytochromes, e.g. the amino acid, trimenthyl-lysine, a common constituent of trypanosome cytochromes, is absent from vertebrates.

Table 4.2. Classification of naturally occurring lipids.

Fatty acids		
	saturated	
	unsaturated	
	branched chain	
Glycerol derivatives		
	glycerides (mono, di and tri)	
	phosphatides	phospholipids
	plasmalogens	phospholipids
Sphingosine derivatives		
	sphingomyelins	phospholipids
	ceramides	phospholipids
	cerebrosides	
	gangliosides	
Sterols (steroidal alcohols)		
Terpenes		
Waxes (long chain alcohols esterified to fatty acids)		
Conjugated lipids		
	lipoproteins	
	lipopolysaccharides	

Intraerythrocytic stages of the malarial parasites produce a haemoprotein-containing pigment within the parasitized red cell. The structure and function of this unique pigment is at present unknown.

Lipids

Lipid composition

Lipids are a highly diverse and heterogeneous group of compounds, unified only by their common extraction from animal tissues by the use of organic solvents, such as methanol, chloroform and acetone. A simplified classification of naturally occurring lipids is given in table 4.2.

Fatty acids. Saturated (with no double bonds) and unsaturated (with varying numbers of double bonds) fatty acids occur in glycerides and phospholipids, and small quantities of free fatty acids can be extracted from all animal tissues. Some nematodes may possess unusually large pools of these free fatty acids e.g. *Nippostrongylus brasiliensis,* 60% of total lipid; *Strongyloides ratti,* 67% of total lipid. More usually, the free fatty acid pool represents less than 10% of the total lipid content, and this will often include glycerides as well. In general, parasite fatty acids are no different from those found in non-parasitic animals.

Glycerides. In mammals, glycerides-composed of a glycerol back-bone and either one, two or three fatty acid-form the major storage element of lipid. In parasites, stored lipid is rather less abundant than in other animals. The parasitic protozoa possess only small amounts of triglyceride, though it is interesting to note that the culture forms of some trypanosomes contain consistently more triglyceride than the equivalent bloodstream forms; the reason for this is not clear. Triglycerides may act as an energy store in the encysted stages of some protozoans, such as the occysts of *Eimeria,* whose metabolism is probably aerobic. The glyceride content of helminth parasites has not been adequately investigated for more than a small number of species. Approximately 30% of schistosome lipids are composed of glycerides and free fatty acids, while somewhat higher figures have been obtained for some nematodes, cestodes and acanthocephalans.

Phospholipids. Phospholipids are the major lipid components of cell membranes and are derivatives of phosphatidic acid (figure 4.4). In the phospholipids, a variety of polar groups are found, e.g. choline in phosphatidylcholine, ethanolamine in phosphatidylethanolamine, serine in phosphatidylserine, inositol in phosphatidylinositol and glycerol in phosphatidylglycerol and diphosphatidylglycerol or cardiolipin. It is likely that most parasites contain phospholipid in their membranes but only a few species have been examined in any detail. The most commonly occurring phospholipids are phosphatidylcholine, phosphatidylethanolamine, phosphatidylserine, phosphatidylinositol, cardiolipin, sphingolipids and plasmalogens.

Sterols. The most abundant sterol in parasite tissue is cholesterol-it is the predominant sterol in trypanosomes, *Entamoeba, Plasmodium* and *Trichomonas*. An alternative sterol, ergosterol , is found in place of cholesterol in the culture forms of some of the trypanosomes (*T. cruzi, T. lewisi* and

(a)

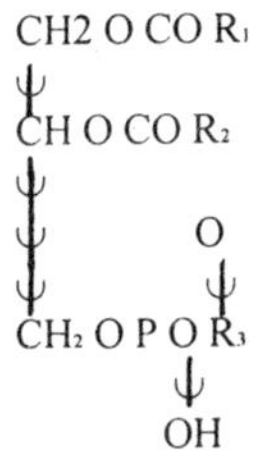

(b)

Phosphatidic acid: R_3 = H; R_1 and R_2 are long aliphatic chains.
Phosphatidylglycerol: $R_3 = CH_2CHOHCH_2OH$
Cardiolipin (diphosphatidylglycerol);

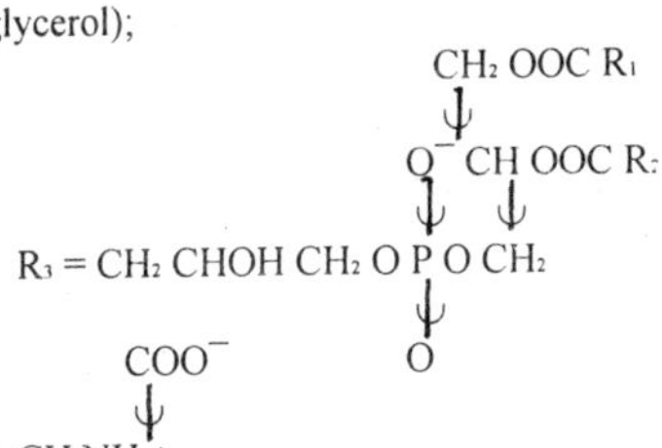

Phosphatidylserine: $R_3 = CH_2\,CH\,NH_3+$
Phosphatidylethanolamine: $R_3 = CH_2\,CH_2\,NH_3+$
Phosphatidylcholine: $R_3 = CH_2\,CH_2\,N + (CH_3)_3$

Fig. 4.4 : (a) The general structure of phospholipids. In the parent molecule, phosphatidic acid, R_3 = H. (b) Some common examples.

rhodesiense). The reason for this switch in sterols during the life-cycle is not known. All the helminths that have been examined possess cholesterol, as well as a variety of cholesterol esters, whose distribution is somewhat uneven. It appears that sterols in general are not synthesized by parasites and are therefore essential dietary requirements.

A small number of parasites contain pigmented carotenoid material, which is derived from cholesterol. This is not uncommon in the Acanthocephala, e.g. the cystacanth larvae of *Polymorphus minutus* are characteristically surrounded by a bright orange capsule which is rich in carotenoids. It is often speculated that this pigmentation makes the parasite more obvious to the final host, a duck, when it is feeding on fresh water shrimps (the intermediate hosts for the parasite). A few digeneans and nematodes also contain carotenoids.

Waxes. Wax esters have been extracted from the tissues of only a small number of parasites, such as acanthocephalan larval stages. There is some doubt whether waxes occur in the parasitic Protozoa.

Lipid biosynthesis

Parasites in general appear to posses only a limited ability to synthesize long chain fatty acids *de novo* and therefore rely to a great extent upon the host to provide their essential requirement. Mammals can synthesize long chain fatty acids by a reversal of the β-oxidation process, using acetyl-CoA as the basic two-carbon foundation. Although this *de novo* pathway may occur in some trypanosomes, the majority of parasites (Protozoa and helminths alike) are restricted to conservative chain lengthening of absorbed fatty acids by the addition of acetate units. Chain elongation is a feature of lipid biosynthesis in a number of parasites including *Leishmania,* some trypanosomes, trichomonads, *Crithidia, Plasmodium, Schistosoma* and *Hymenolepis*. The nematode, *Ascaris lumbricoides,* possesses a limited facility for lipid synthesis using malonyl-CoA. As a consequence of the overall inability of parasites to synthesize their long-chain fatty acids, it is not surprising to find that the fatty acid content of a parasite closely resembles that of the host in which it resides. This is exemplified by *Hymenolepis diminuta,* whose fatty acid any glyceride content may be altered by feeding the host experimental diets with either an excess or a deficiency of various lipids. Both palmitic acid ($C_{16:0}$) and linoleic acid ($C_{18:2}$) proved to be highly variable under such experimental conditions.

Schistosoma mansoni and *Fasciola hepatica* are both capable of chain elongation by the addition of acetate units to absorbed fatty acids. Comparison of rat liver fatty acid content with that of *Fasciola* grown in rats shows considerable similarity, with the exception of the occurrence of two acids, eicosenoic ($C_{20:1}$) and eicosadienoic ($C_{20:2}$), which appear in the worm pool but are absent from the host. *Fasciola* has a high rate of phospholipid synthesis from the precursors phosphate and glycerol, which are readily incorporated into lysolecithin, phosphatidycholine, phosphatidylinositol, phosphatidylserine, phosphatidylethanolamine, ethanolamine plasmalogen and cardiolipin.

Threonine is an important precursor for acetate and acetyl-CoA in *Trypanosoma* and *Crithidia*. The postulated biosynthetic pathway for these compounds involves the intermediate, α—amino—β-oxybutyrate, and is inhibited by tetraethylthiouram, a powerful growth inhibitor of some trypanosomes.

There is little information on the mechanisms employed by parasites for the introduction of double bonds (desaturation) into long-chain saturated fatty acids, not is much known of the processes involved in triglyceride synthesis. *Hymenolepis* rapidly incorporates radioactively labelled glycerol, palmitic acid and oleic acid ($C_{18:0}$) into triglycerides and phospholipids. This synthesis is thought to occur via steps involving α-glycerophosphate, phosphatidic acid and diglyceride. Additionally, studies on *Fasciola* have demonstrated the rapid incorporation of serine, glucose ad oleic acid into phospholipids and glycolipids, including cerebrosides. Phosphatidylethanolamine syntheses in *Hymenolepis diminuta* proceeds via the decarboxylation of phosphatidylserine. Rapid phospholipid synthesis is a characteristic feature of many protozoans.

Table 4.3. The structural components of the nucleic acids, DNA and RNA

	Synthetic sequence	*Components of DNA*	*Components of RNA*
bases	*de novo* synthesis, salvage, or dietary requirement	*purines* adenine guanine *pyrimidines* cytosine thymine	*purines* adenine guanine *pyrimidines* cytosine uracil
nucleosides	base+sugar	*deoxyribonucleosides* (sugar=deoxyribose) deoxyadenosine deoxyguanosine deoxycytidine deoxythymidine	*ribonucleosides* (sugar=ribose) adenosine guanosine cytidine uridine
nucleotides	nucleoside+phosphate	*deoxyribonucleotides* deoxyadenosine-monophosphate deoxyguanosine-monophosphate deoxycytidine-monophosphate deoxythymidine-monophosphate	*ribonucleotides* adenosine-monophosphate guanosine-monophosphate cytidine-monophosphate uridine-monophosphate
nucleic acids	polymerization of nucleotides	DNA	RNA

The ability of parasites to synthesize sterols is apparently limited, and most rely on their hosts to provide these essential compounds. Exceptions to this include *Crithidia fasciculata* and some trypanosomes.

Lipid catabolism

Mammals catabolise lipids aerobically by β-oxidations but, since many parasites are facultative or even obligate anaerobes, we might justifiably expect limited use of such oxidative metabolism. This expectation is borne out, and very few parasites make use of lipid for energy metabolism. Those that do include *Trypanosoma lewisi, Leishmania brasiliensis, Litomosoides carinii, Echinococcus granulosus, Taenia hydatigena, Ascaris* and the free-living stages of some larval nematodes. Other parasites, like *Hymenolepis,* do not possess the full complement of β-oxidation enzymes. Anaerobic catabolism of fatty acids has been recorded in some of the rumen-dwelling ciliate Protozoa, but this is mot unusual.

Nucleic acids

Nucleic acid composition

The two ubiquitous nucleic acids in all animal tissues are deoxyribonucleic acid (DNA) and ribonucleic acid (RNA). Most of the DNA is found within the cell nucleus and it constitutes the basic genetic information or genome. DNA is a double helical molecule with specific pairing of bases between each strand of the double helix so that guanine and cytosine are always linked, as are adenine and thymine. It is the sequence of base pairs in the DNA of any organism that represents foundation for its heritable characteristics. Small quantities of DNA may also be found outside the nucleus, often associated with mitochondria (mtDNA). Mitochondrial DNA probably codes for ribosomal RNA synthesis. RNA occurs in three basic forms, messenger RNA (mRNA), transfer RNA (tRNA) and ribosomal RNA (rRNA). Structurally, RNA is similar to DNA but contains ribose in place of deoxyribose, uracil in place of the closely related pyrimidine, thymine, and is frequently single rather than double stranded. The sequence of base pairs in nuclear DNA determines the base sequences of RNA (transcription), and the latter forms the template for protein synthesis (transcription), and the latter forms the template for protein synthesis (translation). In this sequence of events mRNA is involved in transcription and tRNA in translation. The final step, protein synthesis, is ribosomal in location.

The field of molecular biology, with its core in the biochemistry of the nucleic acids, has developed at a truly dramatic rate and parasitic organisms have not been overlooked in this modern approach to biology. However, the trypanosomatid flagellated have received considerably more attention than any other group of parasites. This is, in part, easily attributable to the presence in these protozoans of an unusual form of extranuclear DNA, known as kinetoplast DNA (kDNA).

The cell nuclei of all parasites are presumed to contain DNA-considerable evidence for this comes from Feulgen staining and, more recently, from direct chemical determinations. There is no reason to believe that parasite DNA contains any unusual bases determination of the base composition, using a range of techniques, including buoyant density determination in caesium chloride gradients in the analytical ultracentrifuge, melting point determination and direct chemical analysis, reveals the expected range of purines and pyrimidines. However, there is some doubt as to whether any of these methods would expose the presence of any minor components, such as methylated bases.

The base composition of DNA is usually expressed as the percentage of guanine and cytosine relative to the total base content, i.e. the G+C content. Figures for the G+C content of parasite nuclear DNA range from 18 to 61%, while the G+C content for extra nuclear DNA is more restricted, lying between 18 and 41%. These determinations have been made on less than 40 parasite species so that any generalisations are, perhaps, inopportune.

Little is known about the RNA of parasites and studies are restricted almost exclusively to the parasitic protozoans and ascarid nematodes. RNA has been characterized from the chromatoid bodies (composed of symmetrically arranged ribosomes, made up of 30S and 50S subunits) of *Entamoeba invadens,* where it is associated with a protein component. Ribosomal RNA and transfer RNA have been isolated from 3 small number of protozoans, but very little is known about messenger RNA.

Nucleic acid synthesis

Purine and pyrimidine synthesis. Most of the available information on the synthesis of nucleic acids comes from the protozoans and rather little from the helminths. Mammals synthesize the purine ring either *de novo* or from salvage pathways, involving preformed bases or nucleosides. Many parasitic protozoans differ from mammals in that they rely almost exclusively on a dietary source or salvage pathway, since they are apparently incapable of *de novo* synthesis of the purine ring. At least two species, *Crithidia oncopelti* and *Trypanosoma lewisi,* are exceptions to this generalisation and synthesize purines from glycine and serine. Purine salvage pathways have been investigated in *Plasmodium* and *Trypanosoma* and it is apparent that hypoxanthine, adenosine and adenine are freely converted to purines.

In direct contrast to the purines, pyrimidines are synthesized *do novo* by many parasites. All of the species studied so far, most of which are protozoans, synthesize pyrimidines from precursors such as aspartate, bicarbonate and glutamine. Additionally, salvage pathways are important sources of these bases in many of the Protozoa. The enzymes involved in the *de novo* synthesis of pyrimidines by *Plasmodium* have been identified and include orotidine-5 monophosphate pyrophosphorylase, deoxythymidilate synthetase and dihydrofolate reductase.

It is not certain whether helminth parasite can synthesize either purine or pyrimidine bases *de novo* or whether they simply rely on the host to be their sources of supply. Several species of cestodes, digeneans and acanthocephalans are freely permeable to a variety of bases, but the mechanisms of acquisition have been examined in one species (*hymenolepis diminuta*) only (see chapter 2). The difficulties in maintaining helminths in defined culture media for any length of time are undoubtedly responsible for the basic lack of knowledge. All of the data on purine and pyrimidine synthesis in the Protozoa derive exclusively from organisms grown in defined or partially defined media containing radioactively labelled precursor molecules.

Nucleic acid synthesis Synthesis of nucleic acids involves a series of enzymes termed polymerases and virtually nothing is known of these polymerases in parasitic tissues.

Nucleic acid catabolism

Nucleases, such as DNA - are and RNA - are responsible for catabolising nucleic acids to nucleotides, and phosphorylases, phosphatases and hydrolases catabolise these nucleotides to nucleosides and free bases. A variety of these enzymes have been isolated from parasites, including *Crithidia oncopelti, Plasmodium, Trypanosoma, Entamoeba, Ascaris* and *Trichuris*. The tissues of *Hymenolepis diminuta* contain the β-amino acids, β-alanine and β-aminoisobutyric acid, which are thought to originate from uracil and thymine. Uric acid, a breakdown product of purines, is excreted by a number of helminths, implying active nucleic acid catabolism.

The extranuclear DNA of trypanosomes

Extranuclear or mitochondrial DNA is a common feature of most organisms, and that of parasites shows no fundamental difference from that of free-living animals. The kinetoplast DNA of trypanosomes was that first extranuclear DNA to be described for eukaryotes.

In caesium chloride gradients, kDNA characteristically appears as a rapidly-banding peak which can readily be differentiated from the nuclear DNA (figure 4.5) is no exception. Kinetoplast DNA can be isolated from the

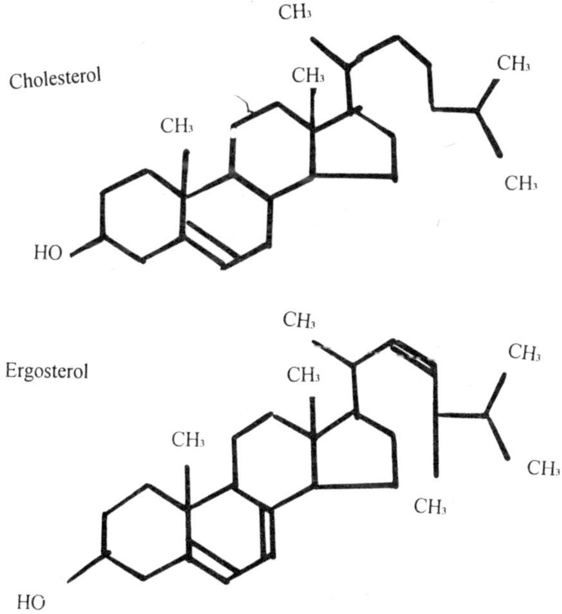

Fig. 4.5 : The sterols, cholesterol and ergosterol.

trypanosome mitochondrion intact and visualised under the electron microscope, where its circular nature is clearly revealed. In fact, kDNA is composed of two distinct circular elements, referred to as mini-circles and maxi-circles. The diameter of the mini-circles ranges from 0.3-0.8μm while the maxi-circles have a mean diameter of 11μm. The smallest mini-circles are typical of the mammalian trypanosomes, while the larger mini-circles occur in the insect-parasitic species. The application of restriction endonucleases (enzymes which cleave nucleic acids in specific places) reveals that there is some degree of nucleotide heterogeneity in terms of sequence in the mini-circular kDNA but the maxi-circles are made up of non-repeating nucleotide sequences. On visualisation with the electron microscope, kDNA is seen as a network of circular components, of which less than 10% is the maxi-circular element; maxi-circles can be completely removed by enzyme treatment without loss of the integrity of the DNA network. In *Crithidia* there are approximately

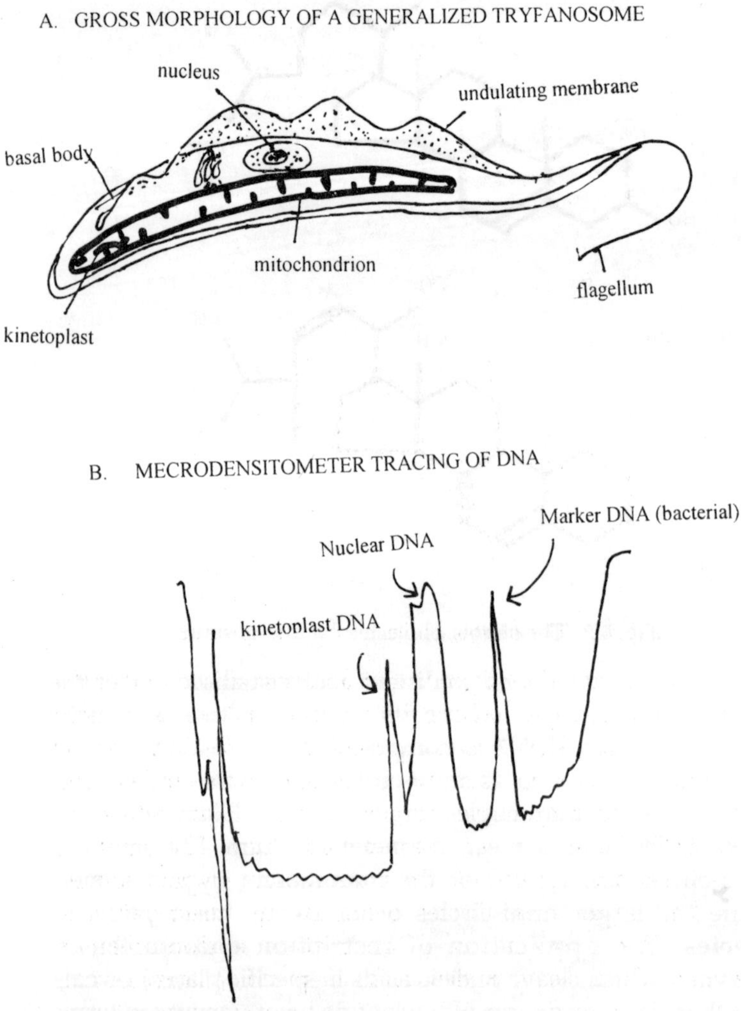

Fig. 4.6 : The kinetoplast and kinetoplast DNA of trypanosomes. Kinetoplast DNA appears as a rapidly-banding peak distinct from nuclear DNA; marker DNA is an added internal standard of bacterial or mammalian origin. The tracing in B. is derived from the analytical ultracentrifuge and is recorded under U.V. light (260 nm) at equilibration in a caesium chloride gradient.

14000 mini-circles and 50 maxi-circles in kDNA. The elements of the kDNA network are linked by covalent bonds are the mini-circles are thought to be catenated or fused together, possibly into long chains.

Although the precise function of kDNA is far from clear, it is thought that maxi-circles represent true mitochondrial DNA and therefore code for ribosomal RNA. The function of the mini-circular component of kDNA is even less clear, but it may have a supportive or structural role. Dyskinetoplastic trypanosomes, occurring either naturally or produced by laboratory manipulation, tend to lose their kDNA and, with it, the ability to synthesize mitochondria. Some dyskinetoplastic forms, however, like *T. equiperdum* and *T. evansi,* seem to possess normal kDNA. There is much to learn from this area of research, particularly concerning the role of kDNA in parasitism by trypanosomes.

Chapter—5
Physiology of Osmoregulation

The excretory system of all animals is responsible for maintaining a relatively constant internal environment in the face of an ever-changing external environment. It follows, therefore, that a parasite inside its host's tissues will be inhabiting an environment that is itself rigorously controlled by the excretory and osmoregulatory systems of the host animal. Parasites disperse themselves from host to host and, during this transmission, they will be exposed to a variety of physico-chemically different environmental conditions. Many parasites have a direct life—cycle with one host and a single transmission stage. At the other end of the spectrum are parasites, e.g. the strigeid digeneans, which, during a single life-cycle, infect three different hosts and have two distinct transmission stages. We would thus expect the excretory system of parasites to be able to remove the waste products of their metabolism and carry out an osmoregulatory role when the prevailing conditions demand.

In general terms, excretory systems perform several major physiological functions: they remove toxic end-products of metabolism; they maintain ionic balance; they control the body water content; and they remove toxic foreign substances. The evidence for these functions occurring in parasites is rather limited and a great deal of important work remains to be done on the excretory and osmoregulatory systems of parasitic animals.

Contractile vacuoles in Protozoa

The organelle concerned with water and ion balance in the Protozoa (and in the Porifera) is the contractile vacuole. It plays a significant role in ion and water regulation in freshwater protozoans, but its functional significance to either marine or parasitic species is questionable—both these groups of protozoans inhabit environments that are isosmotic to their internal fluids.

Amongst the parasitic Protozoa, contractile vacuoles are present in a small number of flagellates and many ciliates but they are not typical features of either the sporozoans or the amoebas. The excretion of nitrogenous waste material is probably not a function of the contractile vacuole. Excess nitrogen, in the form of ammonia or urea, is removed by outward diffusion across the cell membrane.

The protonephridial system of platyhelminths

Protonephridia, comprising a blind-ending system of tubules which open via a nephridiopore, are present in a

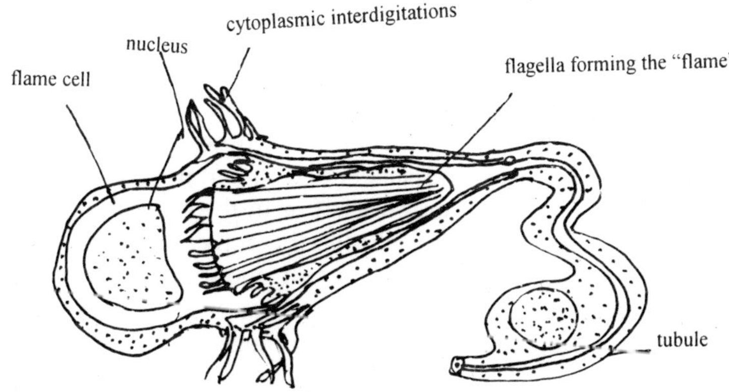

Fig. 4.1 : The flame cell terminal organ of a digenean.

wide range of invertebrate groups. In the protonephridial system of the platyhelminths, a single " flame cell" is located at the terminus of each excretory tubule. The flame cell derives its name from the large number of flagella that beat rhythmically, resembling the flickering of a flame. The protonephridial systems of monogeneans, digeneans and cestodes show a fundamental morphological similarity. The ultrastructure of the excretory system has been described for only a small number of platyhelminth species, including *Fasciola hepatica, Himasthla quissetensis, Moniezia expansa, Echinococcus granulosus* and *Diphyllobothrium latum*.

The flagella of flame cells arise from a circle of projections of the cell arising from the first cell of the excretory tubule (figure 5.1). The excretory tubule itself is extracellular, the walls of which are formed by a sheet of cytoplasm. There are usually between 50 and 100 flagella in the flame; these are sometimes called cilia, but functional evidence suggests that the term flagella is more appropriate. Each flagellum contains microtubules in a 9+2 arrangement, and is attached to the cell by a basal body. The flagella beat rhythmically in an undulatory wave motion which is thought to produce the flow of tubule fluid and, perhaps, draws solutes from the surrounding parenchymal tissue into the terminal organ of the excretory system. A nephrostome, consisting of a ring of pores through which solutes might pass, has been described in the tapeworm *Moniezia*. There is no evidence for such a structure in other platyhelminths. Filtration of solutes and suspended material may take place either at the terminal organ itself or in the epithelium of the excretory tubule, a metabolically active region rich in mitochondria, Golgi apparatus and endoplasmic reticulum. The tubules leading from the terminal organ are extremely small and are extracellular, though they were once regarded as intracellular. The tubules are lined with a highly folded cytoplasm in the Digenea and with microvilli in the Cestoda.

The excretory system in the Digenea (figure 5.2) tends to be simple in larval worms and increases in complexity during development, e.g. the miracidium of *Fasciola hepaticas* has been two terminal organs, each opening to the outside separately, but in the adult liver fluke, there are many hundreds of terminal organs, each with a tubule that enters the larger collecting ducts which then empty into a single excretory bladder. The cestodes are well supplied with terminal organs and excretory tubules and, in addition, there are large ventral and dorsal excretory canals which run the length of the strobila, connected by transverse canals.

Although we refer to the protonephridial system of the platyhelminth parasites as an excretory system, there is remarkably little evidence to support its status as such. Furthermore, it has been debated, at considerable length, whether this system is involved in osmoregulation—we shall examine the evidence for this later. A functional excretory system must carry out selective filtration and reabsorption of solutes in order to regulate the chemical content of the excreta. It has been postulated that filtration in the protonephridial system is effected at the region of interdigitation between the flame cell and the excretory tubule. The flagellar beat may draw solutes into the flame cell and move them along the tubule. It is calculated that a sufficiently large filtration pressure could be developed across the interdigitating membranes for filtration to occur, but there is no direct evidence as yet. There is limited evidence that nitrogenous waste material is transported within the protonephridial system. In *Moneizia expansa,* ammonia is thought to be removed from the body by diffusion through the tissues and across the tegument, while there is positive evidence for the removal of waste nitrogen by the protonephridia of *Hymenolepis diminuta*. The latter tapeworm has been subject of one of the few analytical studies on the excretory system of parasites. Wilson and Webster developed a technique of micropuncture that allowed them to withdraw small volumes of protonephridial fluid from the longitudinal canals. Chemical analysis of volumes less than 10µl revealed the presence of organic ions (Na^+, K^+, Cl^-, CO_3) equivalent to 1.15% NaCl, amino acids

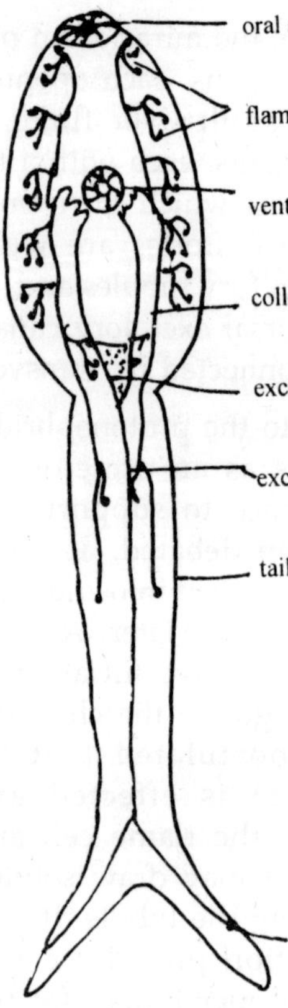

Fig. 5.2 : The digenean excretory system, illustrated in a fork-tailed cercaria.

(12 mg/ml), ammonia (0.04 mg/ml), urea (0.5 mg/ml) and lactic acid (0.7 mg/ml). The occurrence of these substances in the protonephridial system of *Hymenolepis* is an excretory system. However, it remains to be established that filtration takes place at the flame cell and that selective reabsorption and secretion (both prerequisites of a true excretory system) occur within the protonephridial ducts. Glucose, lactate and urea are all reabsorbed in the canals of *Hymenolepis* has been shown to depend upon worm body movements and the excretion rates of the major waste products, in mg/mg dry weight/h, are: lactate 19.4; urea 1.02; and ammonia 0.08. The significance of the protonephridial system in the

excretory physiology of cestodes cannot be fully understood until further critical and analytical studies have been made.

The excretory physiology of the Digenea has been inadequately investigated, and we are left unclear as to the precise function of the protonephridial system. Lipid droplets are present in the excretory ducts of *Fasciola hepatica, Diplostomum spathaceum, Apatemon gracilis, Holostephanus luhei* and *Echinostoma revolutum,* whereas no lipid is observed in the protonephridial system of tapeworms. Alkaline phosphatase activity is associated with the excretory system of several digeneans, but this may be involved in the hydrolysis of sugar phosphates in the tubules rather than any direct excretory function. Ther is no experimental evidence to suggest that nitrogenous or other metabolic waste products are actually excreted by the digenean protonephridial system and it is likely that these substances are lost by diffusion across the tegument. Osmoregulation may be a more important function of the system, particularly in the free-living larval stages.

Excretion in the Acanthocephala

The Acanthocephala are unusual in that the majority possess no discrete excretory organs or system, with the exception of the members of the order Archiacanthocephala, which have a protonephridial system comparable in its structure to that of the platyhelminths. It must be assumed that metabolic waste products are removed by diffusion across the membranes of the cuticular pores.

Excretion in the nematoda

The excretory system of the parasitic nematodes (in no way different from free-living forms) is not protonephridial, but consists, in the promotive condition of gland cells that open to an excretory pore via an excretory vesicle. In the more advanced condition, a canal system, often excretory vesicle. In the more advanced condition, a canal system, often H-shaped, opens to the outside at a single pore; gland cells are commonly associated with these canals. Experimental evidence points to this system functioning primarily in an osmoregulatory, rather than an excretory role.

End-products of nitrogen metabolism in parasites

The major source of excess metabolic nitrogen derives from proteins, amino acids and, to a lesser extent, from purines and pyrimidines. While the majority of parasites excrete ammonia, and thereby resemble free living freshwater animals, a variety of other excretory products have been detected, the most common of which is urea.

The majority of Protozoa examined excrete ammonia and we assume that their small size obviates the need to develop a complex nitrogen detoxication system. Urea may be produced in small quantities, but in those species that have been studied, such as *Crithidia fasciculata,* there is no urea cycle presumably due to the absence of key enzymes like transcarbamylase. Amino acids also appear in the excreta of many protozoans, evidenced by the occurrence of amino acids in culture media used for growing protozoans *in vitro.* It is not certain that these amino acids are true excretory products and they may simply represent counterflow mechanisms involved in the uptake of nutrient amino acids. In the bloodstream-form of *Trypanosoma gambiense,* absorbed glucose is rapidly converted to alanine, which is presumably excreted. Extensive radiolabelling studies are required to clarify the role of the amino acid pool in excretion.

The platyhelminth parasites excrete a range organic molecules derived from nitrogen metabolism: ammonia is the major waste product for some species (*Lacistorhynchus tenuis, Moneizia benedeni*); ammonia and urea are excreted by *Fasciola hepatica* and *Schistosoma mansoni;* ammonia and uric acid are excreted by *F. gigantica* and *Paramphistomum explanatum* and the cestodes *Cysticercus tenuicollis, Echinococcus granulosus* and *Taenia taeniaeformis* excrete urea and uric acid. *In vitro* many helminths, like the Protozoa, release amino acids into the culture media, but the origins of these amino acids remain to be determined.

Ammonia is produced by parasites by a number of well described pathways which include amino acid oxidases (*Plasmodium berghei, Hymenolepis diminuta, Ascaris lumbricoides*), mono-amine oxidases (*Balantidium coli, Schistosoma mansoni*), glutamate dehydrogenase (*Ascaris, Plasmodium berghei, P. lophurae, Microphallus pygmaeus*), and urease (*Lacistorhynchus tenuis*). Urea formation in vertebrates depends upon the urea cycle, but its occurrence in parasitic animals is the subject of some controversy. Many parasites contain some of the enzymes of the cycle and some cycle intermediates, but whether the cycle is operable in any parasite is not known. The enzymes arginase and ornithine transcarbamylase are present in the cestode *Hymenolepis diminuta,* but the worm cannot synthesize the key intermediate carbamyl phosphate. Current opinion is that the functional significance of the urea

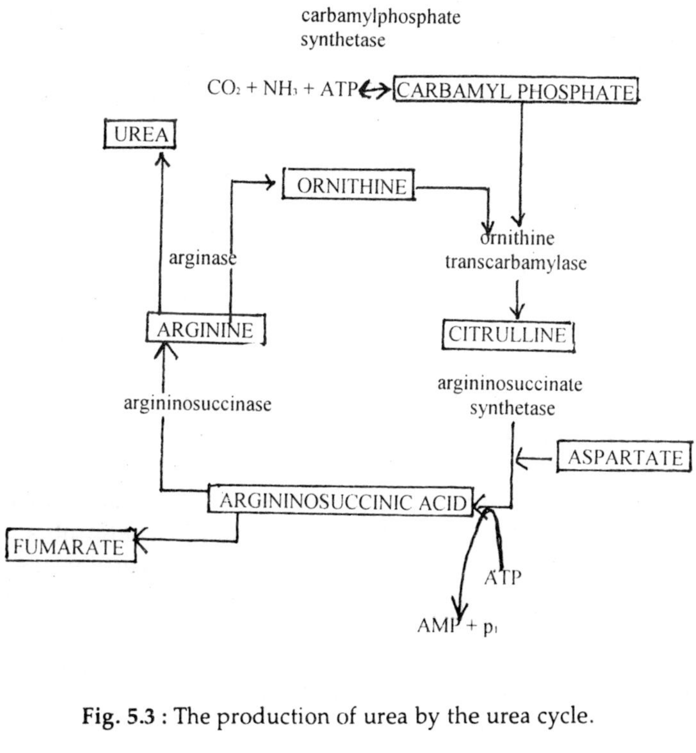

Fig. 5.3 : The production of urea by the urea cycle.

cycle to parasites is, in general, negligible but that those cycle enzymes that do occur may be involved in alternative biosynthetic activities.

Virtually nothing is known of the excretory products of acanthocephalans. *Moniliformis dubius* contain only one enzyme of the urea cycle, ornithine transcarbymylase, so the cycle probably does not function. Almost all of the parasitic nematodes excrete ammonia, but urea is excreted by some species.

Ionic regulation and water balance

Parasites, unlike free-living animals, are not normally confronted with the major physiological problem of maintaining their water content or the ionic composition of their body fluids at levels very different from the external environment. The greatest osmotic problem faced by parasites will occur during transmission between hosts-transmission stages of parasites may inhabit a marine environment, a freshwater environment or a truly terrestrial environment for varying lengths of time. Inside the host, the parasite inhabits an environmental that is approximately isosmotic with its own tissue fluids. Ectoparasites, such as protozoans, almost all of the monogeneans, a single digenean (*Transversotrema patialense*) and the many arthropods, inhabit the external surfaces of aquatic animals. It the host is a fresh water animal the parasite will be faced with a severe water and ion balance problem since the external environment will be hypotonic to its tissue fluids. A similar problem will be shared by the larval stages of platyhelminths, such as eggs, miracidia, oncomiracidia, cercariae and coracidia, which are dispersed in freshwater. The need to resist dessication is a related problem common to protozoans, platyhelminths and nematodes parasitic in terrestrial or aerial vertebrates. In the eggs of these parasites this function is effected by the protective coverings of the eggs, rather than by the osmoregulatory system. In the parasitic nematodes the ability to withstand desiccation (cryptobiosis) is well developed but the physiological mechanisms involved are not understood.

There is only a small literature concerning ion and water balance in parasites, probably due to the small size of most parasites and their inability to survive satisfactorily in culture. In the parasitic Protozoa, contractile vacuoles occur in some, but not all groups. It is not certain whether they carry out an osmoregulatory role since there are practical difficulties in determining with any accuracy very small fluxes of water and ions across the cell membrane. Some degree of osmoregulation has been observed in a few species, e.g. *Balantidium*, *Crithidia fasciculata*, *Entodinium*, *Ichthyophthirius multifiliis*, *Trypanosoma gambiense*, *Vahl kampfia calkensis*. Some of the intestinal amoebae develop contractile vacuoles when placed in hypotonic growth media and can remain viable over a wide range of osmotic pressures. The white-spot organism of fishes, *Ichthyophthirius multifiliis*, although unquestionably endoparasitic in habit, possesses many contractile vacuoles, but their function is far from clear.

The osmoregulatory role of the helminth protonephridial system has been examined in a small number of species, usually by adopting the procedure of measuring the flagellar beat rate of the flame cells when the worm is maintained in media of different ionic strengths and osmotic pressures. In general, these studies suggest that the protonephridial system is not involved in osmoregulation. Platyhelminths appear to be osmoconformers, so that the composition of the protonephridial canal fluid reflects the composition of the external environment.

In the miracidia of some digeneans, the beating of the flagella in the terminal organs is initiated, upon hatching, by light. In *Fasciola hepatica* no further change in the beating rate of the flame is discerned under different external conditions of osmotic pressure or ionic strength. In adult *Fasciola gigantica* that are placed into a hypotonic medium, the excretory bladder becomes distended and frequently empties its contents. This worm can tolerate a wide range of osmotic pressures without changing its weight through water fluxes. In other helminths, weight changes are a

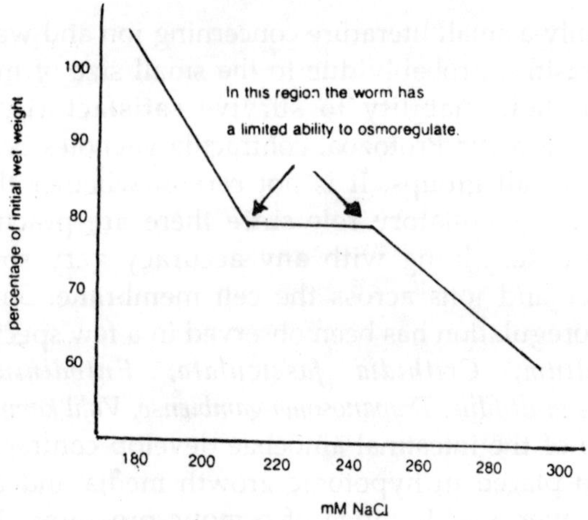

Fig. 5.4 : Weight changes in the cestode *Lacistorhynchus tenuis* related to the external concentration of NaCl.

common response to the environmental conditions, e.g. *Cysticercus tenuicollis* loses weight in hypertonic media and gains weight in hypotonic conditions; *Hymenolepis diminuta* loses weight in hypertonic conditions and gains water, but loses salts in hypotonic salines. The protonephridial system of *H. diminuta* cannot control internal levels of Na^+, but does exert limited control over the concentration of K^+, although it is possible that this control resides with the surface membranes of the worm. Other tapeworms that have been examined (*Calliobothrium, Moniezia* and *Schistocephalus*) are also osmoconformers. It has been suggested that tapeworms with a free-living larval stage, such as the pseudophyllidea and the Trypanorthyncha, would have a particular need to osmoregulate in the larval condition and this ability may or may not be maintained by the adult. *Lacistorhynchus tenuis* can osmoregulate over quite a wide range of osmotic pressures (figure 5.4), as shown by weight changes in media of varying strength. Urea plays an interesting role in ormoregulation in the dogfish tapeworm *Calliobothrium verticillatum*. This worm is freely permeable to urea and the addition of urea to a culture medium will prevent the inevitable uptake of water. This role of urea may not be

Table 5.1. Free amino acid concentrations is selected platyhelminth parasites.

Amino Acid	1 *Hymenolepis diminuta*	2 *Schistosoma mansoni*		3 *Dictyocotyle coeliaca*	3 *Entobdella soleae*	3 *Entobdella hippoglossi*	3 *Discocotyle sagitta*	3 *Diclidophora merlangi*	3 *Diclidophora denticulata*
		♀	♂						
proline (% of total free pool)	4.3	6.6	5.4	63.9	37.3	72.5	8.0	72.3	41.3
glycine (% of total free pool)	10.9	8.3	8.4	2.0	31.8	9.7	17.5	2.0	4.8
total free pool (µmol/g dry weight)	56.9	45.3	40.8	1271.7	1441.1	1343.6	371.4	561.7	243.4

surprising in view of its importance of elasmobranch fishes in maintaining a hypertonic blood. Urea may also play an important osmoregulatory role in other parasites of elasmobranchs.

The little information on osmoregulation in the Acanthocephala suggests that they are also osmoconformers—they do not possess any free-living transmission stages. The parasitic nematodes may posses limited powers of osmoregulation as shown by their ability to alter their body length in hypotonic or hypertonic media. Ionic regulation has been determined only for *Ascaris* which can control the pseudocoelomic fluid concentrations of calcium, magnesium, potassium and chloride ions to some extent. Juveniles of *Nippostrongylus* and *Ancylostoma possess* an excretory ampulla which pulses at a rate that is inversely proportional to the osmotic pressure of their immediate environment.

Free amino acids and osmoregulation

Free amino acids in the tissue fluids are known to play a part in osmoregulation is some free-living invertebrates, proline and glycine, in particular, having a special role. Free amino acids may also be involved is the osmotic relationships of bacterial cells, and here again, proline features in a unique manner.

There is a virtually no direct evidence that the free amino acid pool of parasitic organisms is concerned with osmoregulation. It is interesting to note, however, that the free pool of many ectoparasitic monogeneans is considerably higher than that of endoparasitic helminths. Monogeneans contain rather large amounts of free proline, and some also possess a large clycine pool. It is interesting to speculate that these free pool acids may be involved in osmoregulation; unfortunately there is no information on the size or contents of the free amino acid pool of freshwater monogeneans for comparison.

Chapter—6

Physiology of Nervous System

This section contains a brief description of the nervous systems and sense organs that are typical of helminth parasites. No consideration of the parasitic protozoans will be given, though receptors are present on the surface of these parasites.

Morphology of parasite nervous systems

The platyhelminth nervous system consists of a cerebral complex of ganglia and commissures in the anterior region of the body, proximal to the pharynx. Entending anteriorly and posteriorly form the cerebral ganglia are bilaterally symmetrical nerves that innervate the tissues and organs, and which are joined by numerous lateral branches. The major deviation from the standard pattern that occurs in the parasitic platyhelminths is associated with the innervation of the organs of attachment, the posterior haptor in the Monogenea, the oral and ventral suckers in the Digenea and the scolex in the Cestoda.

The nervous system of monogeneans is basically comparable to that of the free-living Turbellaria. In some monogeneans an apparently primitive condition occurs whereby the cerebral ganglia are situated dorsal and anterior to the pharynx (e.g. Dactylogyridae and Tatraonchidae), but, more usually, the brain, composed of a pair of ganglia joined by a single dorsal commissure, lies immediately adjacent to the pharynx. In a few species, a pair of commissures join the cerebral ganglia, and completely encircle the pharynx. Of the three pairs of major nerve trunks that run posteriorly from the brain, (i.e. the ventral, dorsal and lateral nerves) the ventral pair are the best developed and innervate the posterior haptor, often developing ganglia in the haptor itself. Anteriorly, three or four pairs of nerves supply the head.

The basic pattern of the digenean nervous system differs little from that of the Monogenea; there is an additional nervous supply to the anterior and ventral suckers, which is often extensive. In the Cestoda, paired cerebral ganglia, joined by commissures, give rise to a pair of lateral nerve trunks, that extend the length of the strobila and are connected by commissures, which are particularly evident at the junctions between proglottids.

The nervous system of the parasitic platyhelminths is in the form of a nerve net, and contains unmyelinated nerve fibres, which are usually demonstrated, at the light microscope level, by the histochemical localization of acetylcholinesterases. There is little ultrastructural information on the platyhelminth nervous system, but there is no justification for regarding it as a degenerate system.

The nervous system of the Acanthocephala has been examined in detail in only a few species, e.g. *Polymorphus, Macracanthorhynhus, Bolbosoma* and *Hamanniella*. It consists of a cerebral ganglion, situated ventrally in the proboscis receptacle, and varying numbers of both paired and single nerves which supply the body wall, the musculature and the somatic organs. In male acanthocephalans, the lateral nerves enter the reproductive system, forming a pair of genital ganglia connected by a circular commissure.

Parasitic and free-living nematodes possess essentially identical nervous systems, consisting of a nerve ring of circum-oesophageal commissure, on which are located varying numbers of ganglia, one of which is dorsal, one ventral and two or more situated laterally. Anteriorly, six nerves extend to supply the mouth and head, while between six and eight main posterior nerves serve the body. The major nerve trunk, the ventral posterior nerve, is ganglionated at intervals and is paired along some of its length. At the posterior extremity the dorsal and ventral nerves divide and, on either side of the body, fuse with the lateral nerves to form a pair of lumbar ganglia. Sensory function is associated with the lateral and ventral nerves and motor function with the dorsal and ventral nerves.

Simplified, outline diagrams of nervous systems of helminth parasites are given in figure 9.1.

Sense organs

Monogenea

The major organs in the Monogenea consist of photoreceptive eyes, ciliated, superficial or epidermal structures (sensillae) possibly concerned with chemoreception, tangoreception and rheoreception, and haptoral knobs, that may be touch receptors. While the morphology of these structures has been examined in some detail with both light and electron microscopes, relatively little is known of their true physiological role, due to the many difficulties incumbent in an experimental approach.

Although many monogeneans possess eyes, these organs tend to be more a feature of the oncomiracidium and are often absent in the adult. This may be indicative that photoreceptors are important in host location, but are not necessary to the adult worm once it is established on its host. Primitively (in the Monopisthocotylea), there are two pairs of dorsal eyes, each with a crystalline lens; the anterior pair are small and are directed posterolaterally, while the larger, posterior pair are directed anterolaterally. The arrangement in the Polyopisthocotylea is regarded as more advanced, there being only a single pair of eyes, either lacking a lens or with an oil droplet lens. In all monogeneans the body of the eye consists of a pigment cup, possibly containing melanin, and a nervous supply. The functional significance of eyes in the larval monogeneans is not clear,

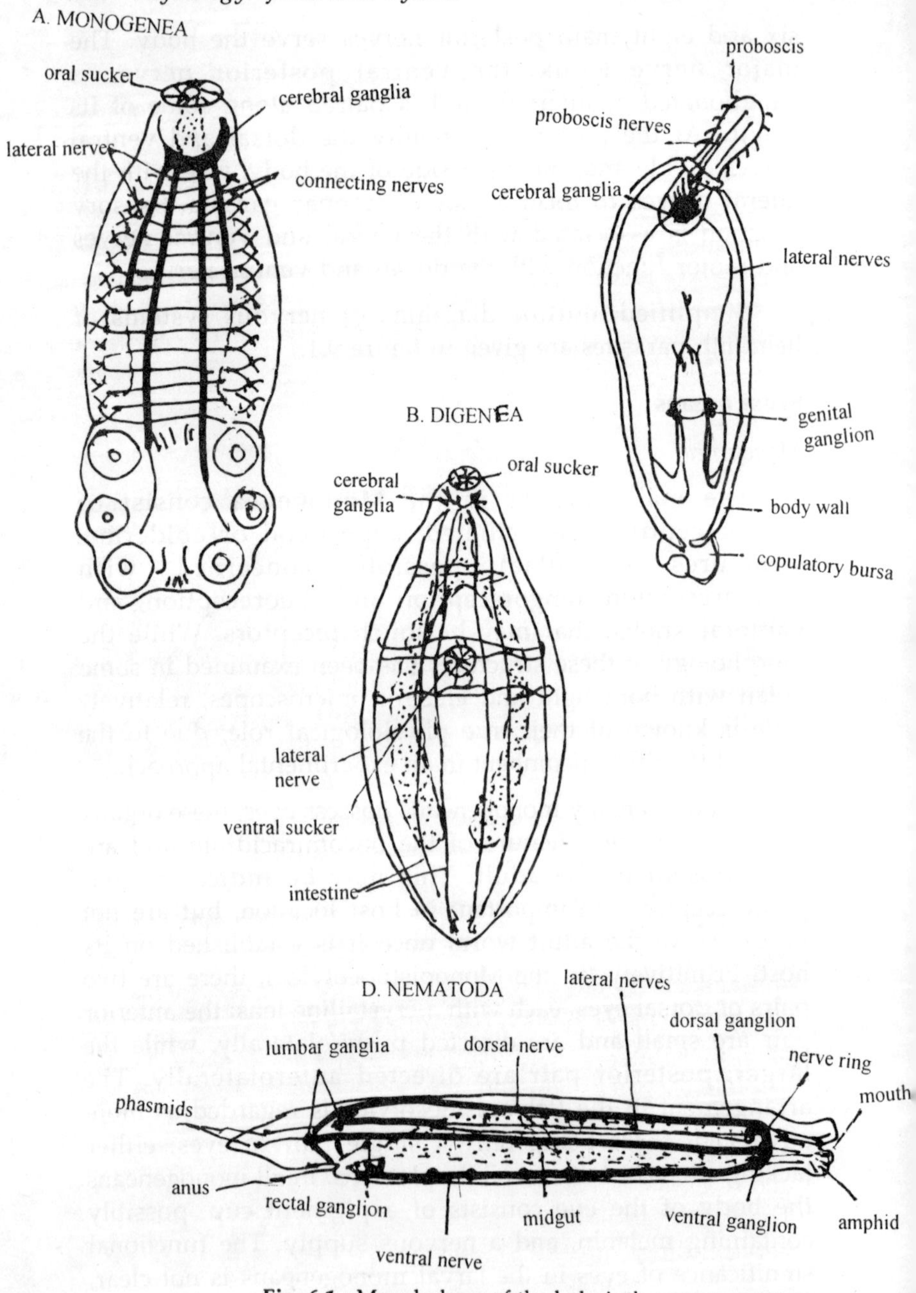

Fig. 6.1 : Morphology of the helminth nervous system

since there is little information of their behavioural responses to light. There is evidence of positive phototaxis in some oncomiracidia, but this may be a short-lived response (e.g. Dactylogyridae). Most larval monogeneans swim with their dorsal surface uppermost, so that dorsoventral orientation may be one function of the eyes, at least during daylight swimming. There is evidence for the occurrence of additional putative photoreceptors in monogeneans, based exclusively upon ultrastructural studies of larval *Entobdella soleae.* These photoreceptors are paired structures, located on the head of the oncomoracidium, close to the eyes. They contain many modified cilia, lamellate bodies and are richly provided with mitochondria. These receptors are morphologically similar to ciliary photoreceptors described in other invertebrates, but convincing evidence for their precise function has yet to be provided.

There is a variety of possible sensory structures associated with the monogenean surface, usually taking the form of papillae containing modified cilia. Their sensory function is deduced from histochemical techniques, particularly silver staining, but this deserves cautious interpretation. These sensory organs are located over much of the body surface and are concentrated in the head region and on the haptors. Electron microscope studies have revealed three types of sensory structure: single receptors (nerve bulb plus a cilium); compound, uniciliate receptors (several nerve endings, each with a single cilium); and compound, multiciliate receptors (one or several nerve endings; each with many cilia). The single sensilla, the most common type of receptor, has been identified in a number of species (*Gyrodactylus, Entobdella soleae, Amphibdella flavolineata, Leptocotyle minor* and *Diclidophora merlangi*) and is thought to be a touch receptor. Compound receptors have been described in *E. soleae, Gyrodactylus, Acanthocotyle elegans* and *A. lobianchi,* and these are thought to be chemoreceptors. The haptoral knobs of larval *E. soleae,* with a rich supply of nerve endings but lacking in cilia, may be mechanoreceptors.

Digenea

Miracidia and cercariae are the free-living digenean larvae concerned with host location and penetration. We have already discussed these processes, in general terms, and examined the controversy concerning active host seeking as opposed to accidental discovery. Most of the studies on digenean sense organs have been made on these two larval forms.

Miracidia The morphology of miracidial sense organs is known for only a small number of species (*Schistosoma mansoni, Fasciola hepatica, Allocreadium lobatum, Heronimus chelydrae, Philophthalmus megalurus* and a species of *Spirorchis*). These sense organs are of several types: pigmented eye-spots, ciliated pits, ciliated nerve endings on the apical papilla, and ciliated nerve endings, both on the lateral papillae and between the ciliated body plates. The eye-spots consist of pigment cells surrounding microvillous rhabdomeres and they show a degree of bilateral asymmetry in their size. It seems probable that these structures are photosensitive, serving to orientate the larva, and perhaps assisting in locating the snail host. Associated with the eye-spots of at least one larval digenean (*Diplostomum spathaceum*) are ciliated cells whose membranes show marked evaginations. On circumstantial evidence alone, these are thought to be photoreceptors. The functions of the other miracidial sense organs are not known for sure, but they include tangoreception (ciliated nerve endings), chemoreception (ciliated pits and ciliated papillae) and mechano- or rheoreception (lateral papillae).

Cercariae Relatively little is known about the sensory structures in the cercaria, but the presence of eye-spots, papillae and sensory hairs suggest that these larvae are capable of receiving a range of types of information from the environment. The morphology of the eye-spots closely resembles that of the miracidia, i.e. they consist of pigmented cells, rhabdomeres and microvilli. In many strigeoid cercariae, the eye-spots lack pigment, but the functional significance of this is not understood. The sensory papillae of cercariae are thought to be tangoreceptors, but no function has yet been ascribed to the hair-like projections that cover the body surface and which are particularly prevalent on the tail.

Adult digenea Sensory structures in adult digeneans are associated with the tegument; schistosomes possess ciliated sense organs over the entire body surface, but these are concentrated, in the male worm, in the anterior region of the gynaecophoric canal. Although similar in basic morphology to cercarial sense organs, in the adult, the cilia are completely enclosed by the membranes of the tegument, suggesting that they are mechanoreceptors.

Cestoda

Sense organs associated with the tegument have been described in a small number of species, using standard histological techniques at both light and electron microscope levels. The sense organs are tegumental protrusions, extending from bulbous swellings in the syncytial region. Their function is not known, but it seems probable that adult tapeworms possess chemoreceptors, that provide information on the chemical nature of the environment, and which may allow them no select the optimum region in the host gut for habitation. It has been suggested that proprioreceptors are present in the tapeworm scolex, associated with stretch reception within the organs of attachment. Histological evidence for their occurrence is available only for *Acanthobothrium*.

Acanthocephala

The only information on acanthocephalan sense organs concerns the presence of small pits with a nervous supply, located on the proboscis and around the male copulatory apparatus.

Nematoda

In contrast with the other groups of parasites, there is an extensive literature on the sensory structures of

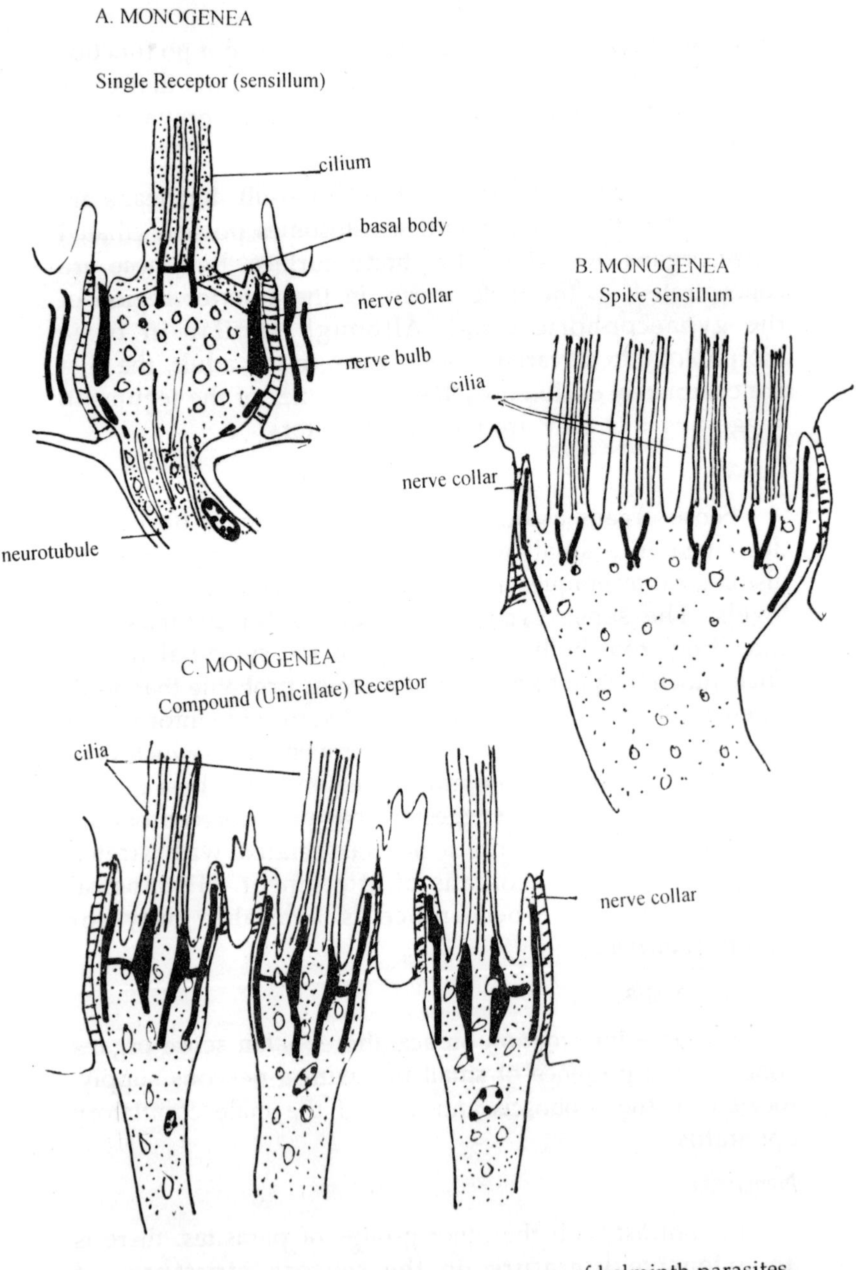

Fig. 6.2 : Some common sense organs of helminth parasites.

D. DIGENEA

Eye-Spot of Miracidium

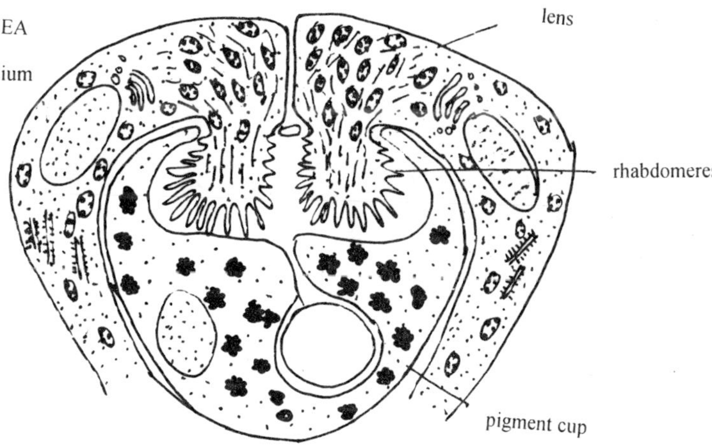

E. DIGENEA

Adult (schistosome) Sense Organ

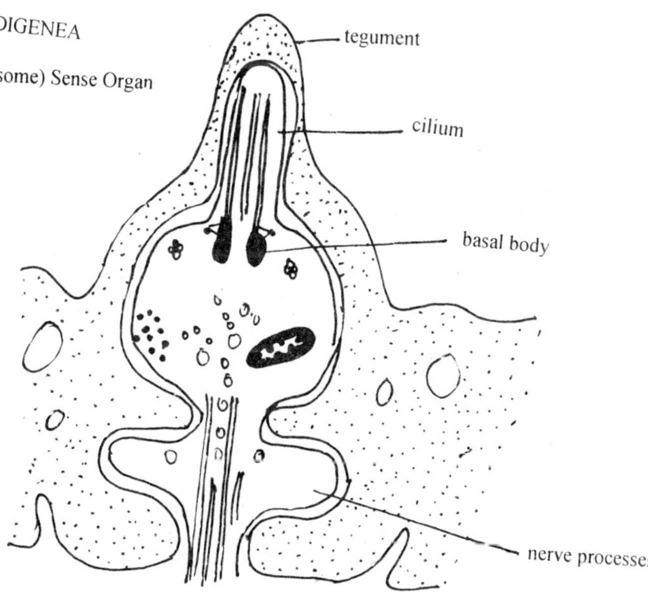

Table 6.1 Nematode sense organs and their functions.

Organ type	*Description*	*Function*	*Examples*
Head			
labial and cephalic cilium, chemoreception	3 concentric circles of papillae *Dipetalonema, Necator,* covered by cuticle	mechano- and	round mouth; single *Nippostrongylus,* Nematospiroides
amphids	contain several modified cilia; open by a pore	chemoreception; photoreception (?) associated with secretory gland	many species
cephalids	anterior and posterior glands	unknown	plant parasites
Neck			
deirids	paired, papillate, lateral organs	mechanoreception (?)	*Syngamus, Ascaris*
hemizonid	refractile band, between cuticle and hypodermis	unknown	*ancylostoma, Trichinella, Haemonchus*
lateral, cervical organs	contains single modified cilium	proprioreception (?)	*Ancylostoma, Necator*
bacillary bands	hypodermal cords; sensory cells with gland cells	ionic and osmotic regulation	Trichuroidea
anterior ventral body	contain 4 cilia; open	chemoreception (?)	Trichuroidea

wall organs	through pore		
photoreceptors	usually absent in parasitic forms		*Mermis*
Posterior body and tail			
caudal papillae	around male cloaca	copulatory chemoreception	*Dipetalonema, Syphacea*
spicules	nervous tissue within spicule of male	copulatory mechanoreception	*Heterodera, Heterakis, Nippostrongylus*
phasmids	lateral, paired, with cilia and a pore	sensory and glandular role	*Dracunculus, Necator, Oesophagostomum*
caudalids	hypodermal commissures in anal region	unknown	*ascaris*
Other sensory structure			
Proprioreceptors	intestino-cloacal junction	stretch receptors	*Heterakis*
setae	cuticular bristles over surface		not in parasites

nematodes, and a wide variety of different sense organs have been identified and described (table 6.1). In general, nematode sense organs are of two types; one with bundles of nerve axons, and the other with nerve endings in the form of modified cilia (amphids, phasmids and papillae). The latter types of sense organs are commonly associated with secretory activities and enzymes, including acetylcholinesterase and other esterases, have been identified, but the sensory role of these organs is not clear.

Nervous transmission, neuromuscular junctions and neurosecretion

The neurophysiology of parasitic animals is an area that has largely been neglected. The probable reasons for this are twofold: firstly, parasites are frequently quoted as being degenerate animals, and hence their sensory physiology would seem to be a poor area of research interest; and, secondly, because of their small size and the difficulties of laboratory maintenance, parasites have proved not to be the best models for neurophysiological investigations.

Acetylcholinesterase activity has been demonstrated in a number of parasites including *Schistosoma mansoni, Fasciola hepatica, Diplostomum phosini, Diphylloobothrium latum, Taenia taeniaeformis, Hymenolepis* spp., *Echinococcus granulosus, Dipylidium caninum* and many nematodes. Adrenaline stimulates muscle contraction in *Hymenolepis nana,* and both catecholamines and 5-hydroxytryptamine (5-HT) are present in the nervous system of schistosomes. In *S. mansoni* there is pharmacologcial evidence that acetylcholine inhibits neurotransmission, while 5-hydroxy-tryptamine acts as a stimulator. It is proposed that catecholamines may act as interneuronal transmitters and stimulate 5-HT release, so transmitting motor nerve impulses to effector organs.

Adult *Ascaris lumbricoides,* being both large and readily obtained from the local slaughterhouse, is a convenient model for studies on neuromuscular physiology. Isolated nerve-muscle preparations have been examined and found to respond to various pharmacological agents rather differently from other animals, e.g. acetylcholine and nicotine stimulate muscle contraction, but adrenaline and histamine have no effect. Excitation of action potentials in the neuro-muscular junction is brought about by acetylcholine and inhibited by y-aminobutyric acid. Acetylcholinesterase is present in varying amounts in the tissues of *Ascaris* and many nematodes, occurring particularly at nerve-muscle junctions, in many of the sensory structures and in the nerve ring. The muscle cells of nematodes are unique in comprising a contractile portion and a non-contractile portion, which makes contact with the excitatory and inhibitory nervous supply. The muscle cell appears to posses proprioreceptive capabilities, but no receptor has been identified (figure 6.3).

Some of the substances used as commercial anthelminthics affect neuro-muscular transmission in parasites, e.g. bephenium salts, 5-hydroxytryptamine and methyridine either bring about neuromuscular blocking or mimic acetylcholine activity, while piperazine, a commonly used drug against intestinal-dwelling parasites, acts through inducing muscle paralysis. Some of the antischistosome drugs, such as hycanthone, affect parasite motility. More complete information on the neuro-muscular physiology of parasites, in general, will have clear implications in the future development of "logical anthelminthics".

Neurosecretory cells have been identified by histochemical procedures in a small number of parasites, (e.g. *Dicrocoelium dendriticum, Hymenolepis diminuta* and *Echinococcus granulosus,* as well as in several species of nematodes). In many nematodes, there is a clear relationship between moulting and neurosecretory activity, and insect juvenile hormones and their synthetic analogues (farnesol and farnesyl methyl ester) can affect the normal moulting process. Bioassay of a substance isolated from infective larvae of *Haemonchus contortus* reveals insect juvenile hormonre activity.

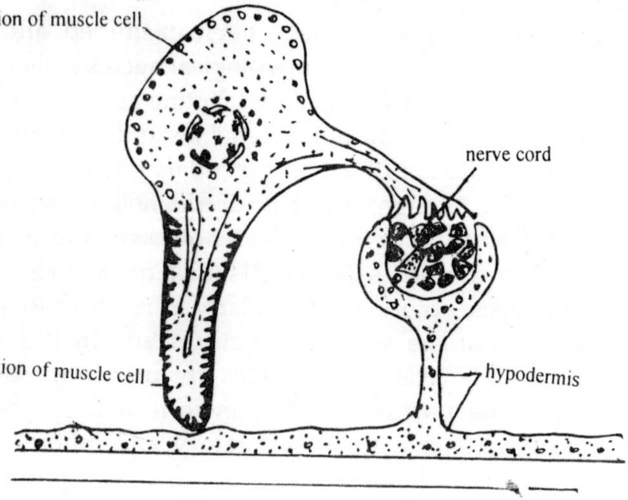

Fig. 6.3 : The nematode muscle cell.

Behavioural coordination

Behavioural studies of helminth parasites are limited to the host-finding activities of larval monogeneans, digeneans and nematodes, and to behavioural analyses of the movements of some adult nematodes. Interest in the latter has only recently emerged and is at an early stage of development.

The mechanisms of coordination of movement have been examined in a few nematodes using a variety of techniques, including agar track analysis, cinematography and closed-circuit television studies, and a neurophysiological approach. The processes and mechanics of locomotion in *Ascaris* are known in considerable detail, and the relative roles of the elastic cuticle, the pseudocoelomic hydrostatic skeleton and the body musculature, in producing the characteristic sinusoidal locomotory movements, are now well understood, However, other distinct patterns of movement can be discerned, associated with activities, such as responses to external stimuli, the act of defaecation, host penetration, mating and feeding. It is becoming apparent that parasitic nematodes are complex and behaviourally sophisticated animals in both their coordination of movement and in their responses to environmental stimuli.

The host-finding behaviour or larval monogeneans and digeneans has attracted considerable attention, but the mechanisms involved are not fully understood. Monogenean oncomiracidia swim freely, but slowly (5 mm sec^{-1}) when compared to the speed of the potential host fish to which they must become attached, and which is of the order of 60 to 300 times faster. It seems improbable, therefore, that successful invasion can take place in open waters, so the behaviour of oncomiracidia must be adapted to facilitate invasion when the hosts are congregated (spawning, feeding or shoaling) and rather less mobile than when swimming freely. The reproductive behaviour of the adult parasite must also be timed so that the release of oncomiracidia coincides with the most favourable time for invasion. Strategies associated with this include the infection, by adult monogeneans, of young fishes only (e.g. *Protancyrocephalus strekovi* on *Limanda aspersa* and *Gastrocotyle trachuri* on *Trachurus trachurus*) or the infection of older fishes exclusively (e.g. *Mazocares alosa* on *Alosa* spp.). These strategies are related to the age-related spawning activities of the host.

Oncomiracidia, miracidia and cercariae are all short lived stages in the parasite life-cycle, remaining active for usually no more than 24 hours, and maintaining their ability to invade and settle on a host for perhaps considerably shorter periods of time. It is unlikely that infection of a host can take place at a great distance form the site of larval emergence. The responses of parasite larvae to environmental stimuli, such as phototaxis, chemotaxis, geotaxis and rheotaxis, probably serve two functions; one is to place the larva is the most favourable position for making initial contact with the host, while the second serves a short range function concerned with precise location, attachment and, finally invasion of the host. Unfortunately we have insufficient information on the sensory physiology of larval helminths to make exact statements about their behavioural coordination associated with invasion of the host.

Nematode larvae respond to environmental stimuli, such as light, heat, chemical substances, mechanical events, gravity and electric fields, and their behavioural and locomotory patterns can alter according to the stimuli present in the external environment. Often these behavioural changes are clearly related to host locating activities, e.g. the heat response shown by larval hookworms, which move rapidly towards a heat source, or the positive phototaxes of pastoral nematodes, such as *Haemonchus contortus*. Many of the other responses recorded seem random and have no apparent advantage, e.g. the responses to changes in electric fields seen in *Heterodera, Litomosoides* and *Pelodera*.

Although it is evident that adult helminth parasites respond to environmental stimuli emanating from the host tissues, this aspect of parasite behavioural coordination is poorly understood. Stimuli that bring about site selection and diurnal rhythmicity, and the ways in which parasites receive, and respond to, this information are yet to be determined.

Chapter—7

Physiology of Reproduction

In group nematoda sexes are separate and sexual dimorphism is well distinct. Females and males can be distinguished by their primary and secondary sexual characters. Males can be readily separated from females by the presence of a copulatory apparatus and in many cases by their smaller size. In so me animal parasites, the female is 30 times larger than the male or the male may be so small that it enters the female and lives within the uterus. In some plant-parasitic genera like *Heterodera* and *Meloidogyme* the 2 sexes show a pronounced dimorphism. The female of these genera enlarges to become a reproductive sac, whereas the male remains slender and worm-like. In other nematodes sexual dimorphism may be also attributed to failure of the males to attain complete development. Thus in males of genera *Paratylenchus, hemicycliophora* the and *Criconemoides,* the stylet and esophagus are often not as well developed as in the females.

Reproductive System

The reproductive system is similar in both sexes of all nematodes. It is composed of 1 or 2 tubular gonads which may vary greatly in length and may be straight, reflexed or coiled. Females with 2 complete genital tubes are called *didelphic* whereas females with 1 genital tube are called *monodelphic*. The corresponding terms in the male are *diorchic* indicating the possession of 2 tests and *monorchic* indicating the possession of 1 testis. If the uteri are opposed at origin the females are termed amphidelphic, if they are parallel and anteriorly directed at origin, the females are called prodelphic and if the uteri are parallel and directed posteriorly at origin, the females are called opisthodelphic.

Female Reproductive System

In most nematodes the germ cells are proliferated only in the proximal end of the gonad. This type of germ cell production is termed telogonic. A relatively small number of nematodes possess hologonic gonads in which the germ cells are formed along the entire length of the gonad.

The Ovary

The proximal end of the telogonic gonad in females is occupied by the *ovary,* a tubular sac in which the germinal cells develop. This sac is made up of an epithelial layer and a germinal cord. In general, the ovary can be subdivided into two regions: (1) The *germinal zone:* a region in which rapid division of cells takes place; (2) The *growth zone :* a region of gradual increase in size of the oogonia. The oogonia are attached to the rachis which is a central strand of non-nucleated tissue. The function of the rachis is uncertain. In some nematodes, the oogonia are arranged around the rachis in the growth zone of the ovary farming a cylinder. The growth zone in most free-living nematodes consists of a single chain of cells and the rachis disappears when the oogonia assume the single file. The free end of the ovary may be either reflexed or it may be outstretched.

The Oviduct

The ovary is followed by the *oviduct*. The oviduct is a functional rather than a structural entity of the female gonad. It is a constricted, thick-walled region between uterus and ovary and consists of a narrow tube with a high columnar epithelium. In some nematodes, a slight enlargement in the proximal part of the oviduct serves as a spermatheca or seminal receptacle.

Uterus

The oviduct is attached to the *uterus* which is a broad tube lined with a flat, cuboidal epithelium and covered by a muscle layer of circular and oblique fibers. The beginning of the uterus usually functions as a spermatheca and fertilization occurs in this region. The eggs are stored in the uterus and as they pass distally along it, they undergo fertilization and maturation, shell formation and some degree of embryonic development.

Monodelphic nematodes generally possess a *postvulvar uterine sac* which may functions as a spermatheca. Such a uterine sac may be the remainder of a second genital tube.

Distally the uterus or uteri enter a common tube, the *vagina* which is lined with cuticle and provided with muscles. The vagina is usually quite short and opens through the female gonopore, the vulva.

The *Vulva* is a midventral transverse slit that is usually situated along the middle third of the body, but may be shifted anteriad or posteriad to the vicinity of the anus. The most cases, a specialized musculature is present which dilates the vulva.

If free-living Adenophorea are compared with free-living Secernentea, a tendency may be noted towards the formation of smaller, but more numerous eggs in Secenentea. This is often accompanied by a lengthening of the growth zone of the ovary.

Parasitic nematodes are usually characterized an increase in the size of the entire genital tube. The uteri and the growth zone of the ovary are especially lengthened and muscle development here is greater. This is correlated with higher egg production.

Male Reproductive System

In most of nematode the testes are of *Monarchic* type and male reproductive system has the following organs:

(1) *Testis.* The *testis* of telogonic nematodes can also be subdivided into a germinal zone and a growth zone. The maturation of the sperm takes place at the end of the growth zone. The sperm are of different shape in the various nematodes. They may be amoeboid or rounded as in many Secernentea, or they may be flagellate or conical. The free end of the testis is often reflexed and long testes may be curved several times.

The testis is followed by the *seminal vesicle* which is a dilated portion of the male gonoduct and functions as a storage for sperm. A few nematodes possess a *vas efferens.* This is a specialized tubular part of the seminal vesicle which separates the growth zone of the testis from the seminal vesicle.

(2) *Sperm duct* or *Vas deferens.* The main part of the male gonoduct is the *vas deferens.* It is usually composed of tubular and glandular regions. It may be provided with musculature throughout its length or only in its terminal portion. In some nematodes, the vas deferens is differentiated into an anterior glandular region comprising the greater part of its length, and a posterior non-glandular region or ejaculatory duct. In *Rhabditis strongyloides,* the vas deferens has 2 large lateral pouches which extend anteriad. These pouches are termed ejaculatory glands and probably secrete the adhesive cement deposited on the vulva of the female during copulation.

Nematodes of the class Secernentea characteristically have a single testis. The only exception to this is the genus *Meloidogyne.* These nematodes possess either 1 testis or 2 testis which proceed posteriad and unite in a single vas deferens. Most Adenophorea are diorchic. Typically, the 2 tests extend in opposite directions but are joined to form the vas deferens as a single tubular canal directed posteriad.

In males of parasitic nematodes the reproductive system is increased in length as compared to males of non-parasitic nematodes.

Mode of Reproduction

Bisexual species usually have equal numbers of females and males. Insemination by a male and egg fertilization in necessary for the development of the eggs in such forms. In many free-living and plant-parasitic nematodes, however, males are much fewer in number than females. This scarcity or absence of males indicates a tendency toward hermaphroditism or parthenogenesis. From the bisexual

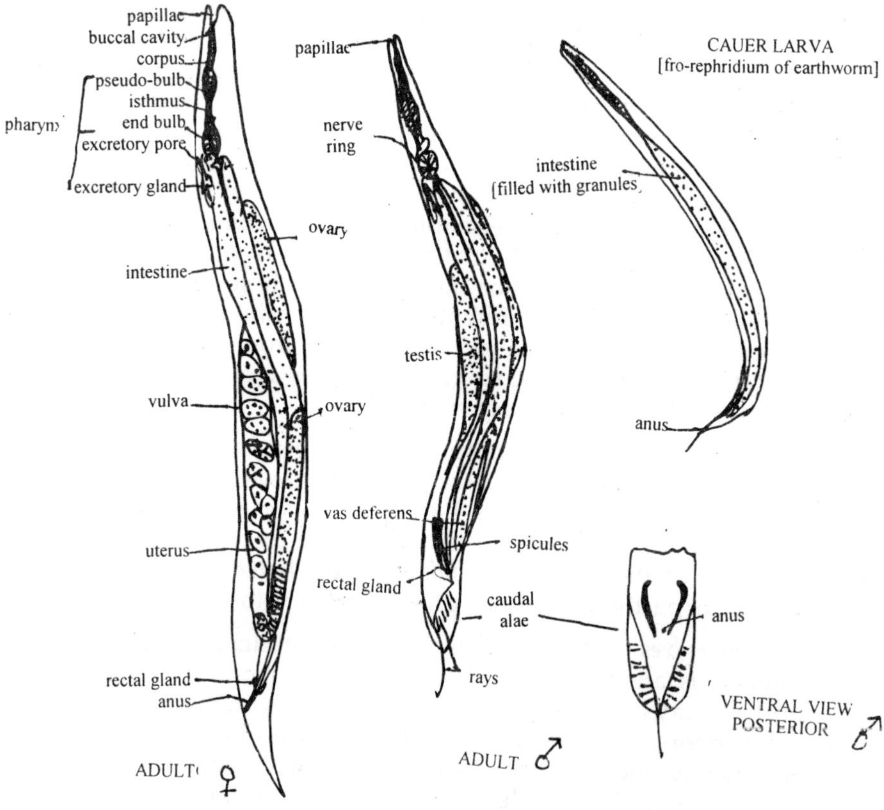

Fig. 7.1 : *Rhabditis maupasi*—morphology of free-living males and females from laboratory culture.

condition with equal numbers of males and females, all intermediate types of hermaphroditic and parthenogenetic species can be found, wherein males have become rare individuals and have mostly lost their sexual instinct, or wherein males may be completely lacking.

In *hermaphroditic species,* reproduction through self fertilization occurs. Hermaphroditic nematodes have the appearance of females. Hermaphroditism is common among the free-living Rhabditida. The hermaphroditism of these nematodes is usually of the syngonic type, that is, sperm and eggs are produced in the same gonad. In most cases, the gonad first produces sperm that are stored and later fertilize the eggs subsequently developed by the same gonad. Males may occur in these protandric hermaphrodites in reduced numbers. They apparently do not function anymore in most instances. A case of digonic hermaphroditism has been reported by Perry (1959) in nematodes of the genus *Helicotylenchus*. Digonic hermaphroditism means that sperm and eggs are produced in different gonads of the same individual. Examination of the gonads of these nematodes revealed that the sperm present in the spermatheca originated in a sphaeroid structure located dorsally from the oviduct near its junction with the spermatheca. This spermproducing organ was termed "*spermagonium*".

Males are lacking completely in most *parthenogenetic species*. In parthenogenesis, the eggs develop without fertilization. Parthenegensis has been proved for a number of free-living nematodes. It is also known to occur in plant parasites of the genera *Meloidogyne* and *Heterodera*. A full account of parthenogenesis is presented in a following chapter.

Intersexes are known in some nematode genera. An intersex is an individual that exhibits a blending of male and female characters. All intersexes reported in Adenophorea such as Mermithidae and other Enoplida, and Secernentea such as *Ditylenchus triformis* were females which showed secondary male characters. They possessed well-developed ovaries with uteri, vagina, and vulva. In addition, they had developed to some extent male sexual openings, spicules, bursal muscles, and male genital papillae. In many cases, these female intersexes copulated with males and were fertile. The only case of male intersexuality has been reported by Chitwood (1949) in *Meloidogyne javanica*. Here, males possessed normal testes, spicules and gubernaculum and in addition exhibited various degree of vulvar development.

Sex Determination

Sex determination has been shown in some nematodes to be dependent on a sex chromosome mechanism. In free-living as well as parasitic Secernentea, there is an XO mechanism or a multiple sex chromosome mechanism of the Xn O type. In some nematodes, however, sex determination has been suggested to be purely environmental. Environmental conditions greatly influenced the sex ratio in the insect-parasitizing Mermithidae and may have a similar effect on the sex ratio of the plant-parasitic general *Heterodera* and *Meloidogyne*.

Reproductive Habit

Little is known about the reproductive habits of nematodes. Some aquatic and terrestrial nematodes lay a small number of relatively large eggs, 50 or less per female, over a long period of time. In saprophagous forms such as *Rhabditis, Diplogaster, Panagrolaimus* a single female may lay more than 200 eggs within one week . Plant-parasitic nematodes, if they are endoparasitic, deposit their eggs inside the plant tissues. Animal-parasitic nematodes discharge their eggs inside the host body. Parasitic nematodes lay more eggs per female than free-living forms.

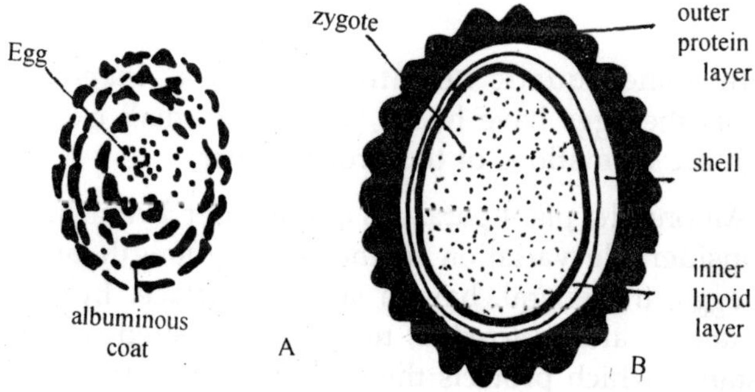

Fig. 7.2 : *Ascaris,* A—Entire egg, B—Egg in section.

The female may lay unsegmented eggs or they may undergo segmentation in the uterus. If the eggs are discharged unsegmented or in various degrees of development, the condition is termed oviparous. In many of the animal-parasitic nematodes, the eggs develop completely while in the uterus and the larvae hatch therein. This is called ovoviviparity. In certain Rhabditida, a peculiar phenomenon has been observed. Old females are unable to discharge the eggs. The larvae thus hatch in the uterus and after the death of the female, break through the uterine wall into the coelom where they soon consume the body contents of the female. The larvae remain enclosed in the cuticle of the mother individual for a period of time until they escape and continue their development normally. This phenomenon has been described by Seurat (1920) as "Endotokia matricida" and has been mentioned later by other investigators.

Nematode Eggs

Eggs of most nematodes are covered by 3 distinct membranes: (1) an external protein coat which is a secretion of the uterine wall; (2) a chitinous membrane of true shell which secreted by the egg itself ; and (3) a vitelline membrane beneath the chitinous membrane. These 3 layers are not present in all nematodes. In some groups, such as the Rhabditoidea and Tylenchida, the protein coat is absent in most of the species. In general, the morphology of the egg varies according to the life habit of the nematode concerned. Usually the eggs have a rounded, oval or elliptical shape.

In some aquatic nematodes, such a *Anaplectus* and *Trilobus* the eggs have hooks, spines, or small excrescences which are formations of the protein layer.

Among the plant parasites of the order Tylenchida, there is considerable variation in the relative length of the egg. The eggs, in general, have a smooth surface. In species of *Meloidogyne* and *Heterodera,* they are deposited in a mucoid substance which protects the eggs. Chemically, this egg sac resembles secreted collagens. It contains phenolic compounds and the enzyme polyphenol oxidase which causes it to change from colorless to orange-brown.

Eggs of animal-parasitic nematodes may bear various specialized structures. Mermithid eggs are characterized by branched polar cords, the byssi, by which the eggs are attached to the surface of plants. From here, they are eaten by the insect host. Terminal, polar or equatorial filaments may be present in other animal-parasitic species. These filaments may function in a similar manner as the byssi or may serve to hold the eggs together in clusters and thus insure better infection of the hosts. The egg shell may be mammillated or may exhibit opercula. Opercula are slightly sclerotized zones in the egg membranes. They are usually located at the egg poles and make it easier for the larvae to escape from these thick-shelled eggs. Mammillation refers to the rounded excrescences over the surface of the protein coat as exhibited by the eggs of *Ascaris lumbricoides*.

Development

Oogenesis—Oogenesis has been studied in free-living, animal and plant-parasitic nematodes. The same is true for nematode embryology which has been the subject of several classical studies in cytology.

Parascaris equorum bivalens, a large animal parasite, shall be selected here to review the maturation cycle and embryology of a bisexual nematode species.

After insemination by a male has taken place, the sperm travel up the female reproductive tract and fertilize the ripe egg cell as they proceed distally. As soon as one sperm has entered the egg cell, a fertilization membrane appears around the egg and the egg protoplasm contracts. The shell begins to form by an endogenous process immediately after sperm penetration. A thin lipoid membrane, the vitelline membrane, is produced on the inner side of the shall. As the egg passes along the uterus the protein membrane is added outside the shell. While the egg membranes are being formed, 2 maturation divisions occur by means of which the normal diploid female chromosome set of the egg cell is reduced to a single or haploid set. *Parascaris equorum bivalens* has 2 pairs of chromosomes ($2n$=4). After the sperm has entered the egg, the first maturation division of the egg nucleus initiates. Early in the prophase the homologous chromosomes pair and then each chromosome is doubled, thus two tetrads are formed. The tetrads take the usual position at the equatorial plate of the cell and the first division of the egg nucleus follows. Two chromatids of each tetrad get into the first polar body which separates from the egg. The 2 chromatids of each tetrad that remain in the egg rotate and the second division follows. The second polar body is formed and the egg nucleus contains now the haploid number of chromosomes (n=2). During the second maturation division, the sperm nucleus increase its volume and the chromosomes appear as in a resting nucleus. The same change takes place with the egg nucleus soon after its second maturation division is completed. These nuclei are the male and female pronuclei. Without fusing each pronucleus again develops 2 chromosomes which immediately, take their position at the equatorial plate of the cell. The first cleavage spindle is formed. Each female and male chromosome doubles and the first cleavage division takes place. This is a mitotic division.

Embryonic Development- The first cleavage occurs transverse to the egg axis and produces 2 cells of about equal size. The second cleavage results in 4 cells which are at first arranged in a "T" shape, but later shift to the rhomboid shape characteristic of the four-cell stage of nematodes. Subsequent mitotic cleavages result in a coeloblastula which is generally dorso-ventrally flattened and elongated in the antero-posterior direction. The coeloblastula is followed by the gastrula stage Gastrulation usually occurs by epiboly. The ectodermal sides of the embryo grow down and push the ventral surface into the interior. The invaginated material forms and oval mass and extends antero-posteriorly in the blastocoel. By the process of gastrulation, the previously flattened embryo is altered to a cylindroid shape. At this stage, the 3 germ layers of the nematode are clearly defined. In cross section the embryo consists of an ectodermal epithelium enclosing a central column consisting of endoderm in the middle with a strand of mesoderm to either side. The embryo then continues to grow and is soon recognized as a larva inside the egg membranes.

The embryonic development of nematodes has a highly determinate character. The fate of each cell can be definitely followed during development of the embryo. There is little if any evidence of regulatory power. If one blastomere is destroyed in the two-cell stage, the other develops as it would have in a normal embryo. As a result the embryo dies at an early developmental stage. Furthermore, the segregation of the germ cell line from the somatic cells takes place at an early stage in embryonic development. In *Parascaris equorum,* as well as in other nematodes, the germ cell line can be distinguished by the behaviour of its chromosomes. During the early cleavage divisions in the embryo, a peculiar fragmentation of the chromosomes takes place in these nematodes. In the second to the fifth cleavage mitoses in all cells which will form somatic tissue, the middle part of each chromosome breaks up into a number of portions. These short segments behave as entirely separate chromosomes in the somatic cells. The remaining club-shaped end portions of the broken chromosomes are left in the cytoplasm where they eventually degenerate. This process of chromatin loss from the nuclei of somatic cells is termed chromatin diminution.

During the late embryonic development cell multiplication ceases in all organs except the reproductive system. The number of cells or nuclei present a that time remains constant throughout the life of the nematode. This condition is known as cell constancy. Subsequent growth of the nematodes is accomplished by enlargement of cells. There is also a tendency toward the development of syncytia.

Post-embryonic Development- Having completed their development within the eggs, the juvenile worms hatch. These juvenile worms or larvae are usually fully developed except for body size and gonad development. The larvae grow and undergo some changes inform especially of the anterior and posterior ends and the gonads develop gradually. A remarkable feature of this post-embryonic development is the occurrence of molts. Periods of growth are separated by molts in which the entire cuticle, including the cuticular lining of stoma, esophagus, vagina, rectum and cloaca, is shed. There are probably always 4 molts. The nematode emerges from the fourth molt as a fully developed adult. The adult may continue to grow to some extent.

In some cases, 1 or 2 molts occur within the egg so that the second or third stage larva hatches. The first molt occurs within the egg in most species of the Tylenchida. In other cases, the larva undergoes the second molt but does not shed the cuticle. The third stage larva remains in the cuticle of the second stage larva. This cuticle serves as a cyst and such larvae may resist adverse conditions. They are termed ensheathed larvae or "Dauer" larvae.

It is possible to determine the sex of the future nematode during the early stages of the post-embryonic development. In *Heterodera* and Ditylenchus, males and females can be distinguished in the third larval stage. In *Tylenchulus semipenetrans and Meloidogyne,* females and males can be distinguished in the second larval stage.

Parthenogenesis and the Role of the Male in Reproduction

Parthenogenesis

Parthenogenesis is the development without fertilization, *i.e.* the egg of an animal develop without fertilization by the sperm. Parthenogenetic reproduction occurs in the absence of the sperm and is of two types; diploid and haploid parthenogenesis. The diploid type is commonest and occurs in some invertebrates including nematodes. Haploid parthenogenesis occurs in some Hymenoptera and Homoptera.

The Haploid Parthenogenesis

The *haploid parthenogenetic* egg is an asexual egg which undergoes complete reduction and is capable of fertilization but may develop without any process of syngamy. Such eggs most commonly develop with the haploid number of chromosomes and in this case, as far as known, always produce males which themselves are of haploid constitution.

The Diploid Parthenogenesis

In *Diploid parthenogenesis* the egg typically undergoes no general reduction and hence develops with the diploid number of chromosomes and so far as is known is incapable of fertilization and even if same took place the product would be a triploid and hence an abnormal zygote. Paula Hertwig in her examination of *Rhabditis pellio* found a mutant which produced only females and formed only one polar body without chromosomal reduction and developed with the diploid number of 14 chromosomes. However, development failed unless the sperm entered the egg although it took no part in the subsequent processes. In may opinion this was not a case of fertilization but rather impregnation of the egg by the sperm and therefore the result was not a triploid.

The general opinion is that diploid parthenogenetic animals were originally bisexual with sexual eggs capable of fertilization. Normally in diploid parthenogenesis there is no reduction in the diploid number of chromosomes and only one polar body is produced. The diploid type of parthenogenesis, even in the sexual egg after completion of reduction, is possible. In a species of starfish the polar nucleus of the second polocyte unites with the egg nucleus subsequent to maturation and the diploid number of chromosomes is present. There seems to be no doubt that the doubling of the haploid number due to fusion of the egg nucleus and the second polar body occurs in some forms of parthenogenesis. In diploid parthenogenesis the offspring may be either male of female as demonstrated by experiments carried out with gall flies, aphids and daphnids while in haploid parthenogenesis only males are produced.

Without exception, as far as known, female-producing parthenogenetic eggs develop with the diploid number of chromosomes. However, recently workers have uncovered new aspects a to the male-producing parthenogenetic eggs of this type. In some invertebrates, males produced through the diploid parthenogenetic process, have one or two fewer chromosomes than the females. Therefore, reduction apparently occurs in the case of one or two chromosome pairs. This occurs in insects such as aphids. A species of *phyllozera* during maturation of the male-producing parthenogenetic egg shows two chromosomes lagging behind and failing to enter the eggs nucleus and are probably cast out in the polar body.

In general, parthenogenesis is followed eventually by sexual forms, or in essence a parthenogenetic series followed by a sexual series, etc.

Parthenogenetic Evolution

Huxley once remarked that comparative evolution is destined to become as important a part of biology as comparative anatomy. Therefore I thought it advisable to consider briefly the evolution of parthenogenesis because in nematodes this method of reproduction is present in some species of economic importance.

Diploid parthenogenesis consists of two kinds, (a) complete parthenogenesis where a species consists of exclusively females or females with rare males and every individual arises from an unfertilized egg, and (b) cyclic parthenogenesis or heterogamy where one or more parthenogenetic generations alternate with a bisexual one. Many parasitic worms have the latter type.

Complete parthenogenesis which may be successful for a time will sooner or later become extinct or perhaps return to sexuality. Bisexual and parthenogenetic races have been found within the same species. However, these should be regarded as "sibling species". White (1954) claims that a parthenogenetic species is really not equivalent to a bisexual one since it consists of an indefinite number of biotypes, between which the exchange of genes is not possible and will continue to diverge in evolution as one mutation after another establishes itself in the different lines of descent. Thus, parthenogenetic species which have been in existence for some time are likely to show a great deal of variability which will not necessarily be correlate with geographical distribution in the same way as variability of a bisexual form.

Since each individual in a parthenogenetic species is capable of reproducing its prolificity is much higher than that of a bisexual form.

Comparison of parthenogenetic animals with related bisexual forms reveals if these polyploid in constitution. In Artemia this is clearly shown. The bisexual race has a somatic number of 42 chromosomes. Meiosis is normal and at metaphase I twenty-one bivalents are present. In the diploid parthenogenetic race there is no reduction in the 21 bivalents and therefore at cleavage I there are 42 chromosomes present. In the tetraploid parthenogenetic race there is no pairing of the chromosomes during oogenesis and therefore the single maturation division takes place with 84 univalents. Triploid parthenogenetic races have been found.

A very unequal sex ratio dies not imply parthenogenesis. Belar found that in *Rhabditis monohystera* where males form only 6-7 per cent of the population, unmated females do not produce offspring. The egg undergoes a single maturation division and then is stimulated into further development into a female by the penetration of the sperm which eventually degenerates. If an egg undergoes two maturation divisions it become capable of being fertilized and after fusion of the male and female pronuclei a male nematode is produced. In some species of this genus there are no maies but the females form a certain quantity of sperm in their ovaries. Sperm from this source stimulates the egg to begin development and then it degenerates. In a unnamed species referred to as *Rhabditis* XX the eggs only undergo a single maturation division, the chromosomes do not pair during the growth stage and at metaphase each univalent splits as in mitosis. The meiosis in the spermatocyte follows a somewhat different course; two meiotic divisions occur but since pairing seems incomplete and highly irregular the chromosomes passing into these sperms is variable. Since these sperms only function for a short time and as stimulating agents and play no part in the genetical make up of the nematode this is probably unimportant. In an evolutionary sense the degeneration of the sperm function has rendered a possible degeneration of meiosis and the regularity of pairing and segregation has disappeared.

In *Rhabditis monohystera* which is bisexual although males are in the minority there is only one maturation division in the egg but pairing of the chromosomes is normal and the 20 chromosomes from 10 bivalents. The somatic number is restored by splitting of the chromosomes which takes place during telophase of the single meiotic division. The degeneration in this species has not proceeded as far as than in *Rhabditis* XX where no pairing occurs in oogenesis. Oogenesis in *Rhabditis* XIV form of Belar's follows the same course as *R. monohystera* but in "XIV" there are no males and the eggs are capable of developing spontaneously without pseudoreduction.

Parthenogenesis in cyst-forming and root-knot nematodes. The possibility of parthenogenesis in *Heterodera* was first fully investigated by Ellenby (1957). He concluded, from the results of his experiments with the golden nematode, *H. rostochiensis,* that parthenogenesis does not normally occur in this species. Fassuliotis (1957) and Williams working independently arrived at the same conclusions as Ellenby with respect to parthenogenesis in the golden nematode. The only instance of parthenogenesis in the genus *Heterodera* was shown by my experiments in 1957. Reproductive in the absence of males has also been observed in *H. trifolii.*

Parthenogenesis is of the diploid type in which the diploid number of chromosomes is not reduced during meiosis and only one polar body is produced. All the offspring were females. There appears to be no pairing of homologous chromosomes.

Reproduction of this species in the greenhouse follows a regular pattern of a generation being produced every forty days. Three pure lines of the clover cyst nematode have been propagated each of which originated from a single unfertilized second-stage larva, to the tenth generation. A single male among the thousands of nematodes produced from these pure lines has been reported.

Activatin of the egg by the sperm

One of the most important processes of fertilization is "karyogamy" which consists of union of the sperm and egg muclei (or pronuclei) to form a cleavage or fusion nucleus. Generally in animals the entire sperm enters the egg. This phenomenon may be clearly seen in *Ascaris megalocephala bivalens* wherein we seen a large sperm composed of a nucleus or chromosomal area and an acrosome. The nucleus is generally surrounded by mitochondria. The fate of the nucleus is generally known but that of the acrosome and the other bodies is still in dispute. The sperm nucleus slowly enlarges after entery into the egg, approaches the egg pronucleus and sooner or later units with it. In *Ascaris* the two pronuclei lie side by side but do not fuse during first cleavage.

In *Ascaris* and various after nematodes the sperm enters the egg before the polar bodies have been extruded or sooner. The sperm nucleus pauses within the egg until the meiotic division are completed. During this time the sperm enlarges, assumes a vesicular structure and finally becomes as large as the egg nucleus.

However, in some instances the sperm enters the egg but does not take any physical part in the maturation process. According to Nigon (1949) in *Rhabditis belari* development can only proceed if the egg is activated by penetration of the sperm. Some eggs undergo two reductional division and then their pronuclei fuse; these produce males and females. Other eggs which produce only females give off a single polar body but the bivalents divide twice; in these pseudogamous eggs development proceeds from the diploid female pronucleus the sperm remaining inert. As previously mentioned Paula Hertwig found a mutant in *Rhabditis pellio* which produced only females, formed only one polar body without chromosomal reduction and developed with the diploid number of 14 chromosomes. However, development failed unless the sperm entered the egg although it remained inert.

Chang (1951) and Chang and Graham (1955) report that in *Strongyloides papillosus* the eggs of the parasitic females may produce one, or two, or no polar bodies while the eggs of the free-living female undergo a single maturation division with no reduction in the chromosome number which results in three morphological forms; free-living males and females and filari-form larvae which will under proper conditions become parasitic females. These workers believe that the sperm of the free-living forms contribute to the genetic constitution of the offspring. However, in contrast to this view Nigon and Roman (1952) hold that pseudogamy takes place in the free-living *Strongyloides ratti.*

Hybrid fertilization

Fertilization within invertebrates of one species by another sometimes occurs. According to Wilson (1925) the somatic number of the hybrid, as might be expected, is typically equal to the sum of the parental gametes or haploid number. This is shown when *Ascaris megalocephala bivalens* is mated with *A. megalocephala univalens* giving a hybrid with three chromosomes or the sum of the two gametes.

White (1954) maintains that comparison of genetic systems of related species is generally facilitated if interspecific hybrids are obtained. In such instance a few hybrids are produced in nature. Many species which never seem to produce hybrids, when kept under crowded conditions in the laboratory and are not supplied with mates of their own species, can be induced to produce hybrids. The failure of chromosome pairing during meiosis is generally the cause of hybrid sterility. Species with the same chromosome number are more likely to produce functional hybrids than those with different numbers.

Moore, Woodroffe and Sanderson (1956) carried out some very interesting experiments with two species of the Ptinit beetle *Ptinus latro and P. hirtellus* are separate sexually dimorphic species. *P. hirtellus* is a bisexual beetle in which males and females are abundant. *P. latro* is parthenogenetic but reproduces in a peculiar manner. The authors mentioned carried out cross-breeding experiments with these two species as follows:

Virgin females of *P. latro* when tested produced no progeny while females mated with *P. hirtellus* males produced larvae. Females of this generation where again mated with *P. hirtellus* resulting in a very large culture; all were female *P. latro*. Further experiments with virgin females failed to produce progeny.

The authors conclude that this is a peculiar type of parthenogenesis in which mating of the female *P. latro* with the male *P. hirtellus* is essential. The male and female *hirtellus* have a diploid constitution of 2N=18 and the female *latro* has 27 chromosomes and is therefore a triploid. In *hirteleus* reduction occurs during meiosis while in *latro* the triploid number is retained. Insemination of the *latro* female is essential for development but since the egg shows no reduction in chromosome number the sperm can play no part in regulation of the chromosome number.

Giant egg or larvae formation

The occasional occurrence in several groups of animals of giant eggs of double the usual size has long been known. These may arise in several ways of which the following two are common; (a) failure of protoplasm cleavage in the final oogoinal division; (b) formation of a monster accompanied by doubling of the chromosome number and subsequent growth of the egg to twice its normal size. In either case it is evident that they are bivalent in respect to nuclear and cytoplasmic content.

When fertilized they often produce double monsters of various types but they may give rise to larvae twice the normal size as in the case of *Ascaris* and in sea-urchins.

In 1958, I successfully carried out impregnation between two species of cyst-forming nematodes. Previous experiments had shown that the clover cyst nematode is of the diploid parthenogenetic type while the sugar-beet nematode is a bisexual species and males are numerous. Mixed cultures of these two nematodes revealed that as many as thirty per cent of the clover cyst nematode were impregnated with sperm of the sugar-beet nematode. The offspring from the clover cyst nematode population were all females. However, some of this second generation produced giant size eggs. Both normal and giant eggs were found in several females. Giant larvae averaging 660 microns were produced. These were alive and active. Seventeen of the giant eggs averaged 178 microns long and 45 microns wide. Two meiotic spindles with many chromosomes were observed in one giant egg and another contained a large spindle with two sets of chromosomes.

Onions (1953) reported the occurrence of a number of giant first—stage larvae from a single cyst of the potato eelworm *H. rostochiensis*. Normal sized larvae were also present in this cyst and therefore he concluded that gene mutation or polyploidy, rather than environmental factors, were responsible for these giant larvae. In the case of *H. trifolii*, poly-ploidy has taken place. According to Goodey (1941) two workers Debray and Maupas found large and normal adults in populations of the stem eelworm *Ditylenchus dipsaci*. These workers suggested that the large forms were due to environmental factors. Goodey (1941) found a wholly giant race of *D. dipsaci*, approximately the same size as the large forms of Debray and Maupat, attacking broad beans in Portugal and Great Britain. He showed, in his case, that the larger size was not due to environmental factors.

In many experiments on impregnation of the clover cyst nematode I has been observed that the *H. schachtii* sperm frequently penetrates the *H. trifolii* oocyte (primary) and proceeds some distance in the egg cytoplasm. Its chromatin becomes resolved into discrete chromosomes. However, the behaviour of the egg nucleus, which in *H. trifolii* has 27 chromosomes and the sperm nucleus was not determined.

From these rather limited observations we might conjecture that since the sperm shows activity it may in some manner be uniting with the egg nucleus producing a tetraploid. In this connection if is observed that *H. trifolii* may have originated from perhaps a bisexual nematode such as *H. schachtii* in the following manner: *H. schachtii* has a diploid complement of 18 chromosomes in the germ line and perhaps during maturation these were not reduced. The sperm with its 9 chromosomes would unite with the egg nucleus to form a triploid with 27 chromosomes or the same number as *H. trifolii* possesses. Furthermore, if sperm now entered *H. trifolii* oocytes there is the possibility that a tetraploid or perhaps twins with 18 chromosomes each would be produced.

Polyploidy

White (1945) maintains that polyploidy in animals is rare. According to Muller (1952) this may be because polyploidy will inevitably upset the sex-chromosome mechanism since tetraploidy hexaploidy, etc. will automatically abolish the heterogamety upon which sex depends. But the sex-determining factor perhaps does not exclude the possibility of tetraploidy in mammals especially tetraploidy arising from hybridization. Darlington (1953) cites the case of the golden hamster which appears to have its origin in the Caucasian region where two geologically older species, its presumed parents, overlap. The evolution may be diagrammed as follows:

Cricetus cricetus (Europe eastwards) 22 chromosomes	X	*Cricetulus griseus* (China westwards) 22 chromosomes
	Mesocricetus auratus Syrian or golden hamster 44 chromosomes.	

How could a species arise as above. Perhaps like plants wherein through frequent production of infertile diploid hybrid whose sole means of making effective germ cells would be through nonreduction, the failure of the first of second meictic division. The egg would then have all XX chromosomes, the sperm XX, XY and YY. If the last of these gave less fertile zygotes the others XXXX and XXXY might be efficient enough to start a new species with sexes in equal numbers. A second possibility is that tetraploids arise by non-reduction in both parents in the original crossing of the two species.

Muller (1925) believes that polyploidy should be effectively confined to those groups in which reproduction is hermaphroditic or parthenogenetic. Polyploidy however is not present in Cestoda or digenetic trematodes at least in those species studied. Earthworms show a number of polyploid species and Turbellarian flatworms have a few probable polyploids. According to White (1954) triploids result from the fusion of a haploid and a diploid unreduced gamete and tetraploids are rare. Generally the triploid arises from a haploid sperm and a diploid egg nucleus which is not reduced.

Muldal (1952) considers that triploidy is a necessary step between diploid and tetraploid forms. In *Drosophila* 3N females are fertile and vigorous and when mated to diploid males they produce, apart from the diploids of both sexes, females like themselves and inviable or unfertile unbalanced individuals. Thus the number of triploids will tend to decrease in each succeeding generation.

Chapter—8

Transmission of Parasites

To the student of parasitology, the life cycle of parasites can often present a bewildering array of detail, with no obvious pattern appearing at first glance. Nevertheless there are certain events that are common in the life-cycles of all parasites, one of which is transmission to the next host. Transmission may occur more than once during a single life-cycle, should the parasite develop in one or more intermediate hosts. The adult of mature parasite completes its development and reproduces in the final or definitive host, releasing eggs, larvae or other infective stages that must be transmitted to another host, whether it be of the same species (as in a direct life-cycle) or to hosts that are of different species (if the life-cycle is indirect)

The transmission of a parasite between consecutive hosts in the life-cycle can take place in one of three ways. First, transmission can occur when the potential host feeds on the egg, larval stage or on the intermediate host harbouring the larval parasite. Secondly, an intermediate host, or vector, may acquire the parasite and subsequently reinfect the definitive host, during feeding on host blood or tissues. Thirdly, transmission can often involve the active penetration of the host by free-living larval stages of the parasite. In an individual species of parasite, more than one of these modes of transmission may be employed at different stages of the life-cycle.

Direct life-cycle

e.g. many Protozoa (coccidia, ciliates), almost all Monogenea, some Cestoda, many Nematoda.

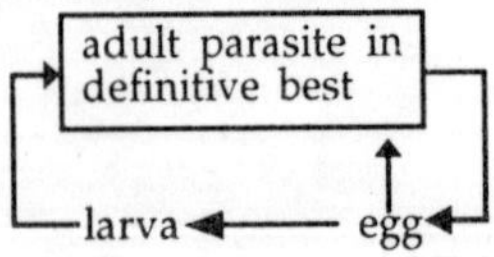

There is only one host in a direct life-cycle. The egg (oocyst in Protozoa) is released from the host, usually in the faeces, and is acquired by being eaten by the new host, or, if it hatches into a larva, either by ingestion or by penetration of the host by the larva.

Indirect life-cycle

Two hosts

e.g. many Protozoa (Sporozoa, flagellates), many Digenea, many Cestoda, most Acanthocephala, many Nematoda.

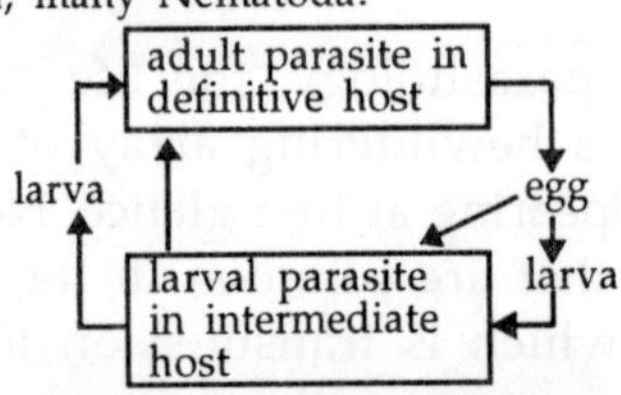

Passage through both hosts is mandatory for completion of the life-cycle. The egg may either be eaten by the intermediate host or it may hatch into a larva which actively penetrates or is eaten by the host (in Protozoa the transmission does not involve eggs but various other stages of the life-cycle). The definitive host become infected either by eating the infected intermediate host or when the intermediate host feeds on host blood or tissues; penetration by the larva also occurs.

Three hosts

e.g. some Digenea (Strigeoidea), some Cestoda (Pseudophyllidea), some Acanthocephala.

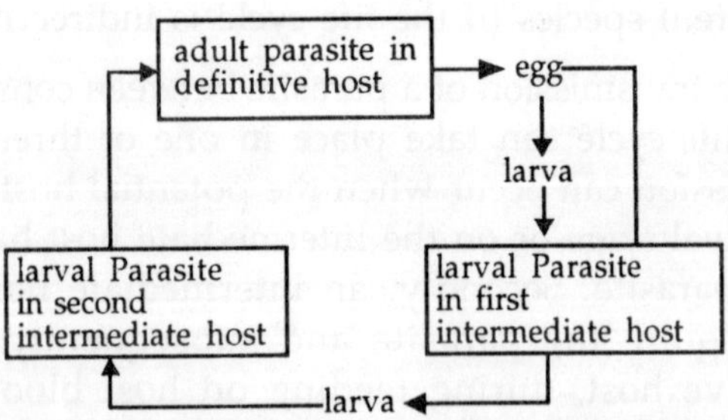

The addition of a third host to the life cycle merely amplifies the position outlined above. Usually both intermediate hosts become infected by active penetration of the parasite larvae; the final host normally ingests the infected second intermediate host for the life-cycle to be completed.

Fig. 8.1 : Simplified schemes of parasite life-cycle.

To help clarify the enigmas of parasite life-cycles, simplified schemes are shown below. In these diagrams, the stage outside the boxes are responsible for transmission.

A parasite may inhabit widely differing environments at various stages in its life-cycle, and it will be adapted for life in each of these environments, however different they may be. Rather surprisingly, little attention has been paid by parasitologists to this ability of parasites to accommodate, both morphologically and physiologically to several quite dissimilar environments, differing perhaps in ambient temperature, oxygen availability, salt and water content, pH and food content. The DNA content a parasite will not alter during a single life-cycle, yet the parasite can perhaps adopt many forms and develop physiological attributes each unique to a particular phase in its life-cycle. Parasitic animals demonstrate this phenomenon to a much greater extent than do free-living animals. Little is known of the factors that bring about the "switching on" of some parts of DNA while other parts are "turned off". We presume that environmental stimuli are responsible for initiating such switching mechanisms via the action of operator genes which can activate the appropriate areas of DNA and inactivate the inappropriate areas. When we consider the factors that control the establishment and development of parasites. Suffice it to say here that these putative environmental triggers, which operate throughout the life-cycle of a parasite, will play a particularly important role during the transmission process, because a parasite must be able to establish in a new host once that host has been located, i.e. its DNA must be switched to the appropriate regions once successful transmission has been accomplished.

Parasites typically large numbers of offspring in the form of either eggs or larvae, and this high degree of fecundity is usually regarded as an adaptation to the peculiar hazards of the life-cycle, involving time spent both inside and outside the host or hosts. Transmission between hosts probably represents a period of considerable loss in the life of most parasites, being a time when a great many individuals either fail to locate and enter a a suitable host, or fail to establish successfully within the right host.

	Monogenea	*Digenea*	*Cestoda*	*Acanthocephala*	*Nematoda*
Direct Life Cycle	egg				egg ↓
	oncomiracidium or viviparous larva ↓				1 to 4 larval stages ↓
	adult worm				adult worm
Indirect Life Cycle		egg ↓ miracidium ↓	egg ↓	egg ↓	egg ↓
1. *two hosts*		sporocyst ↓ sporocyst or redia	oncosphere ↓ cysticercoid or cysticercus or hydatid	acanthor ↓ acanthella ↓ cystacanth	1st, 2nd & 3 rd larvae ↓
		↓ cercaria ↓ metacercaria ↓	↓	↓	4th larva and adult worm
		adult worm	adult worm	adult worm	
		egg ↓ miracidium	egg ↓ coracidium ↓	egg ↓	
2. *three hosts:*		sporocyst ↓ sporocyst or redia	procercoid ↓	acanthor ↓ acanthella ↓ cystacanth	
			plerocercoid (sparganum)	cystacanth	
		↓ cercaria ↓	↓	↓	
		metacercaria ↓			
		adult worm	adult worm	adult worm	

Table 8.1. Various stages in the life cycles of metazoan parasites. The stage represented outside the boxes are the free-living forms that are responsible for transmission to the next host. Each host is represented as a box within which the parasite may pass through a number of developmental larval stages, accompanied often by a dramatic increase in parasite numbers. Although there will be exceptions to the situations portrayed above these patterns of life cycle and the nomenclature for each stage are valid for the great majority of parasites. Protozoan parasite, however, are excluded since they often have complex life cycles than do not readily fit the above scheme and the names of the various stages are not readily comparable.

Mechanisms for locating the host

A variety of different stages are involved in the act of host location. These include oncomiracidia and adult worms in the Monogenea, miracidia and cercariae in the Dignea, coracidia in the Cestoda and various larval stages (usually the third larval stage) in the Nematoda. All of these individuals are active, free-living organisms that usually penetrate the host at its surface, or, in the case of the Monogenea, simply attach to the host surface. Research into the mechanisms employed both by parasite larvae and adults for host location involves behavioural studies, attempts to identify chemo-attractants and various biochemical approaches.

Monogenea

The mechanisms by which monogeneans transfer from host to host have fascinated parasitologists for a very long time, but rather few species have been studied critically. Probably the best known example is *Entobdella soleae,* parasitic on the body surface of the flatfish *Solea solea* (the common European sole), through the work of G.C. Kearn almost a decade ago.

Adult *Entobdella soleae* attach to the surface of the sole by broad, cup-shaped *opisthohaptor;* the worm is quite mobile and capable of limited excursions, such as might be required for mating to be effected. Eggs are laid into the surrounding seawater and these hatch after 3-4 weeks, releasing the free-swimming larva, the oncomiracidium. This larva is small, about 0.25 mm long, and is ciliated in three distinct regions of its body surface. There are four pigmented eyes at the anterior end, each covered by a lens. Information on the host specificity of *E. soleae* strongly suggests that, under natural conditions, the parasite is quite definite in its preference for the sole and rarely occurs on other fishes. It this brought about by active selection on the part of the parasite or do the oncomiracidia attach, but fail to develop in a wide range of fish species? These questions have been answered by the ingenious laboratory studies of Kearn, using isolated fish skin and scales.

When the oncomiracidia of *E. soleae* are offered a choice of isolated scales of its normal host in the presence of scales of related flatfish, such as *Microhirus (Solea) variegatum* (the thick-backed sole), *Byglossidium luteum* (the solenette), *Pleuronectes platessa* (the plaice) and *Limanda limanda* (the dab) (all of which co-occur around the British Isles) *Solea solea* scales are selected almost exclusively. It was noted that the parasite larvae settled, almost in every case, on the remnant of growing skin attached to the excised scale, rather than on the bony element of the scale itself, suggesting that fish skin might possess a specific attractant for oncomiracidia. Attachment to isolated scales occur in complete darkness, ruling out visual perception as a means of host location. The transfer of *E. soleae* from heavily infected sole, placed in a crowded aquarium tank with other species of flatfish, was limited primarily to recruitment of parasites by other sole and not by the other species of flatfish. Additional experiments showed that oncomiracidia will actively select and settle upon agar circles that have been impregnated with extracts of the skin of sole and this finding clearly implicates chemoreception as, perhaps, the major means by which *E. soleae* larvae detect their host (figure 8.2). It is evident that the mucoid secretions, from glands in the fish skin, are the principal attractants, since oncomiracidia are not attracted to sole skin lacing mucus glands, e.g. the cornea. It has been suggested that host-finding by chemotaxis in the Monogenea may be related to their evolution from a carnivorous, free-living ancestor, that found its prey by chemical identification.

Many other species of monogeneans are profoundly host-specific, but it is not known whether this phenomenon is brought about by the recognition of specific secretions of a suitable host or by other means.

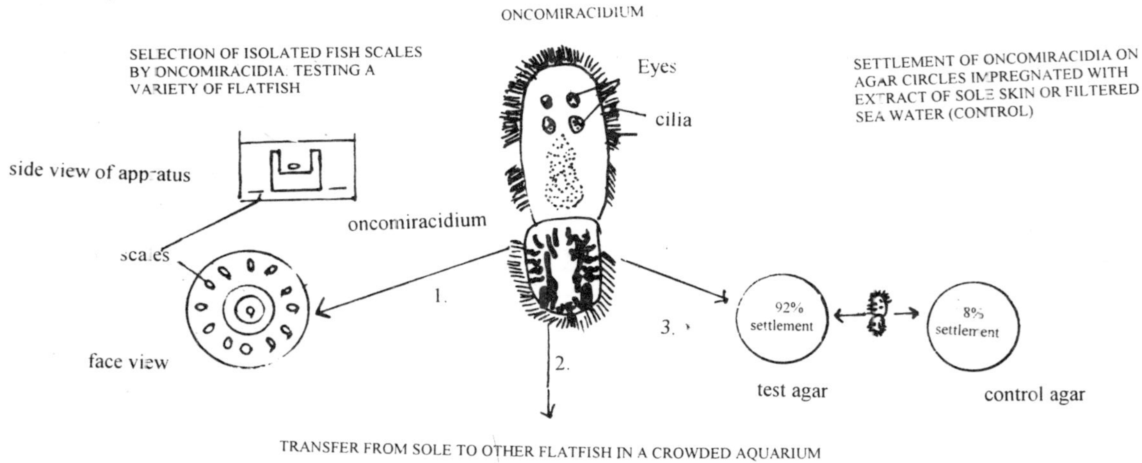

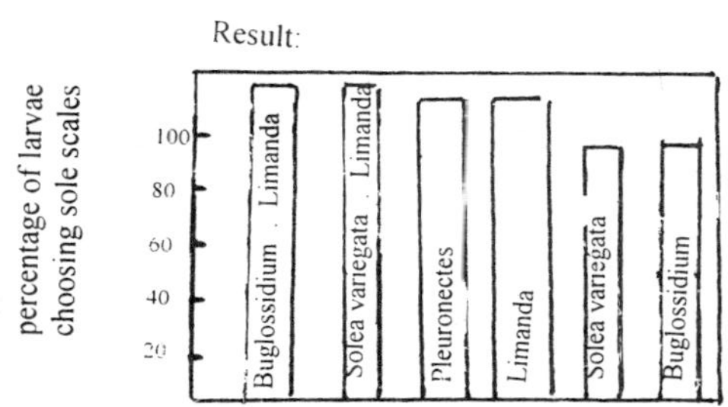

infected Host	Recipient Fish	Percent Transfer
sole	sole	84
	Buglossidium	13
	Limanda	2
	Pleuronectes	0.5
	Raia	0.5

Fig. 8.2 : Host location and host specificity in the monogenean parasite, *Entobdella soleae,* of the flatfish *soleae*. The omcomiracidium is responsible for locating and setting on the most suitable fish, in this case exclusively the sole.

Monogeneans will also transfer from host to host as adult worms when there is physical contact between host individuals and this is probably a common means of transmission in parasites of gregarious fishes. In fishes that do not shoal, parasite transmission many occur when the hosts make contact during their reproductive activities. These are areas where out knowledge is limited and they deserve further attention.

Digenea

The stages of the life-cycle concerned with transmission of digeneans are the miracidia, which hatch from the egg and penetrate the molluscan intermediate host (usually a gastropod), and the cercariae, which are released from the mollusc after a period of several weeks development and penetrate the next host, which may either be another intermediate host or the definitive host. The cercaria may, alternatively, encyst on vegetation as a metacercarial stage and reach the final host by the ingestion route, e.g. *Fasciola hepatica*. The majority of miracidia actively penetrate the mollusc, but some species must be eaten by the snail for continuation of the life-cycle, e.g. *Haplometra intestinalis* and *Glypthelmins quieta* are eaten by the snail *Physa ayrina,* then they penetrate the snail gut wall and invade the tissues.

The miracidium is a ciliated larva with sense organs that consist of eye spots and a variety of lateral papillae, which contain a nervous supply. Most miracidia respond emphatically to the physical elements of their environment, and they are thought to resemble their mollusc hosts in this regard. They are usually phototactic, thermotactic, thigmotactic and geotactic and will often respond to alteration in pO_2, pCO_2 and pH. Responses such as these probably function to bring the miracidia into close physical contact with molluscs. Once in the vicinity of a suitable snail, chemical attraction may facilitate the final stages of host location, although chemotaxis in miracidia is a controversial issue of long standing. While there is a substantial body of evidence implicating the release, by snails, of chemical attractants that are specific for individual species of Digenea and possibly mucoid in orgin, there exist in the literature a great many reports that fail to offer support for the hypothesis of chemical attraction and alternatively suggest that location of snails by miracidia is simply a process of trial and error.

Considerably varied experimental approaches have been adopted to test the chemical attraction hypothesis and apparatus employed includes four-armed mazes and Y-shaped choice vessels, collectively called chemotrometers. Other workers have studied miracidial locomotion under the flying spot microscope and yet others have preferred a strictly biochemical approach, using pure chemicals as potential attractants. Despite this wealth of experimental data, the controversy remains unresolved. Support for chemical attraction hypothesis comes mainly from studies on the miracidia of *Schistosoma mansoni* and *Paragonimus ohirai*. In experimental chambers, the miracidia of *S. mansoni* appear to recognize and swim towards whole snails, crushed snails and snail extracts that have been impregnated in agar blocks. It is not completely certain, however, whether the parasite larvae recognize only those snails in which development can continue of whether unsuitable snails are also detected and approached. Schistosome miracidia detect specific chemicals, such as short-chain fatty acids, certain amino acids and sialic acid impregnated in agar blocks, and will alter their swimming behaviour according to the nature of the attractant. Unfortunately, this approach to the study of chemotaxis, as pioneered by A.J. McInnis, has not been thoroughly pursued using other species of digeneans, yet it would appear to be a fruitful line to adopt. Short-chain fatty acids may be recognized by the miracidia of *Fasciola hepatica*. There is, however, no unequivocal evidence that snails release specific substances that will attract only those digneans which can grow in them. The chemical evidence available at present suggests that snail mucus contains a number of distinct components, any which might be involved in attraction.

There are many reports in the literature that tend to refute the hypothesis of chemical attraction by snails, including studies on *Fasciola hepatica, Fascioloides magna, Schistosoma mansoni, Austrobilharzia* and *Zoogonoides*. The age of miracidia can affect their ability to locate the host, but the mechanisms involved are not known, e.g. studies on *Megalodiscus temperatus* have demonstrated that 2-5 hours old larvae are more successful in locating than snail (*Helisoma trivolvis* the either younger or older larvae.

Table 8.2 Selected references of miracidial location of snail hosts.

Species of digenean Miracidia	*Host snail Species*	*Authors*
Data supporting the chemical attraction hypothesis		
Schistosoma japonicum	*Oncomelania nosophora*	Leiper and Atkinson (1915)
Schistosoma mansoni	*Biomphalaria glabrata*	Kloetzel (1958, 1960), and Etges and Decker (1963)
	Helisoma anceps	
	Bulinus spp	
Paragonimus ohirai	*Assiminea parasitologica*	Kawashima *et al.* (1961)
	A. japonica	
	A. latericea muyazakii	
Megalodiscus temperatus	*Helosoma trivolvis*	Ulmer (1970)
Data refuting the chemical attraction hypothesis		
Zoogonoides laevis	*Columbela lunata*	Stunkard (1943)
Schistosoma mansoni	*Biomphalaria boissyi*	Abdel-Malek (1950)
Schistosoma mansoni	*Biomphalaria glabrata*	Stirewalt (1951), Chernin and Dunavon (1962)
Austrobilharzia variglandis	*Littorina pintada*	Chu and Cutress (1954)
Gigantobilharzia huronensis	*Physa gyrina*	Najim (1956)
Trichobilharzia elvae *T. Physellae* *Schistosomattium douthitti* *Schistosoma mansoni*	A variety of normal and abnormal hosts	Sudds (1960)

On the precise nature of a possible chemical attractant for miracidial host location, snail mucus has been put forward most frequently as the likely candidate. Yet, as we have seen there is no unequivocal evidence to support this notion.

It may be, therefore, that other snail exudates, e.g. reproductive or excretory material, faeces or haemolymph, could be involved, but there is little experimental evidence either to support or refute this. It seems probable, yet unproven, the miracidial responses to the physical environment bring them close to snails and that chemotaxis only operates, if operate it does, at close quarters. What is clear is that most miracidia are more strongly attracted by light than by any other stimulus.

Host locating mechanisms employed by cercariae, the other free-living digenean larva, are also surrounded by controversy. These larvae swim by means of a tail and do not appear to use their cilia for locomotion; some cercariae do not swim, at all but crawl along the substrate to find the host. In the swimming forms the tail is lost once contact with the host is made and penetration commenced. Emergence of cercariae from the snail in which they have develop appears to be under the control of such factors as light, temperature, humidity, pO_2 and pH. In the laboratory, stress (induced by crowding snails) will also stimulate cercarial emergence. Once emerged from the snail, the cercaria exists as a free-living larva for only a short time, up to 36 hours as a rule. The behaviour patterns of swimming cercariae, which represent the most common type, are characterized by alternating periods swimming and resting activity; during the rest period they tend to sink, rising in the water on resuming swimming. This is aided by their positive phototaxis, mediated through pigmented eyespots. Some cercariae do not possess eyespots and are negatively phototactic. These are commonly crawling (rather than swimming) cercaria that infect bottom-dwelling animals. Superimposed upon the primary response to light are additional responses that assist in the process of host location, e.g. the cercaria of *Posthodiplostomum cuticola,* which penetrate and encyst in a variety of European freshwater fishes, respond to shadows, such s might be produced by a passing fish, and swim spontaneously in a swarm. Spontaneous swimming is also initiated by increased water turbulence and by contact with a moving object. There is no evidence or chemotaxis in *P. cuticola,* however.

In general, cercariae are positively phototactic and negatively geotactic, with the noteable exception of crawling forms, but there is doubt as to whether chemotaxis plays any significant role in host location. Early studies on *S. mansoni* and *Clonorchis sinensis* indicated that chemotaxis was unimportant, but more recent work suggests that chemical attraction may play a not insignificant role in host location by cerariae. Chemotaxis may occur in a small number of species, including *Gorgodera amplicava, Schistosoma mansoni, Schistosomattium douthitti, Trichobilharzia, Austrobilharzia,* and *Gigantobilharzia huronesis.* Although it is not clear whether schistosome cercariae actually locate the host skin chemotactically, there is ample evidence that the specific site for cercarial penetration is selected by responses to chemical components in the skin itself, perhaps released by the sebaceous glands. Fatty acids may attract some schistosome cercariae to the host, since it has been demonstrated experimentally that *Schistosomattium douthitti* cercariae fail to penetrate excised mouse ears if the skin is extracted with ether to remove the fatty acids. Penetration resumes if ether-extracted ears are rubbed with fatty acids. The bird schistosome, *Austrobilharzia terrigalensis,* requires contact with sterols, particularly cholesterol, for successful penetration, but these substances do not appear to be attractants for host location by the cercaria.

Cestoda

The only active, free-living larval stage in the cestodes is the coracidium of the pseudophyllidea. This is a ciliated larva, which hatches from the egg after its release into freshwater within the faeces of the bird or mammal definitive host. The coracidium is infective to copepod crustaceans. It differs in two ways from miracidia and cercariae; first, it appears to swim exclusively by random movements, and secondly, it is eaten by its intermediate host, a copepod. As it presents itself as food to its intermediate host, transmission is a passive process.

Transmission in all other cestodes also occur exclusively by passive mechanisms involving the ingestion of eggs or larvae by the host.

Nematoda

Many plant-parasitic and some animal-parasitic nematodes, such as the hookworms of vertebrates, posses active, free-living larval stages that are concerned with locating a new host. Chemical attraction may play a particularly important role in host location by many of the plant parasites. Parasitic nematodes, in general, are will endowed with sense organs, including the antero-lateral amphids and the postero-lateral phasmids, which are thought to be chemoreceptors. Little direct evidence is available to support this hypothesis, however.

Host location in plant-parasitic nematodes is accomplished by random movements that operate when the parasite larva is some distance from the target plant, and by chemoreception when the plant is near. Mechanical stimulation may also be important for successful host location. Considerable exploratory behaviour characterized plant nematodes when searching for a suitable entry site once the plant is reached. By contrast with the initial randomness that is a feature of the plant nematodes, the hookworms possess behavioural patterns that are clearly adaptation for locating a warm-blooded (homoiothermic) animal in a terrestrial environment. Larval mammalian hookworms, such as *Ancylostoma* and *Necator,* tend to be inactive for much of the time as are only stimulated into activity by he increased temperature due to a passing homoiotherm. *Nippostrongylus* larvae seem to search actively for their rat hosts by characteristic waving movements of the head and once a host is located the larvae spiral rapidly down the hairs following a thermal gradient.

Nematode larvae are more sophisticated in their behaviour patterns than the equivalent transmission stages of platyhelminths. They respond to a wide range of stimuli, including light, heat, gravity, various chemicals, mechanical events and electrical fields.

Mechanisms for penetrating the host

Information on the processes of active penetration of the host epidermis by larval parasites is rather scanty and tends to be concentrated on a small number of well studied species. The stage concerned with host penetration are miracidia, cercariae, nematode larvae and some parasitic insects, the latter remaining outside the scope of this book.

Digenea

The miracidium, as we have seen, locates and penetrates a snail and then migrates through the host tissues, finally setting in the digestive gland (hepatopancreas) where it develops into either the sporocyst or redia stage. There is some indication that miracidia search the snail epidermis for a suitable penetration site, but most appear to penetrate at the site of initial attachment. The larvae adhere to the snail surface by means of the apical papilla, which is a mobile and suctorial organ. Penetration is thought to be effected by secretions from the glands that open at the papilla. *Fasciola hepatica* miracidia possess three types of papillary glands. There is a single large, multinucleate apical gland and several pairs of lateral, uninucleate glands, termed accessory glands; additional glands may open at the base of the papilla. When the miracidium a firmly attached to a suitable host the secretions of these glands are extruded by muscular contractions of the entire body of the larval parasite. Miracidia of *S. mansoni* and *S. mattheei* possess similar papillary glands. It is postulated that the secretions of miracidia contain a mucoid lubricant to aid both attachment and penetration and that lytic enzymes are responsible for breaking down the host epidermis. At (or perhaps just prior to) penetration, some miracidia, e.g. *Fasciola hepatica* and

Fascioloides magna, shed their ciliated epidermal plates, whereas other species keep these plates intact during penetration, e.g. *mansoni.* The relevance of these events of obscure at present.

Penetration of mammalian skin by the cercariae of digeneans, most notably the schistosomes, has received

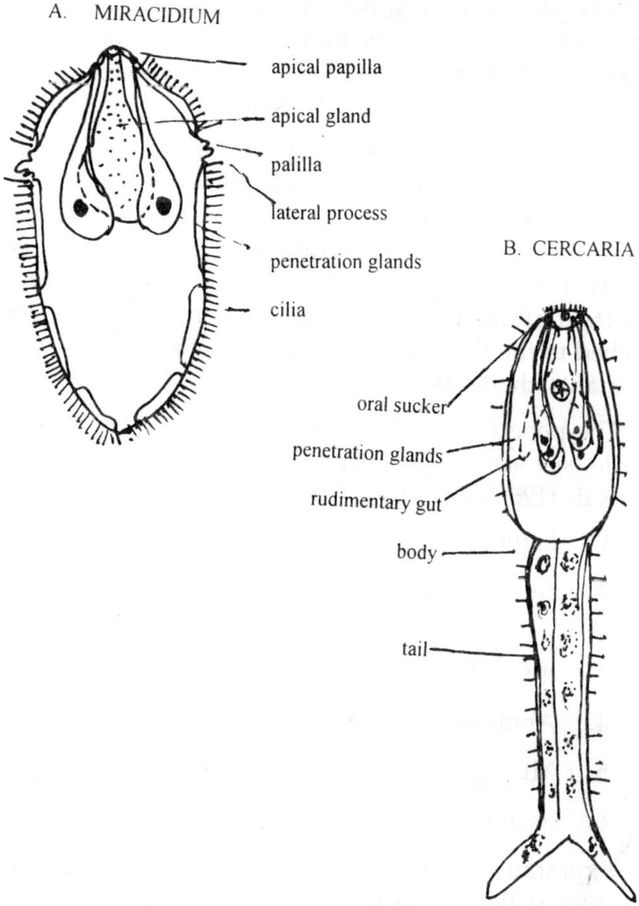

Fig. 8.3 : The gross morphology of skin-penetrating digenean larvae.

considerable attention. Mammalian skin is a complicated structure made of three distinct regions: an outer layer of keratinized dead cells (the *stratum corneum*), below which are living layers, the epidermis and the dermis, separated by a basement membrane. The layers of the skin vary in thickness and composition between different species of mammals, between individuals of the same species and even between different regions of the surface of the same individual. These factors may play an important role in determining both host specificity and the success of parasite establishment in any particular host.

There is some evidence that skin lipids are essential for stimulating penetration by schistosome cercariae, e.g. cholesterol is the major stimulant for the avian schistosome, *Sustrobilhrzia terrigalensis,* while mammalian schistosomes require free fatty acids. These lipids are thought to be important stimulants only after the cercariae have actually attached to the host skin, and they are probably not chemical attractants themselves.

The penetration of mammalian skin by the cercariae of *Achistosoma mansoni* has been categorized into six stages by Stirewalt (1966), and these may be summarized as follows:

1. Exploration of the skin.
2. Secretion of mucus for attachment.
3. Piercing and penetrating the *stratum corneum* through the muscular action of the cercarial body.
4. Abrasion of host tissue.
5. Mucus secretion to soften the *stratum corneum.*
6. Secretion of (putative) lytic enzymes.

Exploration of the host surface s much more a feature of cercarial than miracidial penetration, and it seems that cercariae, although capable of entering at any site, expend energy in locating the most suitable region, such as wrinkles, ridges, areas where dead cells are flaking off, or even hair follicles (gland-cell ducts are not used as sites of entry). Exploration is accompanied by the secretion of mucus from the cercarial post-acetabular glands, and this aids adhesion of the larva to the skin surface. The process of entry itself is part mechanical, brought about by the vigorous and abrasive actions of the cercaria, and part enzymic, due to the secretions of the acetabular glands, which are expelled by the body movements of the penetrating larva. On penetration, the cercarial tail is shed and the migratory larva is now called a *schistosomulum,* a term reserved for species of the genus *Schistosoma*. The cercariae of *S. mansoni* can penetrate a wide of mammals but they do not survive to maturity in all the species entered. Thus, while the skin forms a distinct barrier, it is not the sole location in the host body at which host specificity may be determined.

The transformation from free-living cercaria to tissue-dwelling schistosomulum is a physiologically dramatic one, both in terms of temperature acclimatization (from 25°C to 37°C) and adjustment to a medium with a considerably increased ion content. A schistosomulum that has transformed from a cercariae for only a very short time will die if returned to fresh water at 25°C. Transformation in a facet of skin penetration, but it can also be accomplished experimentally, *in vitro,* by using a biphasic culture medium and also by allowing cercariae to penetrate excised mouse skin. The latter is a routine laboratory technique in the culture of schistosomes.

Nematoda

Some members of the order Strongylida, such as the hookworms and related worms (Ancylostomatoidea, Strongyloidea, Trichostrongyloidea and Heligmosomatoidea) are transmitted to the vertebrate host, usually a mammal, by actively penetrating the skin as a free-living stage. The earliest evidence of skin penetration by hookworms was accidental, when in 1897, the famous parasitologist, Looss, became infected with *Ancylostoma duodenale* in error. Fortunately this mishap has no lasting effect on the pioneering studies of Loss in the field of parasitology. A variety of more controlled techniques has been subsequently employed to study the methods of skin penetration by larval nematodes. These include histological studies, histochemistry, tissue lysis studies, penetration though excised skin and behavioural approaches.

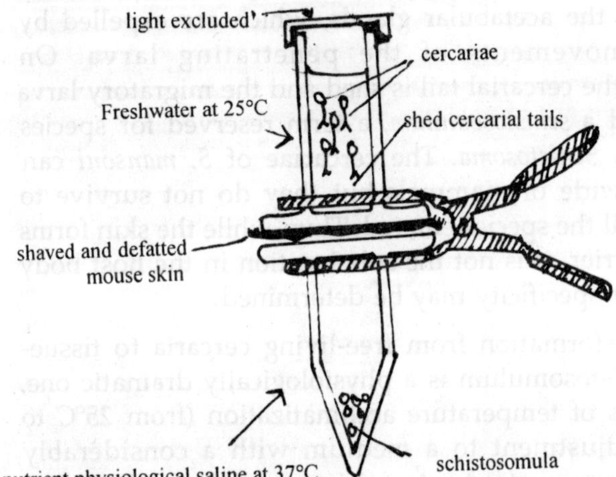

Fig. 8.4 : Apparatus used to transform schistosome cercariae into schistosomula in *vitro*. The all-glass flanged apparatus is incubated in a water bath for approximately 60 min. with excised mouse skin separating the two halves of the apparatus. Alternative methods of obtaining this transformation in the laboratory include mechanical shaking of cercariae to remove their tails or culture of cercariae in a biphasic medium.

Early work on skin penetration by larval nematodes was carried out primarily *in vivo* and led, in general, to the conclusion that enzymes were secreted to effect entry across the skin barrier. More recent work has involved *in vitro* studies, using modifications of the Goodey "floating raft" technique, in which excised skin is stretched across a window in a sheet of cork that is floated in saline. Studies of this type have revealed that there is some variability in the mechanism of penetration adopted by larval nematodes. Initiation of penetration may come through a response to a thermal gradient, as shown by the hookworm, *Ancylostoma tubaeforme,* and entry occurs most frequently between dead cells of the *stratum corneum.* Some larvae exsheath (shed the cuticle of the previous larval stage) before penetration can commence, e.g. *Ancylostoma caninum* and *Necator americanus,* while others penetrate without exsheathment. The act of exsheathment, when it occurs, is thought to be initiated by skin lipids. Enzymes of parasite origin have been implicated in skin penetration in a number of species (*Necator americanus* and *Strongyloides fulleborni*), but is by no means certain that all species rely on enzymes to effect entry across mammalian skin; equivocal evidence on this subject exists for *Ancylostoma caninum, A. ceylanicum* and *A. tubaeforme.* It is apparent that penetration may required a strict sequence of signals for the larva and this is exemplified by the fact that *A. tabaeforme* larvae cannot cross revised skin (i.e. skin with the dermis presented outermost), *in vitro.*

Plant-parasitic nematodes posses a sharp anterior stylet which is used to pierce cell walls during feeding. The penetration of plant tissues, such as roots, appears to be effected exclusively by the mechanical action of the stylet, although there is some evidence that enzymes might assist with the process.

Circadian rhythms associated with transmission

Circadian rhythms, based on 24-hour cycle of activity, are characteristic features of number of parasites and are frequently associated with transmission either to the final host, if the life-cycle is direct, or to the final or an intermediate host, if the life-cycle is indirect.

Hawking (1975) has classified the circadian rhythms of parasites as follows:

1. Rhythms associated with synchronous cell division, e.g. malarial parasites.
2. Rhythms associated with the discharge of infective forms:

a. from a definitive host, e.g. coccidia, pinworms, schistosomes.

b. from an intermediate host, e.g. schistosomes.

3. Rhythms associated with parasite migration, e.g. trypanosomes, malarial parasites, microfilariae (larval filarial nematodes).

Circannual, or yearly, rhythmicity may represent yet another aspect of parasite transmission, particularly when associated with annual cycles of host occurrence.

Many species of malarial parasites are strictly periodic in their asexual reproductive cycle in avian or mammalian blood, which cell division, or schizogony, occurring every 24 hours (*Plasmodium knowlesi*), every 48 hours (*P. cynomolgi*) every 72 hours (*P. malariae*). The circadian rhythmicity in malarial schizogony is closely related to the production and viability of the gametocytes, the stage involved in the process of transmission to the mosquito intermediate host, or *vector*. Male gametocytes of several species of *Plasmodium,* including *P. cynomolgi, P. knowlesi, P. berghei, P. chabaudi* and *P. cathemerium,* become ripe (shown by the occurrence of exflagellation) during a limited period of the day, which coincides with the feeding activities of the appropriate mosquito on the final host's blood, e.g. the male gametocytes of *P. knowlesi* are immature between 09.00 and 13.00 hours, mature between 21.00 and 05.00 hours and subsequently become moribund and die. It has been suggested that both schizogony and the time required for gametocyte maturation are, in each species, related to the biting times of the various mosquitoes involved in their transmission. It is not certain whether the female gametocyte possesses a similar circadian cycle of development. Experimentally-induced hypothermia in monkeys infected with malaria, has led to the conclusion that the synchrony of malarial schizogony is related, or *entrained,* to the daily temperature cycle of the host mammals, since prolonged cooling severely disrupts the parasite's circadian rhythm of development.

In malaria, then, circadian rhythms are associated with the requirement for the infective stage (the gametocyte) to be present in the peripheral blood at time when the transmission to the appropriate biting insect would be most likely to succeed. Other parasites possess similar rhythms which involve the active release of infective stages (eggs or larvae) from the host. The release of infective oocysts of coccidian parasites from the gut of birds is timed to coincide with the optimum period for transmission to another bird, e.g. in the case of *Isospora* in sparrows, the optimum time for transmission would occur in the late afternoon and early evening, since this is the time when roosting takes place, and many birds congregate; at other times of the day sparrows tend to be solitary in their existence. The life-cycle of *Isospora* is direct and infection is acquired by birds ingesting the liberated oocysts.

Mammalian pinworms (*Enterobius vermicularis* in humans and *Syphacia muris* in rats) deposit their eggs in the perianal region of the host by migrating from the rectum, their normal habitat, on a circadian basis. Migration of the female pinworms is associated with the daily lowering of the rectal temperature during sleep. The advantage of laying eggs during this period is probably to avoid contamination with host faeces, an occurrence that might reduce the chances of successful transmission, which takes place via scratching and the accidental ingestion of eggs. A similar cycle of egg release has been described for human infections of *Schistosoma haematobium;* a peak on the numbers of eggs present in the urine has been found to occur at midday. Hatching of schistosome eggs may also follow a circadian cycle.

Schistosome cercariae show strong preferences for liberation from the gastropod host at prescribed times of the day and this appears to be related to the optimum availability of the final host *S. mattheei, S. mansoni, S. haematobium* and *S. bovis* cercariae tend to emerge during the early part of the day, while *S. japonicum* and *S. rhodaini* show a preference for emergence during the early evening and night. The emergence of *S. japonicum* cercariae at night suggests that this schistosome may have evolved as a parasite of nocturnal mammals and that man's involvement is perhaps accidental.

Many parasites commonly migrate within the tissues of their hosts and in certain cases these migrations are associated with transmission to an intermediate host; their function is to locate the parasite in the most suitable position within the host body for successful transmission to occur. Circadian periodicity in the distribution of microfilariae (larval filaria nematodes) in the mammalian blood system is a well known example of just such a migration. Hawking (1975) has examined the various microfilariae and recognizes four categories of microfilariae, according to their disposition within the hosts:

1. Numerous in the peripheral blood by night, rare or absent by day, e.g. *Wuchereria bancrofti* and *Brugia malayi*.
2. Numerous in the peripheral blood by day, rare or absent by night, e.g. *Loa loa*.
3. More numerous in the peripheral blood in the evening, e.g. *Dirofilaria immitis*.
4. Present in the peripheral blood over the entire 24 hour period, but more numerous in the afternoon, e.g. *W. bancrofti* in the Pacific region.

Microfilariae are transmitted by various species of mosquito and the feeding habits of these insects coincide with the time of appearance of microfilariae in the peripheral blood, e.g. day-biting mosquitoes transmit *Loa loa* and night-biting mosquitoes transmit *W. bancrofti*. When not present in the peripheral blood, microfilarae accumulate in the lungs, specifically in the pulmonary circulation at the junction between arterioles and capillaries. The process of larval nematode accumulation at this location appears to be related to the difference in oxygen tension (ΔpO_2) between pulmonary venous and arterial blood, e.g. in *W. bancrofti,* a ΔpO_2 of 55 mm Hg is sufficient to induce accumulation of microfilariae in the lungs, and a drop of ΔpO_2 to 44 mm Hg will initiate migration of the nematodes to the peripheral blood. Artificial elevation of the pulmonary ΔpO_2, by taking exercise or breathing oxygen, will also bring about pulmonary accumulation of microfilariae. It is supposed, but yet to be substantiated, that the sense organs of the microfilariae are chemoreceptors, capable of detecting changes in pO_2 in the host blood. It is not known whether microfilariae possess an endogenous rhythm of their own, for, in the above example, they are clearly responding to an endogenous rhythm of host origin.

A similar circadian migration has been described for some species of trypanosomes inhabiting amphibians. One species, *Trypanosoma rotatorium* appears in the peripheral blood of *Rana clamitans* by day and accumulates in the renal blood vessels by night—this trypanosome is transmitted by leeches that feed by day. A second species of trypanosome, parasitizing the same host, shows the reverse pattern of migration and is thought to be transmitted by night-biting insects. It is not clear whether these migrations are due to an endogenous rhythm of the parasites or of the host, but they are presumably related to the light-dark cycle. There is not evidence for this type of migration in the mammalian trypanosomes, but some malarial parasites appear to be distributed differentially within the host blood system according to a 24 hour cycle.

Annual periodicity occurs in many parasites and there is some evidence that this may be related to transmission. It might will be regards as wasteful for a parasite to produce and release infective forms it the potential host animals are not available for infection. Annual cycles, related to transmission, are evident in some malarial parasite's filarial worms, such as *Dirofilaria immitis* in dogs, *Onchocerca gutterosa* and *O. cervicalis* in cattle and horses, respectively, and in other nematodes, such as *Haemonchus contortus* in sheep. It is probable that circannual rhythms are, in fact, common in many parasites, particularly those inhabiting temperate regions of the world, where climate is seasonally demarcated and host occurrence varies considerably with season.

Chapter—9

Growth of Parasites

A parasites must not only be successful in reaching the next host in its particular life-cycle, but must also establish itself, grow and mature if reproductive success is to be achieved. Sexual maturity will normally occur only in the final host. In the intermediate host (or hosts, should the life -cycle be indirect), the parasite may grow in size and may increase its numbers dramatically by asexual multiplication. Alternatively it may develop into a quiescent stage, such as the metacercaria of the Digenea, awaiting transmission to the next host, either by ingestion or death of the intermediate host.

In this chapter we shall consider the array of factors that permit parasite establishment in a potential host animal. A simplified sequence of events, depicting the barriers to invasion and establishment that confront a parasite during its infection of a single host, (Fg. 9.1). This cycle will clearly be repeated twice, or even three times, in parasites that infect two or more hosts during a more complex life-cycle. The first barrier to invasion is primarily an ecological one and is outside the scope of this book. The second barrier represents the physical limitations to invasion. This physical barrier applies both to parasites that penetrate the host surface and to those that enter either the digestive system or hypodermically, via a blood feeding vector, and pertains to the conditions immediately encountered by the parasite once contact with the host has been made. The third barrier is a complex of physico-chemical and biotic factors that the parasite must be able to overcome or tolerate over a more extended time scale once access to the host body has been gained. These include the immune defence mechanisms of the host. Here we shall consider the physico-chemical and interactive biotic factors that may allow or prevent parasite establishment, growth and maturation. Our attention will be focussed primarily on parasite development within the final host, since this has, in general, attracted the most attention from parasitologists.

Hatching mechanisms

Parasites that invade the host a cystic or egg stages enter the host passively during feeding. Such parasites excyst, or

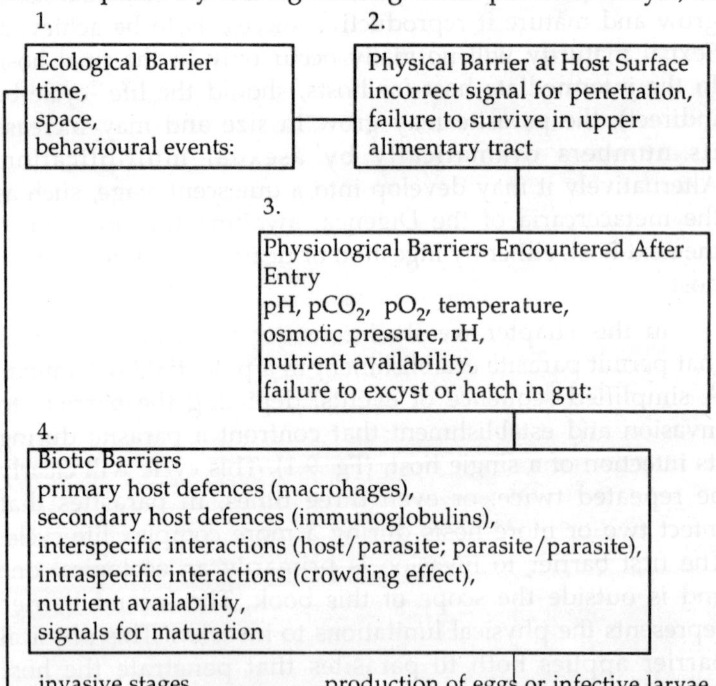

Fig. 9.1 : Principal factors determining the success of invasion, establishment and growth of a parasite.

hatch, into larvae or pre-adults in the alimentary canal of the host, where they may either remain and develop to maturity, of from which they may migrate to other tissues on the host's body. The variety of stages that enter via the mouth are listed below.

Protozoans usually produce infective cystic stages that occur free in the environment, e.g. the oocysts of the Coccidia, and these are acquired when the host accidentally ingests them. Many digeneans from infective metacercariae, which are usually encysted and located within the tissue of an intermediate host; a few metacercariae inhabit the external environment (e.g. *Fasciola hepatica*). The cestodes form a variety of infective stages; these include eggs, that are liberated into the external environment; cysticercoids, cysticerci and hydatid cysts, that are encysted in the tissues of intermediate hosts; and the unencysted parasitic larvae, the procercoids and plerocercoids. Acanthocephalans are infective to the final host as the cystacanth larva, which is encysted in an invertebrate intermediate host. Nematodes are infective either through an embryonated second-stage larva within the egg or an ensheathed third- stage larva, the former occurring free in the external environment, the later encapsulated within an intermediate host or free on herbage (trichostrongyles).

Almost without exception, the infective stage of those parasites that are acquired by being eaten by the final host have a reduced level of metabolism, i.e. they are quiescent. Possible exceptions to this general rule include the progenetic larvae, such as pseudophyllidean plerocercoids (Cestoda) and strigeid metacercariae (Digenea), which grow actively and, as larvae, are very close to the adult stage in their development, lacing only ripe gonads.

On ingestion by the hosts, the infective parasite must be able to withstand the rigours of life in the upper alimentary tract, particularly the stomach, but at the same time, it must be capable of utilizing a variety of stimuli to free itself from any protective covering and activate its hitherto quiescent metabolism. Precocious activation could prove lethal to the parasite, so a very delicate balance of factors must operate to ensure that hatching and activation take place in that region of the gut where the prevailing conditions can be tolerated by the liberated, immature parasite. This location is very often the duodenum in mammals and birds, but there are exceptions to this.

Protozoa

Excystation and activation of infective protozoan cysts has been studied on only a small number of genera, including *Eimeria, Isospora, Toxoplasma, Sarcocystis* and *Entamoeba*. Most of these investigations have involved *in vitro* systems and may, therefore, shed little light on the events that occur naturally. Excystation and activation can usually be accomplished, *in vitro,* by raising the ambient temperature to 37°C for mammalian parasites and to 43°C for avian forms, using a culture medium of neutral pH, containing reducing agents and with a low O_2 and a high pCO_2. Bile salts may be an additional requirement for optimal excystation and activation. There is evidence that excystation and activation of protozoan cysts is a biphasic process, therefore, with *Eimeria* and *Entamoeba hystolytica,* initial activation of the encysted parasite is brought about by the elevation of temperature, the presence of reducing agents and a high pCO_2. Subsequent excystation is due to the activity of proteolytic enzymes, produced either by the parasite itself of by the host gut. Excystation in *Eimeria bovis* requires pretreatment with CO_2 and reducing agents, followed by trypsin and bile treatment. Mechanical disruption of the cyst walls may also be of importance in the initial processes, allowing the permeation of host enzymes. Rather unusually, the infective stages of *H. Histomonas meleagridis* are transmitted unencysted, within the eggs of parasitic nematodes (*Heterakis,*) and in this way they are able to survive the potentially inimical conditions of the bird stomach.

In the Coccidia, the infective stages are sporozoites that are contained within sporocysts, themselves within the oocyst. Rupture of the oocyst micropyle is influenced by treatment with carbon dioxide in mammalian parasites, while in avian forms mechanical damage is thought to be more important. Excystation occurs when sporocysts are released from the oocyst and are exposed to bile and trypsin or chymotrypsin in the duodenum, thereby freeing the sporozoites into the gut. In many species of *Eimeria* and some species of *Isospora*, the sporocyst contains a structure, termed the *Stieda body*, which breaks down under the influence of host bile and trypsin to release the sporozoites. In other species of *Isospora (I. canis, I. arctopitheci, I. endocallimici* and *I. begemina)* the sporocyst lacks a Stieda body and excystation occurs following the general breakdown of the sporocyst wall on exposure to proteolytic enzymes. In general, the process of excystation of coccidian sporozoites begins in the stomach or rumen and is completed in the upper small intestine. Coccidian inhabitants of the large intestine may, however, delay excystment until they reach a more posterior region of the gut, e.g. *Eimeria tenella*.

Digenea

The infective stage for the final host in many digeneans is the metacercaria, a quiescent larva that may be encysted either within an intermediate host or attached to vegetation, or alternatively, unencysted within an intermediate host, as is the case with some strigeid metacercariae, e.g. *Diplostomum spathaceum* and *D. gasterostei* within the eyes of many species of fish, and *D. phoxini* within the brain of minnows. The majority of metacercariae are enclosed within a cyst in the tissues of an intermediate host. The structure of the cyst varies widely with species, and thus the requirements for optimum activation and excystation may be very different. In those species that encyst on vegetation (Fasciolidae, Paramphistomidae and Notocotylidae), the metacercarial cyst wall is a much-thickened structure, capable of withstanding dessication and abrasion. In such cases excystation may be a complex process, requiring elevated temperature, bile salts and high pCO_2 to activate the larva, possibly stimulating it to secrete endogenous lytic enzymes. At the same time host proteases, such as pepsin in the stomach and trypsin in the duodenum, digest the cyst wall from the outside. In the case of *Fasciola*, the larva emerges from the cyst through a digested mucopolysaccharide plug in the cyst wall. Triggers for excystation are quite specific, since many metacercariae can pass through the alimentary canal of an unsuitable host and fail to excyst, yet retain their infectiveness.

The chemical structure of the walls of metacercarial cysts has not been adequately determined. The number of distinct layers that occur varies from one to four, depending upon species. *Parvatrema timondavidi* metacercariae are surrounded by a single layer of unidentified viscous material, and the metacercariae of *Posthodiplostomum* and Xiphidiocercariae and surrounded by two-layered cysts, composed of a layer of glycoprotein external to a layer of protein. Three-layered cysts walls are found in *Holostephanus luhei* and *Cyathocotyle bushiensis* (mucopolysacharide, protein and host tissue), *Parorchis acanthus* (protein, glyco-protein and mucopolysaccharide) and in *Notocotylus attenuatus* (acid mucopolysaccharide and two protein layers). Four-layered cysts walls are a feature of the metacercariae of *Fasciola hepatica* (tanned protein, tow layers of mucoprotein plus mucopolysaccharide, and an innermost keratin layer) and *Psilotrema oligoon* (mucopolysaccharide, protein, mucopolysaccharide and an innermost protein layer). There is a large gap in our knowledge concerning the structure of the metacercarial cyst and the relationship between its structure and the optimal conditions required for activation and excystation of the enclosed larva.

With this wide variation in cyst structure, one would expect that the process of excystation in the vertebrate gut is equally varied in its biochemistry. Limited studies carried out *in vitro* suggest that this is, indeed, the case. In some species, such as *Parvatrema*, elevation of the ambient temperature is sufficient to cause excystation, while other species require pepsin pretreatment, followed by trypsin-bicarbonate or pancreatin treatment. It is not clear what contribution is made by the larval digenean itself in the process of excystation, but it may aid its liberation from the cysts by active physical movements or by secreting lytic enzymes. Host bile salt seem to be prerequisites for excystation in some species (e.g. *Fasciola hepatica*) but their role in general remains to be elucidated.

Cestoda

The infective stage of cestodes include eggs, cysticercoids, cysticerci, hydatid cysts, procercoids and plerocercoids, of which only the latter two are not covered by a protective cyst.

The eggs of many tapeworms hatch in a free aqueous environment, but those of the Cyclophyllidea hatch either within the vertebrate gut (e.g. Taeniidae) or in the gut of an invertebrate host. The cyclophyllidean eggs possesses a thick, multilayered envelop that surrounds the larva (the oncosphere), the outer layer of which is called the *embryophore* (figure 9.2). In non-taeniid cyclophyllideans, the oncosphere is released from the egg, partly by the mechanical action of the host mouthparts, and partly by enzyme action. *In vitro* studies on hatching in hymenolepidid eggs suggest that bicarbonate, free CO_2 a neutral pH and proteases, particularly trypsin, are important factors. The taeniid eggs is characterized by the presence of a considerably thicker embryophore. Hatching in these cestodes involves the enzymic disruption of the embryophore (by pancreatin in *Taenia psiformis*, and by pepsin in *T. saginata*) and activation of the oncosphere, which is then released from the enclosing oncospheral membrane. Bile salts are obligatory prerequisites for activation in these two cestodes.

Excystation of cysticercoids, cysticerci and hydatid cysts takes place in the vertebrate intestine following their ingestion along with the tissues of the intermediate host.

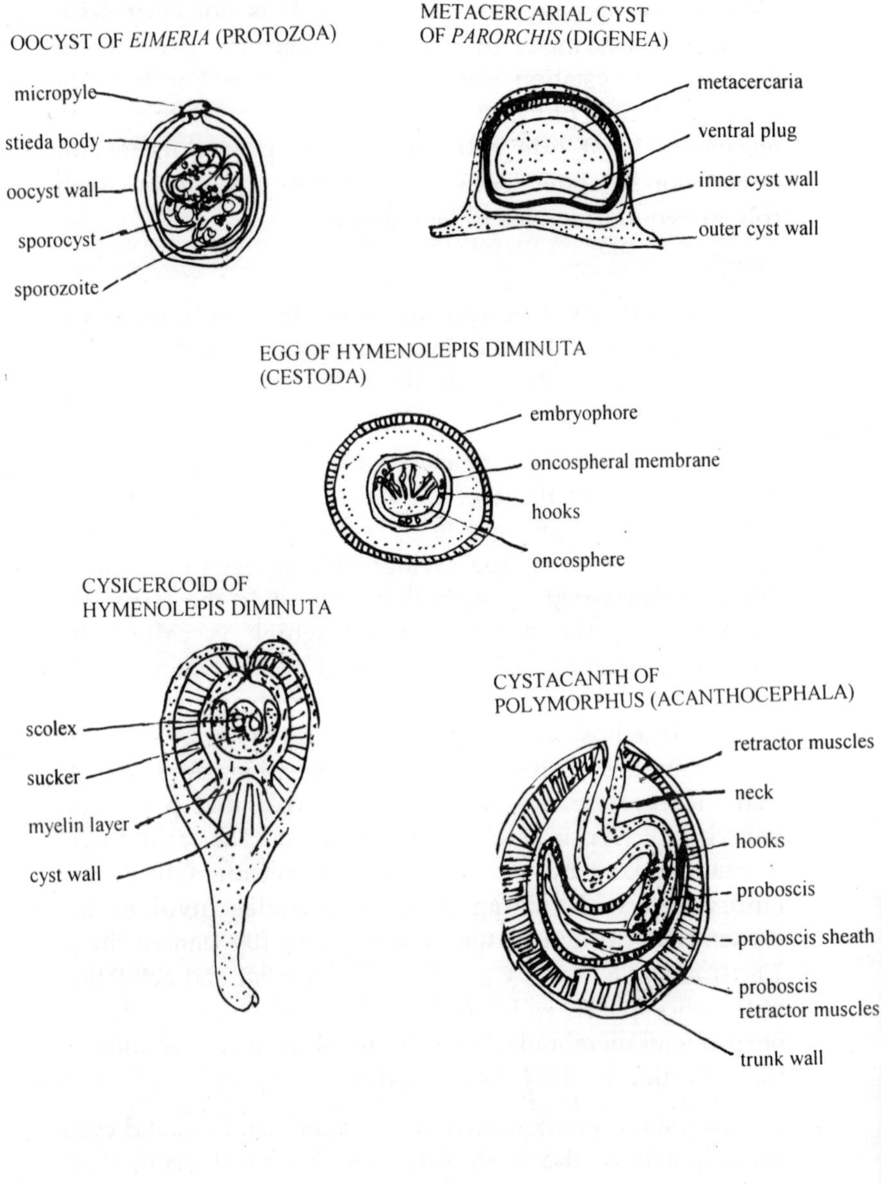

Fig. 9.2 : Some encapsulated, infective parasite larvae.

periods, causing pathological responses in the host during their migration and often becoming encapsulated in organs and tissues. In the correct host these larvae would migrate via a strict route through the body, ultimately reaching the alimentary tract, where they become sexually mature and produce eggs. Another example is that of avian schistosome cercariae, which will penetrate the skin of man but never develop to the adult worm; a mild pathological condition, known as schistosome dermatitis, is produced at the site of

HOT SURFACE

TARGET ORGAN OR PARASITE LOCUS

3.

←——feeding migration——→

1.	2.
ontogenetic migration, and site selection	migration associated with transmission

infective stage of parasite. ←———— release of infective stage

Fig. 9.3 : Migratory patterns of parasites.

1. Ontogenetic migrations. The parasite migrates from the site of entry to its preferred locus or target organ, usually undergoing development (growth and differentiation) during the journey. Ontogenetic migrations are both passive (i.e. via the host's blood) or active (i.e. through the host's tissues). Almost all metazoan parasites migrate ontogenetically.

2. Transmission migrations: Adult, sexually mature, parasites migrate from the target organ in which they reside to release infective stages (eggs or larvae), e.g. pinworm; alternatively the infective stages of the parasites migrate from the target organ occupied by the adult to the peripheral regions of the host body, e.g. microfilariae, schistosome eggs, some trypanosomes.

3. Feeding migrations: Adult or larval parasites migrate actively within their chosen target organ, associated with host feeding patterns; only known for *Hymenolepis diminuta*.

entry. In man the schistosomules of avian schistosomes begin to migrate but this is brought to a half in the dermal layers of the skin.

Ontogenetic migrations by parasites may result in either the adult stage reaching is preferred target organ or in the production of a quiescent larval stage that will await transmission to the next host digenean metacercariae are examples of the latter. It should also be noted that a great many parasites undergo ontogenetic migrations in their invertebrate intermediate hosts, e.g. trypanosomes and malarias in insects, digeneans in molluscs, cestodes in aquatic invertebrates and nematodes in insects. In these instances, successful transmission to the next host can only be accomplished after the completion of the ontogenetic migration, when the parasite is at the correct developmental stage and in a suitable location within its invertebrate host, e.g. tapeworm cysticercoids will be infective to vertebrates only after they penetrate the insect gut wall and complete their development in the haemocoel.

A third type of migration has been characterized in the tapeworm, *Hymenolepis diminuta* and is thought to be related to the feeding patterns of the rat host. In worms that are eight days old, or older, there is a posterior migration in the intestine of the rat during the day and an anterior migration by night. This takes place in both single and multiple worm infections when the rats are feeding by night and fasting by day, which is their normal routine. The pattern of migration can be reversed, experimentally, if rats are fed by day and denied access to food by night. This migration represents a real and considerable movement of tapeworms in the host's intestine, but the factors that initiate this migration have not been isolated. It has been suggested that either the passage of food or the gradually diminishing levels of free glucose in the intestine are responsible for stimulating the posterior phase of the migration, while the subsequent anterior migration might be along a gradient of glucose or some other substance, such as bile. Alternatively, the

INTESTINAL MUCOSA (Hymenolepis nana)

A. MOUTH STOMACH SMALL INTESTINE MIGRATED TO FAVOURED LOCATION WITHIN GUT OR ASSOCIATED STRUCTURES

e.g. Most Digenea and Acanthocephala
Most Cestoda

B. Nematada: INTESTINE MUCOSA SMALL INTESTINE

(Nematospiroides dubius)

MOUTH STOMACH SMALL INTESTINE

BLOOD SYSTEM STOMACH SMALL INTESTINE

LIVER (Ascaris lumbricoides)

LUNGS OESOPHAGUS

C. Nematodes that invade via a skin-penetrating larva: e.g. hookworms of man

SKIN BLOOD SYSTEM HEART LUNGS TRACHEA

SMALL INTESTINE STOMACH OESOPHAGUS

---

2 PARASITES THAT INHABIT HOST TISSUES

A. Adult and larval Digenea that invade via a skin-penetrating cercaria:

MUSCULATURE (Posthodiplostomum in fishes)

BRAIN (Diplostomum phoxini in fishes)

SKIN SUBDERMAL TISSUE BLOOD EYES (Diplostomum spathaceum in fishes)

LUNGS (Paragonimus in man)

LUNGS LIVER

HEPATIC VEINS

MESENTERIC VEINS

(Schistosoma mansoni in man)

B. Cestodes that invade via the mouth:

MOUTH STOMACH SMALL INTESTINE LYMPH BLOOD SYSTEM LIVER OR LUNGS

e.g. Echinococcus granulosus in man

Fig. 9.4 : Common patterns of ontogenetic migrations within vertebrate hosts.

posterior movement could simply be passive, the worms being swept backwards, with the food, by peristalsis; once in the lower small intestine it may be that the worms recognize the unsuitability of their surroundings and actively migrate anteriorly to the duodenum. Certainly, they grow less well in the posterior intestine than in the more anterior regions. It is not known whether this type of diurnal feeding migration is common occurrence in gut-dwelling parasites (it would be most likely to occur in parasites that were not firmly attached to the mucosa of the host intestine).

Site selection and recognition

The great majority of invading parasites do not settle at the site of entry of the host but migrate to their preferred locus. Additionally, most parasites appear to be rather specific in their site requirements and rarely does one encounter ectopic parasitisms in the normal host. These facts strongly suggest that parasite migration from the site of entry to the final location is a directed event, involving the active participation of the parasite itself. The ubiquity of migrations associated with site selection should, ideally, allow us considerable insight into the mechanisms involved in the processes of site recognition and location. Unfortunately, this is not the case, and we known very little about the physiological signals which guide a migrating parasite to its preferred locus.

In the Monogenea, post-settlement migration is commonplace and many of the gill flukes of fishes change their position on the gill once the oncomiracidium has settled. Evidence of this has been obtained for *Tetraonchus monenteron, Diplectanum aequans, Diplozoon paradoxum* and *Discocotyle sagittata*. Analysis of the distribution patterns of *Discocotyle sagittata* and *Diclidophora merlangi* on the gills of trout and whiting, respectively, show clearly a deviation from the expected if settlement and establishment were random processes. These parasites actively select and migrate to specific areas of the gills, usually on the most anterior gill. A rather special case exists for the amphibian parasite *Polystoma integerrimum*. As a juvenile worm, the parasite inhabits the internal gills of the tadpole and, during the metamorphosis of the host, it migrates along the ventral, external surface of the host to the cloaca, from which it enters the urinary bladder, where it resides as an adult worm. Sexual maturation of *Polystoma* coincides with gonadotrophin release by its host, ensuring that parasite eggs are released into the water at a time when larval frogs are abundant. Fertile, neotenic larvae of *Polystoma* are formed it the external gills of the tadpole are invaded, and a secondary cycle of

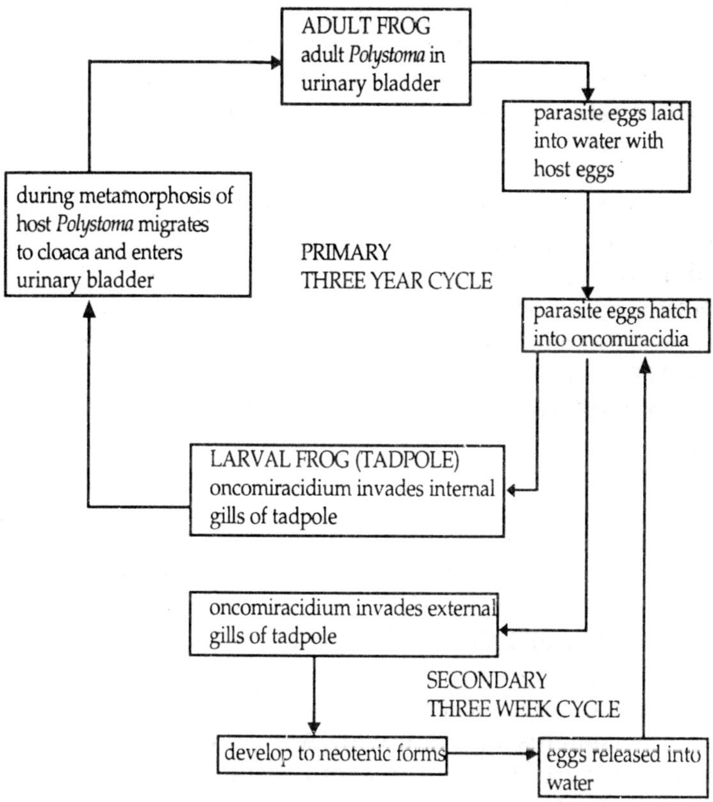

Fig. 9.5 : Site selection and migration in *Polystoma integerrimum* in Amphibia. The secondary, short cycle, leading to the production of neotenic larvae of *Polystoma*, arises only when the oncomiracidia invade the tadpole at its external gill stage; the oncomiracidia produced by this cycle then mature in the normal way, via the primary cycle.

development takes place. The stimulus that controls the migration of juvenile *Polystoma* from the gills to the urinary bladder is probably endogenous, since larvae placed on metamorphosing tadpoles migrate first to the gills and then to the cloaca.

The majority of adult digeneans parasitize the alimentary tract, and its associated structures, of vertebrates. Little is known about their ability to select sites, but it may be assumed that this facet of their physiology is highly developed, as judged by their patterns of distribution within the host. Ingested digeneans will hatch or excyst in the upper small intestine where there is an abundance of proteases and bile salts, and from there they may migrate to other regions of the gut. Preferred sites in the Digenea include the stomach (*Otodistomum* in elasmobranchs, *Hemiurus* and *Derogenes* in teleosts), the duodenum (*Centrovitus, Glypthelmins* and *Loxogenes* in amphibians, *Mesocoelium* in reptiles, *Acanthoparyphium* in birds, *Nudacotyle* and *Alaria* in mammals), the posterior small intestine (*Lasiotocus* in teleosts, *Plagiorchis* and *Echinostoma* in birds, *Echinostoma* in mammals), the rectum (*Stephanostomum, Zoogonides* and *Podocotyle* in teleosts, *Diplodiscus* in amphibians), the intestinal caeca (*Catatropis, Postharmostomum* and *Notocotylus* in birds). Other sites invaded, though less commonly, by adult digeneans include the mouth and buccal cavity (Plagiorchidae), the lungs (*Paragonimus*), the eustachian tubes (*Halipegus*), the pancreatic ducts (*Eurytrema*) and parts of the blood system (*Schistosoma*, Sanguinicolidae). From the above examples it will be obvious that migration to the final site can often be extensive, perhaps requiring the parasite to move against the flow of the gut contents or the blood. Although many of these migrations are well documented, little is known of the physiological mechanisms involved.

Some of the parasitic larval stages of digeneans are known to undergo extensive migrations within their intermediate hosts. This is typical for both the intramolluscan stages (miracidia to sporocysts and rediae) and the stages that inhabit vertebrates (cercariae to metacercariae). Larval strigeids, of the family Diplostomatidae, can migrate over large distances in their fish hosts. The fork-tailed cercariae may penetrate the fish at any region of its body, lose their tails and then migrate to specific organs that can be a considerable distance from the initial site of penetration. The metacercariae of *Posthodiplostomum cuticola* encyst in the outer body musculature, *Diplostomum spathaceum* localized exclusively in the lens of the eye, *D. gasterostei* in the retinal tissue and *D. phoxini* in the ventricles of the brain of their respective hosts. Studies on *Cercaria X,* which is probably *D. spathaceum,* demonstrate that migration can occur via the blood system, the lymphatic system and through the tissues; larvae that do not reach their target organ within approximately 24 hours of penetration are thought not to survive.

Adult cestodes are all inhabitants of the vertebrate gut and their site selectivity within the alimentary tract is known for only a few species. The majority of tapeworms are located in the small intestine, or the equivalent region that lies immediately posterior to the opening of the stomach. Occasionally tapeworms occur in other regions, e.g. *Hymenolepis microstoma* in the bile duct of mice, and *H. fausti* in the caeca of ducks. Within the intestine, a tapeworm prefers a particular region for attachment, through this apparent preference can alter on a diurnal basis, as we have seen for *H. diminuta.* In the chicken gut, *Raillietina cesticillus* excysts close to the opening of the bile duct in the duodenum and, during growth and development of the worm over the next seventeen days, its position of attachment gradually shifts in a posterior direction, so that the fully mature worm inhabits a region 25 cm posterior to the opening of the bile duct. Transplanted *H. diminuta* will migrate to a preferred region of the small intestine regardless of the site of inoculation, but in this species, the notion of a preferred site is complicated by the occurrence of a diurnal feeding migration. The fish tapeworm, *Proteocephalus filicollis,* inhabits the rectum of the three spined stickleback while immature and the intestine when mature. All of the above three cestodes, then, exhibit ontogenetic migrations and a pronounced degree of site selection. In some other cestodes the site of attachment is determined by the morphology of the scolex in relation to the morphology of the host gut mucosa. This is illustrated by two tapeworms that cohabit the intestine of rays (*Raia radiata*): *Pseudanthobothrium,* which has a deep cup-shaped scolex, is restricted to the anterior three tiers of the intestinal spiral valve, where the gut villi are long: *Phyllobothrium* possesses a much flattened and foliaceous scolex and attaches only to the posterior tiers of the spiral valve, which have shorter villi.

ESTABLISHMENT AND GROWTH OF PARASITES

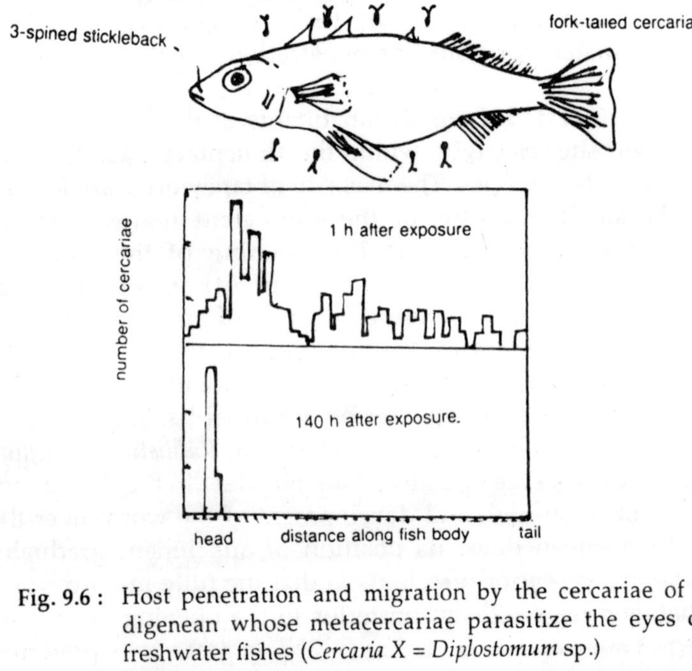

Fig. 9.6 : Host penetration and migration by the cercariae of a digenean whose metacercariae parasitize the eyes of freshwater fishes (*Cercaria X* = *Diplostomum* sp.)

The two histograms record the distribution of migrating parasite larvae within the body of the fish at two times following experimental exposure to cercariae. The numbers of migrating larvae were determined from serial transverse microscope sections of the fish. Larvae that failed to reach the eye within twenty-four hours are thought not to survive.

Larval cestodes may be somewhat less specific in their choice of site, as they often inhabit the body cavity or musculature of the intermediate host. In the case of *Echinococcus granulosus,* the hydatid cysts develop in the first suitable site encountered after penetration through the gut wall, though this may be partly determined according to whether the blood or lymphatic system is used as a migratory route.

Adult acanthocephalans all inhabit the vertebrate gut, in which they are attached to the mucosa and submucosa by the armed proboscis. The proboscis itself is freely retractable so that the potential for movement and change of location within the gut exists. This has not been observed, however, except by implication in concurrent infections of *Moniliformis* with *Hymenolepis,* and *Neoechinorhynchus* with *Proteocephalus.* From observations on the location of acanthocephalans in natural infections it seems probable that each species has a preferred locus in the host's gut. Some species are found in the stomach, particularly in fishes (*Acanthocephalus lucii* in perch, *Acanthorhynchus borealis* in *Lota* and *Silurus*), but a large majority inhabit the upper intestine (*Pomphorhynchus, Neoechinorhynchus, Acanthocephalus ranae, Moniliformis dubius* and *Polymorphus minutus*), or the more posterior regions including the large intestine, the caeca in birds and the rectum in fishes. Larval acanthocephalans inhabit the body cavity of their invertebrate intermediate hosts, but it is not known to what extent they select specific sites within the haemocoel.

Nematodes parasitize a great many organ systems of invertebrates and vertebrates and each species appears to possess its own specific requirements in terms of sites selected. The much quoted study by Schad (1963) clearly demonstrates that several closely related species of pinworms, of the genus *Tachygonetria,* can coexist within the colon of turtles by virtue of their differences in site specificity, along with divergence in their nutrient requirements and feeding methods. Few other studies have examined site specificity in nematodes in such detail. It is evident, from the predilection of many nematodes for complicated patterns of migration around the host's body, that they must be able to recognize and respond to a variety of sequential stimuli concerned with their development, and that they are capable of actively choosing the site in which to reside.

The general subject of site selection in parasites has been largely overlooked by parasitologists, even to the extent that simple descriptions of the sites actually occupied by naturally-occurring parasites are often vague and imprecise. There is a whole field of parasite physiology lying fallow at present, which would be likely to reveal some very interesting aspects of host-parasite interaction. The most fundamental unanswered questions are:

1. What factors permit or prevent the establishment of a parasite at a particular locus in the host's body?

2. How do parasites recognize suitable environments for their continued development?

Invasion of host tissues

It is possible to recognize two separate processes of invasion in the life history of many parasites. The first invasion involves the entry of the parasite into the host's body, while the second invasion occurs then the parasite invades the particular tissues of the host is which it resides. Biphasic invasion, therefore, is typical of tissue parasites, but does not necessarily apply to those gut parasites that inhabit the lumen. Obvious examples of parasites with distinct invasion processes include the malarias and babesias that invade red blood cells, coccidians, nematodes and cestodes that invade the gut mucosa, schistosomes that invade the blood system and trypanosomes that invade nervous tissue.

Recent attention has focussed on the invasion of mammalian and avian red blood cells by malarial parasites, and of intestinal epithelia by coccidians. The accumulating evidence points to a high degree of specificity in this secondary invasion, and this feature may be important in determining the host specificity of certain parasites.

Malarial parasites (*Plasmodium*) invade the vertebrate blood stream as sporozoites, during the feeding of an infected mosquito on the host blood. The sporozoites do not remain in the perpheral blood of the mammal for long, but invade liver tissue where they multiply asexually, by schizogony, resulting in the production of merozoites. These forms are released into the blood stream and invade host red cells (erythrocytes), where a second schizogonic cycle takes place. The mechanisms by which the merozoites locate and then penetrate the erythrocytes are complex and far from completely understood. The merozoite is surrounded by a thick surface coat, which is lost during invasion of the red cell. The surface coat, carries a net negative charge, and is covered with fine filaments, that are involved in the initial attachment of the parasite to the host cell. It can be removed experimentally by treatment with proteases. At the anterior end of the merozoite is an apical protruberance which contains a complex of organelles (rhoptries, micronemes, polar rings and microbodies). These organelles characteristically disappear shortly after entry into the red cell is accomplished. The actual process of invasion is initiated when the merozoite attached to the surface of the red cell by way of the surface filaments, and the apex of the merozoite is appended directly to the host cell surface. This latter event causes local bending of the red cell membrane. Since the surfaces of both the parasite and the host cell are negatively charged, the initial attachment must be regarded as a "captive" process that overrides the natural electrostatic repulsion of the two cells. As entry proceeds, the red cell membrane invaginates, and this is thought to be due to the release of material from the rhoptries and micronemes. There is, therefore, no penetration as such, of the red cell membrane and the engulfed parasite comes to lie within what is termed, a *parasitophorous vacuole,* bounded by the red cell membrane. The entire act of invasion may be very rapid, e.g. occupying as little as one minute in *Plasmodium knowlesi.* The mechanisms by which the *parasitophorous vacoule* is formed are not known, but a histidinerich protein has been isolated from various stages of *P. lophurae,* and this induced cupping of red cell membranes at low concentrations, *in vitro,* and membrane lysis at higher concentrations. Whether such a protein is secreted by the apical organelles of the merozoite to facilitate invasion remains to be clarified. The sloughed-off parasite surface coat appears to plug the opening to the *parasitophorous vacuole* and may also aid the movement of the parasite into the red cell. Once invasion is completed, the merozoite transforms into the trophozoite which remains within the parasitophorous vacuole.

There is evidence that invagination of host cells is a common occurrence during tissue invasion by many protozoan parasites, and it may prove to be the typical method of invasion in both sporozoans and coccidians. *Babesia, Eimeria* and *Toxoplasma* all invade by the invaginative route. In these genera, the parasitophorous vacuole often break down shortly after invasion, resulting in an intimate contact between the parasite and the host cell cytosol. It has been alternatively suggested that *Toxoplasma* invades the host mucosal epithelium by inducing phagocytosis and avoiding subsequent digestion. However, inhibitors of phagocytosis, such as colchicine and cytochalasin B, do not interrupt invasion unless the parasite itself is treated directly, implying active invasion by the merozoite.

In the same way that the merozoites of *Plasmodium* recognize and attach to red blood cells, so *Eimeria,* among the coccidians, exhibits a marked preference for intestinal epithelial cells. So strong is the apparent degree of site specificity that sporozoites of *Eimeria* will develop only in gut cells, regardless of whether the site of entry to the host's body is via intravenous, intramuscular, intraperitoneal or parenteral injection. Little is known about the invasion of gut epithelia *in vivo*, but there is a growing body of information based on *in vitro* studies. In culture, *Eimeria* sporozoites will invade and develop in cells from at least seventeen different hosts. Entry into these cells is rapid and appears to be an active process. The role of the apical

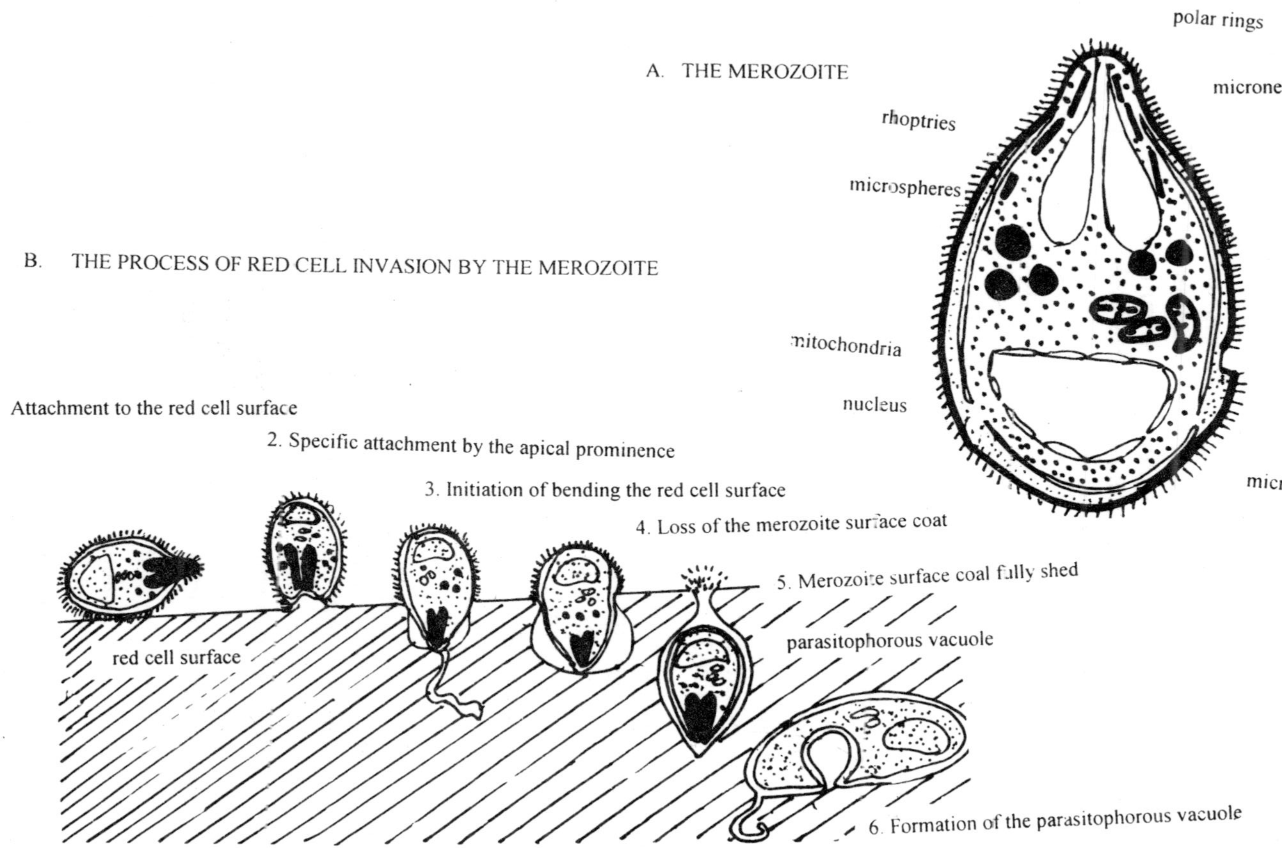

Fig. 9.7 : The invasion of red blood cells by malaria parasites.

organelles in this process is not clear. There is controversy concerning whether the sporozoites of *Eimeria* are phagocytosed by the host cell or whether they actively induce phagocytosis. Treatment of cell cultures with anti-phagocytic drugs suggests that the latter event occurs.

Recognition of the suitable host cell by a tissue parasite will depend upon surface receptors situated on the host cell membrane. There may even by receptors that are specific for attachment and others specific for invasion, since the merozoites of *Plasmodium* will attach to a variety of cell types, but will only invade erythrocytes. Invasion of red cells by malarial parasites is unaffected by neuraminidase treatment, but is reduced by treatment of red cells with proteases, concanavalin A and wheat germ agglutinin. Although the chemical nature of these surface receptors remains elusive, it is evident that they are labile entities that are readily lost or neutralized.

Tissue invasion by helminth parasites has been inadequately investigated and little is known about the processes of cell recognition or invasion. It is not known whether helminths invade host cells by the invaginative route or by direct penetration.

Factors inhibiting parasite growth and development

Crowding effect

Intraspecific crowding of parasites, within host tissues or within the gut lumen, may result in a reduction in their rate of growth, the maximum size attained and their fecundity. Only a small number of examples of the crowding effect have been described, yet it seems likely to be a common phenomenon amongst the helminths.

Probably the best described example of crowding is that of *Hymenolepis diminuta*. When this tapeworm is grown in carefully controlled infection densities in rats, there emerges a clear relationship between the intensity of infection and a reduction in growth rate, maximum worm weight and length, and egg production. Any increase in the number of worms in an infection, however, has no effect upon the rate of maturity of *H. diminuta,* and egg production always commences on the sixteenth day after infection. There is, as yet, no unequivocal explanation for the crowding effect in this cestode. Competition for space, for oxygen and for essential nutrients, such as glucose, have all been proposed as likely candidates. Alternatively, the production of toxins or growth inhibitors by the worms may be responsible, and the immune responses of the host may be directly or indirectly involved. Crowding of *H. microstoma* in the mouse bile duct affects these worms in a similar manner.

The crowding effect has been observed in some species of parasitic nematodes, e.g. the length of adult *Skrjabingylus nasicola,* which infections of increased intensity. *Ancylostoma caninum* infections of dogs are also subject to a crowding effect.

Interspecific interactions

Interactions may take place between cohabiting parasites of different species, and these can result in reduced parasite growth and fecundity, and also be accompanied by alterations in site selection. Concurrent laboratory infections of rats with *Moniliformis dubius* and *Hymenolepis diminuta,* result in a marked restriction in the distribution of each species within the rat's small intestine, as well as a reduction in size of both (the tapeworm shows the greatest size decrease in this relationship). A similar situation can be found in natural infections, e.g. co-occurring *Proteocephalus filicollis* and a species of *Neoechinorhynchus* mutually alter the site of each other in the gut of the three-spined stickleback, so that each species is restricted in its range as a result of the interspecific interaction. This is termed *competitive exclusion,* but the factors that bring it about are no known.

Other classes of interspecific interactions may exist between the parasite and its host, e.g. variations in the size and fecundity of parasites in relation to the size or sex of

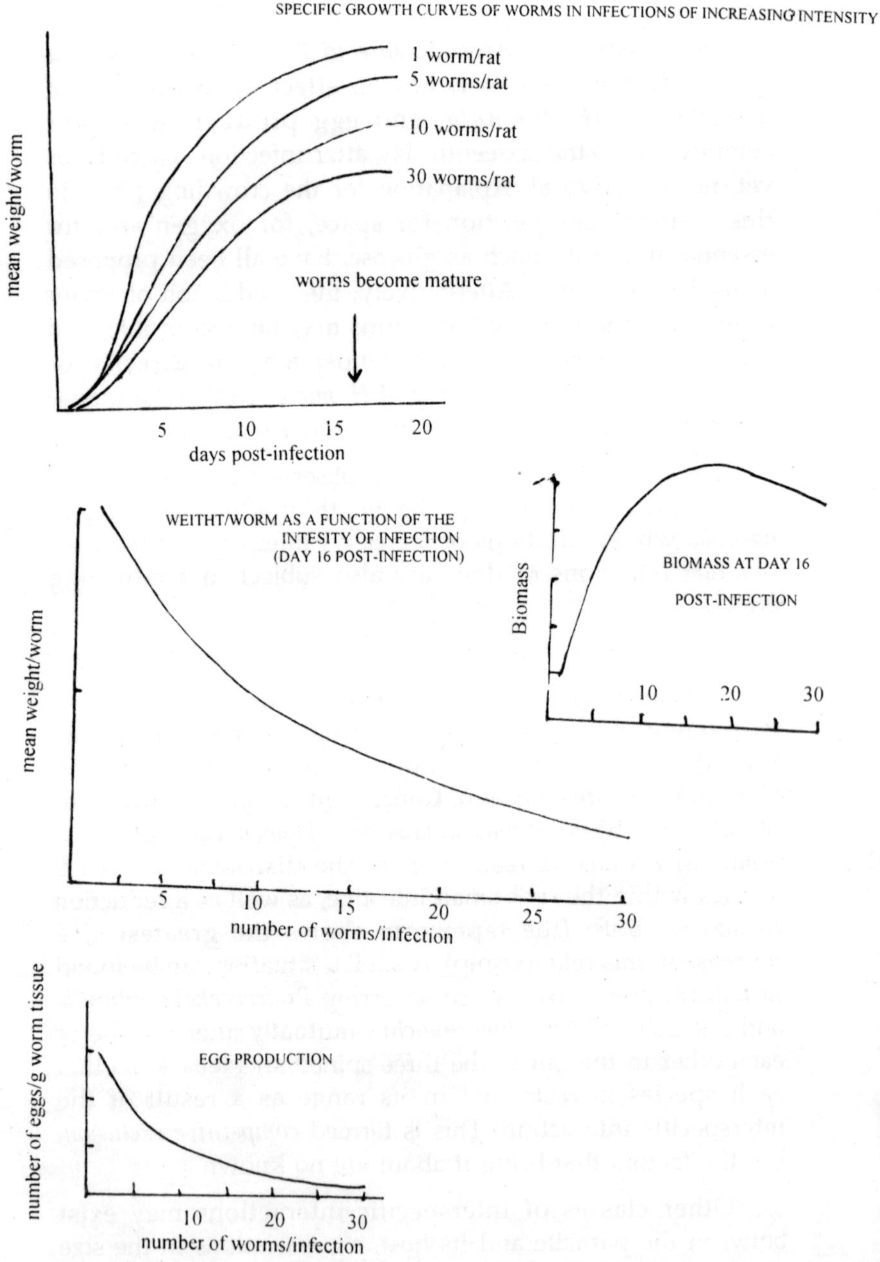

Fig. 9.8 : The "crowding effect" in the cestode *Hymenolepis diminuta*.

their host. Furthermore, a single species of parasite may grow and develop quite differently in a range of host species. Such interactions will involve a variety of physiological factors that include the immune response of the host, its immunological status, as well as the physico-chemical conditions that promote or limit establishment and growth of a parasite.

Labile growth patterns

Members of the phylum Platyhelminthes possess a remarkable ability for degrowth and dedifferentiation under adverse nutritions conditions. This is characteristic of the free-living Turbellaria and the Cestoda, but is known whether it occurs in the Monogenea of Digenea. In the tapeworms, degrowth is termed *destrobilation,* whereby the majority of the proglottids are shed, leaving only the scolex and the neck region attached to the mucosal surface. Destrobilation can be induced in experimental animals by depriving the host of carbohydrate, and this has been demonstrated with *Raillietina cesticillus* and *Davainea proglottina*. Natural destrobilation has been observed in *Dibothriocephalus,* a parasite of bears, when the host's diet changes prior to hibernation. Natural destrobilation may be common in cestodes that inhabit animals which hibernate, where it would serve as an overwintering response by the parasite. Experimental deprivation of carbohydrate may induce a marked reduction in tapeworm size, without the obvious shedding of proglottids. This occurs in *Hymenolepis diminuta, H. citelli, Lacistorhynchus tenuis* and *Oochoristica symmertrica,* and it reflects the dependence of many cestodes on the availability of carbohydrate. Other tapeworms, e.g. *Schistocephalus solidus* and *H. nana,* do not possess such a high degree of sensitivity of depletion of dietary carbohydrate. Cestodes with a typically high rate of growth are more sensitive to a carbohydrate deficiency than those that grow more slowly.

A second feature of labile growth and development is found in the development of the hydatid cysts of *Echinococcus granulosus*. The protoscoleces within the cyst are capable of development of differentiation in one of two

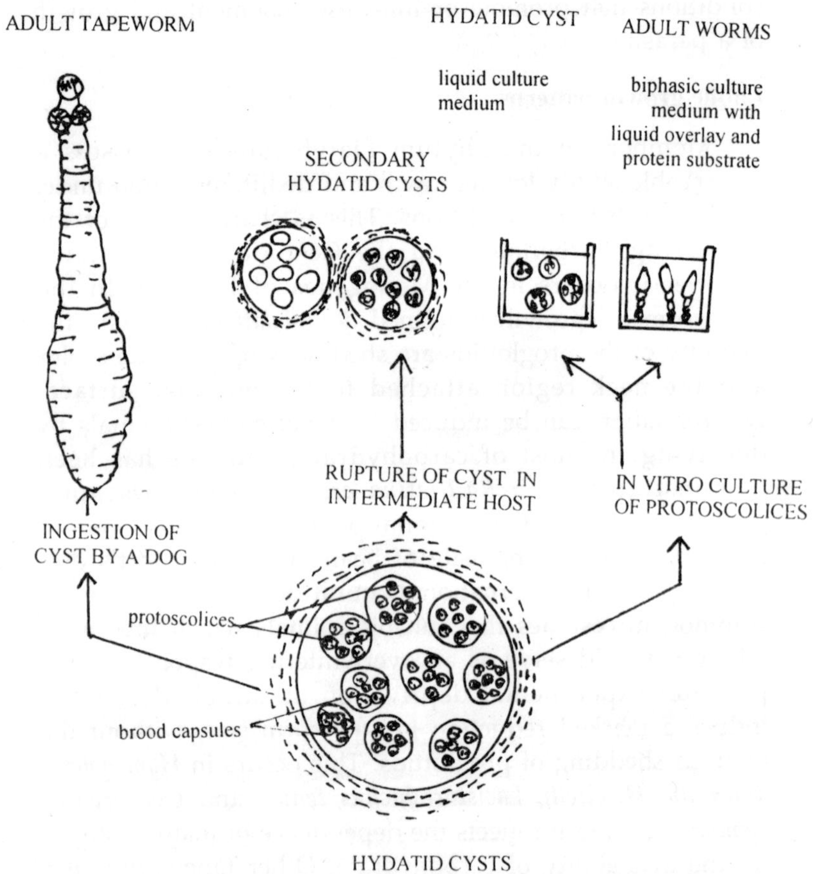

Fig. 9.9 : Development of the protoscoleces of *Echinococcus gramulosus* (Cestoda) *in vitro* and *in vivo.* The protoscoleces will develop into either secondary cysts or adult worms depending upon the environmental conditions encountered. This is though to be controlled at the level of operator genes.

separate directions according to the environmental signals available. Should the cyst burst and release the protoscoleces into the body cavity or the tissues of the intermediate host, each protoscolex can become yet another cyst, in which new protoscoleces will develop; this is termed *secondary hydatidosis*. Protoscoleces that enter the alimentary canal of a suitable host will grow into mature tapeworms. A similar degree of lability in development os seen in the related cestodes *Taenia serialis* and *T. multiceps*. *In vitro* studies on *E. granulosus* suggest that the stimulus for strobilar development to the adult worm is contact with a proteinaceous substrate. Cultivation of protoscoleces in a liquid medium always results in the formation of secondary hydatid cysts, whereas provision of a punctured protein base induces strobilar development (figure 8.9). The ability of protoscoleces to become secondary hydatid cysts represents a particular problem for the surgeon, and he must exercise considerable care during extirpation of cyst to prevent its rupture and the release of vast numbers of protoscoleces. Although it is not fully understood how environmental stimuli control the direction of development in *Echinococcus*, it is evident that a complex interaction between operator and repressor genes will take place in the switch from larva to adult worm, or larva to secondary larva.

Moulting in nematodes

A detailed consideration of growth and the phenomenon of moulting in the parasitic nematodes is outside the scope of this book. Suffice it to stress here that, because of their cuticular covering, all nematodes grow by distinct stages and moult, or shed the cuticle, between each growth stage. Commonly there are five growth stage, with four moults, between egg and adult, and in the parasitic nematodes the third-stage larva is normally infective to the host. In some species, the cuticle of the second stage is retained after moulting, forming a protective sheath around the larva; exsheathment takes place in the gut of the host.

Chapter—10
Influence of Oxygen on Life Processes

All living beings have to depend upon their environment whether it is aerial or aquatic to fulfil their oxygen requirement.

For instance, the composition of the gaseous atmosphere is extremely constant with 20.94 per cent oxygen and 0.03 per cent carbon dioxide. Terrestrial animals, at least when living on the surface of the soil, are hence under constant conditions insofar as availability of oxygen is concerned. However, animals burrowing in the soil are liable to encounter more or less frequently, and for more or less long periods, rather sharp variations in the oxygen of the air available in their immediate surroundings. For example, rainfall will fall the pores of a thicker or thinner layer of soil with water and thus establish a barrier to the rapid downward diffusion of oxygen. Since a certain biological oxygen demand exists in many soil layers with some inorganic oxidations also presumably taking place, an oxygen depletion frequently is observed in the air spaces below the water barrier. This mechanisms is assumed to be responsible for thé familiar reaction of earthworms coming to the surface in great numbers after a rainfall.

Evidently then, the occurrence of near-anaerobic or totally anaerobic conditions in permanently water-logged soils is not surprising. These conditions have been described in swamplands and bottom deposits of many water basins. Especially widespread and severe are the oxygen deficiencies in the bottom deposits of the ocean, if one disregards the most superficial layers. It is interesting, from the standpoint of nematode physiology, that large numbers of unidentified nematodes were found in the oxygen-free bottom mud of the Scottish Clyde Sea area. Their independence of molecular oxygen was demonstrated by Moore (1931) who found them alive and even active after an experimental exposure to anaerobic conditions for 35 days. He gained the impression that these worms can live permanently under anaerobic conditions.

Considerable independence of molecular oxygen is also shown by some fresh-water nematode populations. Several species of roundworms have thus been described as members of the so-called sapropelic fauna, that is, as living in small shallow ponds or bogs which are so rich in decaying organic material that the oxygen is consumed at a faster rate than it can be replaced from the atmosphere. In deeper freshwater accumulations, such as lakes, oxygen deficiencies ranging from minor to complete depletion can occur either during the summer of the winter, or, more rarely, the whole year round. The fundamental mechanism responsible for these oxygen deficiencies is apparently fairly uniform and, in essence, the same as that mentioned above for the soil; namely, biological oxidations proceed in a water layer which is more or less efficiently cut off from the oxygen supply. Such a barrier is set up during the summer by thermal stratification of the water masses, establishing a dividing line between an upper, oxygen-rich, and lower, oxygen-poorer water mass in the so-called thermocline. This thermocline prevents mixing of the upper and lower water strata, the only efficient mechanism for a rapid downward transport of oxygen. Downward diffusion of oxygen of course remains but it is too slow a process to prevent an oxygen depletion of the lower strata if their biological oxygen demand is sufficient. During the winter season, on the other hand, the oxygen depletion may occur throughout the entire water column when a sufficiently thick layer of ice-prevents even the diffusion of oxygen from the atmosphere. In such anaerobic lake habitats various species of nematodes have been found repeatedly.

Many other terrestrial or aquatic habitats exist, exhibiting either transient or permanent oxygen deficiencies of various degree. Since no nematodes has been reported from these as yet, they will be omitted here, with one exception. That is the "oxygen minimum layer" of the oceans, which is of such enormous size that it is without doubt one of the most important oxygen-deficient habitats, although its fauna has not yet been studied experimentally from the standpoint interesting to us here. It has been pointed out that many of the animals living in the oxygen minimum layer especially crustacea and fish, exhibit red or purple colouration. Planctonic animals living above this level are usually colourless, while fish living below it are frequently black or brown. The physiological implications of this curious observation are not obvious at present. The oxygen minimum layer is not found at the bottom of the oceans, but at an intermediate depth which varies from region to region. The oxygen depletion is not very severe (around 50 per cent) in the Atlantic ocean, but in the Pacific large water masses contain only traces of oxygen (around 0.01 ml/liter). The oxygen depletion is undoubtedly due to the oxidation of organic matter which had originated in the photosynthetic upper layers of the sea. Its amount is limited by the amount of plant nutrients present. Redfield (1958), by calculating the phosphorus-oxygen ratio in Atlantic and Pacific waters, has shown a satisfactory correlation of the organic matter synthesized and oxidatively decomposed with the degree of oxygen depletion.

A third group of habitats with widely varying oxygen content are the parasitic habitats. Some parasites, *e.g.* those living in the lungs, especially the bronchi, find in their immediate surroundings oxygen at almost the same pressure as do free-living animals. Another habitat ensuring a plentiful oxygen supply is the blood, while intermediate oxygen tensions ranging from high to fairly low are encountered in tissues. No generalizations are possible, since many factors combine to establish the oxygen level encountered by tissue parasites. Mention may be made here of the size of the host, the method of oxygen supply to an organ (primarily diffusion, circulatory or tracheal system), the state of activity and the temperature (which is more important in the case of cold-blooded than warm-blooded hosts). Still other habitats exist in which the oxygen tension is always extremely low and where it can, at times, be even so low that the usual methods of analysis fail to demonstrate any molecular oxygen. The intestinal canal of larger warm-blooded animals is the best example; another is their bile ducts. It is of course an undisputed fact that the rich vascularization of the intestinal wall leads to oxygen diffusion towards the intestinal lumen. However, the microbial flora of the intestinal lumen apparently has such a high rate of oxygen consumption that a very steep oxygen gradient is established leading to minimal values only a short distance from the intestinal wall. Evidently, then, parasite such as *Giardia* which fix themselves directly to the intestinal cells will have a better opportunity of gathering significant amounts of oxygen than parasites living in the lumen proper, especially toward the centre of the lumen.

Insofar as plant-parasitic nematodes are concerned, no direct determinations of the oxygen concentration in their immediate surroundings have come to my attention. However, indirect evidence seems to indicate that usually they may have access to rather large amounts of oxygen. It has been found that even the roots of rice plants, of the bog bean (*Menyanthes*), or of mangroves (*Avicennia* and *Rhizophora*) contained large amounts of oxygen (usually varying between 10 and 18 per cent, seldom lower), although the roots of these plants are embedded in almost completely anaerobic environments. The observation may also be pertinent that several other plants, such as dog-wood or barely, die when their roots are exposed to anaerobiosis in the soil. Pending direct determinations on parasitized plants a word of caution must nevertheless be sounded, since it is not impossible that parasitization may alter the oxygen supply of the structure involved. In animal tissues at any rate it has been found that inflammatory processes lower the oxygen tension.

Whether nematodes occur in habitats with oxygen tensions higher than air, remains to be seen. It is known that the composition of gases in the swim bladder of different fish species varies considerably, with the oxygen percentages of the secreted gas mixture reaching surprisingly high values in some cases. Although the genus *Cystidicola* occurs in the swim bladder of fish it is not known yet, as far as I am aware, whether or not fish with extremely high oxygen concentration in the swim bladder harbour parasites (nematodes or other) in this organ. In view of the well known toxicity of high oxygen concentrations, a pursuit of this question would seem of general biological interest.

We turn now to the problem of the influence of varying oxygen tensions on the metabolism of invertebrates, where the rate of oxygen consumption is most often used as a convenient yardstick. The work of numerous authors has revealed that the invertebrates can be divided by and large, though without sharp demarcation, into 2 groups: one group in which the rate of oxygen consumption remains constant over a wide range of tensions and a second group in which the rate varies with the tension, the decline often beginning at relatively high tensions, in some cases even far above those met with in nature.

The animals in the first group can maintain an unaltered rate of oxygen consumption even at relatively low tensions because some peculiarity of their organization insures an easy entry of oxygen into their body and/or an efficient distribution of this gas to the various organs and tissues. The structural peculiarity involved varies from group to group. Many very small animals are included in this type; they have a surface/volume ratio favourable to oxygen diffusion and in such cases no special respiratory organs are necessary. The free-living nematodes are in this category. Overgaard Nielsen (1949) showed that their respiration was independent of the tension over the entire range tested, although this range was not very wide. Other animals, such as jellyfish, though large consist mainly of water; their over-all metabolic level is therefore so low that diffusion alone can provide all living cells with all the oxygen needed. In other animals, both large and relatively small, special organs have been developed, such as lung-like structures or gills which specifically serve the oxygen uptake and which usually have a close connection with a more or less efficient circulatory system, the effectiveness of which may in some cases be increased by the presence of a respiratory pigment, such as hemoglobin or hemocyanin. Finally, in many arthropods the air is brought directly in contact with practically all the tissues of the body by means of the tracheal system. In all these later cases, then, there is only a short distance through which the oxygen has to diffuse before reaching the sites of actual consumption.

The classical example for the group of animals that have a respiration dependent on the tension are bulky animals lacking a circulatory system, such as the actinians. According to the explanation first proposed by Henze (1910) and Krogh (1916), the external tissue layers of these animals consume all the oxygen entering the surface by diffusion, while the inner layers must subsist largely on anaerobic processes. If the external oxygen tension is raised or lowered, a thicker or thinner layer of tissues receives sufficient oxygen and the over-all oxygen consumption rises or falls respectively. If this explanation is correct, a reduction of the distance through which oxygen has to diffuse should eliminate or at least materially reduce the dependency of the respiration on the tension; that is, minced or homogenized material should show a different relationship than entire animals. Such experiments were done with actinians by Harnisch (1932) and, on the whole, upheld the above explanation.

It must be emphasized, however, that Henze's and Krogh's theory apparently does not satisfactorily explain all cases of dependency of oxygen consumption on the oxygen tension. It has been shown that the respiration of a Latapie mince of *Ascaris* tissues, a preparation in which no diffusion difficulties should exist, is markedly dependent on the oxygen tension. This is in contrast to similar experiments done with free-living and parasitic flat worms. Since these latter experiments were carried out by the same methods and by the same research group as the *Ascaris* experiments, it seems unlikely that experimental errors should be responsible for the difference in results between *Ascaris* and flatworms.

Before attempting an explanation, another set of observations will be reviewed which seems to be in contrast to the classical explanation. If aerobic animals are exposed to a non-lethal period of anaerobiosis and are then readmitted to an oxygenated atmosphere, they frequently show a phenomenon which has been variously called the repayment of an oxygen debt, respiratory rebound, or respiratory overshoot. It consists in the fact that during the post-anaerobic period the rate of oxygen consumption is increased over the pre-anaerobic rate for a shorter or a longer time before reverting to the original level. No difficulty with the diffusion theory arises from this phenomenon in the case of animals whose respiration is independent of the oxygen tension, but the situation is different in cases where the pre-anaerobic respiration is dependent on the tension. This is true in the case of certain aquatic snails. If their post-anaerobic respiration is studied at oxygen tensions at which the pre-anaerobic rate is sub-optimal, there is a very distinct increase over the pre-anaerobic rate determined at the same tension. Now it is evident that the post-anaerobic respiration could not be raised over the pre-anaerobic level at these tensions if diffusion difficulties were the limiting factor pre-anaerobically unless some mechanism intervened to increase diffusion post-anaerobically. (Such mechanisms are theoretically possible, for example, facilitated oxygen release by the blood due to the Bohr effect, steeper post-anaerobic oxygen gradients due to accumulation of oxidizable metabolic end products in the tissues, and other factors whose magnitude at present are not clearly recognized). Against the assumption of the role of possible auxiliary post-anaerobic diffusion mechanisms is the observation that the respiration of a mince of the aquatic snail *Australorbis glabratus* is even more dependent on the oxygen tension than that of the entire animal. This, of course, is the opposite of what would be expected if diffusion difficulties were responsible for the dependency on the oxygen tension.

It would seem possible that the tentative explanation given by von Brand and Mehlman (1953) in the case of the aquatic snails has a wider applicability. They point out that all difficulties in explaining the observed behaviour of the oxygen consumption would disappear if the importance of the oxygen tension per se were greater for cellular oxidations than is commonly assumed. Contrary to the classical explanation it would be assumed that all tissue layers would receive oxygen by diffusion at the tensions studied, but that some key process, probably one of enzymatic nature, would proceed less efficiently at lower tensions and thus represent the limiting step in the overall rate. It may be significant in this connection that the fluoroacetate inhibition of citrate utilization by pigeon breast muscle is profoundly influenced by the oxygen tension and that certain oxidative enzymes appear to be far more sensitive to lowered oxygen tension than usually assumed. We are dealing here with a complex of problems that deserves a great amount of additional work.

Regardless of what habitat is under consideration, the question arises whether its fauna leads a predominantly aerobic or anaerobic life when the oxygen content of that particular habitat is abnormally low. Since environments with low oxygen tensions are of course encountered much more frequently than truly anaerobic habitats, the above question is of considerable importance. It cannot be answered in a general way because many different factors enter the picture and they vary from case to case. The intestine as habitat of numerous parasites may serve to illustrate the point. It is well known that up to about 20 years ago a controversy existed as to whether intestinal worms should be considered as anaerobic or aerobic animals. Characteristically the protagonists of the former view worked primarily with *Ascaris* or large tapeworms, those of the latter with *Ostertagia* or related species, that is, the two groups studied parasites differing enormously in size. Von Brand (1938) has pointed out that herein lies the clue resolving the controversy.

The above nematodes, like numerous other parasites, have no special structural means of securing oxygen, but rely exclusively on diffusion from the surface. The surface-volume ratio is, however, much more favourable to diffusion in the case of small than in the case of large worms. The respiration of the latter is furthermore dependent on the oxygen tension. These two factors combined suffice to prevent large parasites from realizing more than a very small fraction of their potential consumption at the low oxygen tensions normally prevailing in their immediate surroundings. In small parasites, on the other hand, enough oxygen enters through the surface to satisfy a rather large percentage of their maximal need *in situ*. For instance, *Nippostrongylus muris* and *Nematodirus* spp. may reach 80 to 40 per cent respectively of the *in vitro* rate when studied at the maximal oxygen tension occurring in their natural habitat. The difference between the rates characteristic of these two worms may be due to the fact that only *Nippostrongylus* contains large amounts of hemoglobin, which is characterized by a very steep dissociation curve. It will therefore enable the nematode to extract oxygen even from very oxygen-poor surroundings, but evidently will release oxygen to the tissues only if the latter's oxygen tension is extremely low. It must be emphasized, however, that the presence of hemoglobin alone does not insure an improved oxygen supply, since exchange depends on the properties of the pigment. The hemocele hemoglobin of *Ascaris* or *Dioctophyme* binds oxygen so tightly that it is almost impossible to dissociate it even by chemical means; hence serious doubts exist that it fulfils any respiratory function. Considering the various factors involved, it seems on the whole that size is the most important factor in determining the percentage of the maximal potential oxygen consumption that can be satisfied at a given low oxygen tension. An interesting consequence is that in one and the same habitat some worms may lead a predominantly aerobic and others a predominantly anaerobic life, while actually living side by side. The old controversy of aerobic versus anaerobic life of intestinal worms alluded to above is hence meaningless, and, as is often the case in controversies, both sides were partially correct.

It can be assumed that similar considerations will be valid in the case of free-living invertebrates living in oxygen-poor habitats, although no detailed analysis along the indicated lines seems to have been done as yet. Occasionally the question arises whether an animal leads an anaerobic or aerobic life in a habitat that does not lend itself readily to direct oxygen determinations, for example, the scanty cyst fluid surrounding many encysted parasitic worms. In suitably selected cases a rather clear-cut answer can be arrived at by the following procedure. As mentioned previously, during an anaerobic period many aerobic animals accumulate an oxygen debt which is either totally or partially repaid during a post-anaerobic period in oxygenated surroundings. If the rate of oxygen consumption of such an animal is studied immediately after removal from its natural habitat and shows no signs of an elevated initial rate, it is justified to assume that no marked oxygen debt accumulated in its natural surroundings, indicating a primarily, if not completely, aerobic life. If the accumulation of an oxygen debt should be shown, it can be assumed that the animal had experienced lack of oxygen in its natural habitat, but no conclusions as to the severity of the oxygen deficiency are possible. It could be shown, by the above procedure, that larval *Eustrongylides ignotus* lead an aerobic life when living encysted in their intermediate hosts, several species of fish. It must, however, be emphasized that the procedure cannot be applied indiscriminately. Some invertebrates do not show a respiratory over-shoot under experimental post-anaerobic conditions; in others it is only of very small magnitude. In such cases evidently the above test would be meaningless or could even be misleading.

The reactions of aerobic invertebrates to lack of oxygen vary. Many show very little and others considerable tolerance towards anaerobiosis. An early reaction—not necessarily immediately a sign of damage, but a definite sign of lacking adaptation to life without oxygen—is cessation of mortality which can end in complete paralysis. This state is reached in a few minutes, occasionally even seconds, by especially sensitive animals, for example many terrestrial insects; while other invertebrates, such as worms, may remain motile for several hours to several days in the absence of oxygen. Among the aerobic nematodes, an example for the latter group is the larva of *Eustrongylides ignotus*.

The reasons for the early onset of paralysis are not obvious. It is generally assumed that accumulation of anaerobic metabolic end products, usually acidic in nature, plays a role and it is certainly characteristic that many of the relatively resistant species are able to excrete a greater or lesser fraction of these metabolities. But the nature of the end products accumulating in the tissues also seems to importance in determining the tolerance to lack of oxygen. It has thus been shown that the less tolerant species of aquatic snails produce mainly lactic acid, while the more resistant species form much less lactic acid but more of the evidently less toxic volatile fatty acids. It is possible, though not definitely demonstrated, that extreme sensitivity of the nervous tissues to accumulation of toxic waste products may be responsible for rapid paralysis. It may be significant in this connection that Richards (1941) observed chromatin clumping in the cells of the central nervous system of *Culex* larvae that had been exposed to anoxic conditions.

Even fully paralyzed invertebrates recover completely after restoration of aerobic conditions, provided the anaerobic period does not last too long. If the stress is too severe, the animal may fail to recover motility and die, or as often happens, it partially recovers motility but nevertheless dies sometimes during the recovery period. The time required for resumption of motility, in the case of fully recovering animals, varies considerably from species to species. For example, it takes *Formica cinerea* 10 to 17 hours to revive from an 18-hour exposure to lack of oxygen, while larval *Eustrongylides ignotus* resume movement in about 30 minutes after an anaerobic stress of equal severity.

The fact that the rate of post-anaerobic respiration is frequently increased over the pre-anaerobic rate, that is, the phenomenon of the respiratory overshoot, has already been mentioned. Some of the peculiarities of the post-anaerobic respiration deserve more attention. It is found rather frequently that the respiratory quotient is often exceedingly low during a short interval of the post-anaerobic period, *e.g.* 0.12 in the case of *Zootermopsis nevadensis*. This is explained by the fact that organic acids predominate among the anaerobic metabolic end products, and are responsible for the liberation of carbon dioxide from the bicarbonate or carbonate reserves of the body during the anoxic period. During the post-anaerobic period these reserves are rebuilt by carbon dioxide retention thus leading to a spuriously low respiratory quotient. The period of carbon dioxide retention coincides occasionally, but not regularly, with that of excess oxygen consumption; it may often be either longer or shorter than the latter.

As to the excess oxygen consumption itself, it should be realized that the term "repayment of an oxygen debt," if taken literally, is somewhat misleading. It is of course self evident that an animal exposed to anaerobiosis fails to consume oxygen during this period and the amount of oxygen missed can be calculated from the pre-anaerobic rate and the duration of oxygen deprivation. In some cases, and they were by chance the first ones thoroughly studied, a "complete repayment" took place post-anaerobically, that is, the excess oxygen consumed corresponded rather closely to the amount of oxygen that would theoretically have been consumed during the anoxic period. This led to an overemphasis on the term "repayment." Subsequent work has shown that in many other cases the "repayment" is only fractional and in others a considerable" "overpayment" occurs. As pointed out by von Brand and Mehlman (1953) there is no valid reason to assume a causal connection between the amount of oxygen missed anaerobically and that taken up in excess post-anaerobically. The latter amount will depend upon the level of anaerobic metabolism, the nature of the anaerobic end products and the question of whether an animal stores all the anaerobic end products in its tissues, or whether it excretes a certain fraction during the anoxic period. A related, equally important question is whether all anaerobic end products that have accumulated in the tissues are used oxidatively during the post-anaerobic period, or whether some are excreted during this latter period, as happens in aquatic snails. Finally of great importance is the manner in which the non-excreted end products are utilized. In invertebrates essentially two possibilities seem to be realized; the end products are removed either by complete oxidation, or by partial oxidation and partial resynthesis to carbohydrate.

Almost any animal shows some degree of resistance to lack of oxygen. Those species which normally are dependent on oxygen, but can tolerate lack of oxygen can be called collectively facultative anaerobes, although this term is frequently restricted to forms that can withstand anaerobic conditions for long periods and that do not lose their motility during such exposure. Examples are the trypanosomes among the parasitic protozoa, and the polychaete *Owenia fusiformis* among the free-living worms. Other animals, such as the large intestinal helminths or the trichomonads seem not to be dependent on oxygen; they rather tolerate it. Such forms can be designated as facultative aerobes. True anaerobes, on the other hand, are those animals which are damaged even by low concentrations of oxygen and which carry out all vital functions (mobility, food uptake and digestion, reproduction, etc.) under anaerobic conditions. Only few animals can be assigned to this group. They are, according to our present knowledge, the sapropelic protozoa, the termite flagellates, and the rumen ciliates. Whether the marine nematodes described by Moore (1931) should be placed into the same class cannot be decided at present; a decision will have to wait for studies concerning their reactions to various oxygen tensions.

The question whether the above anaerobic protozoa can utilize oxygen at all, *i.e.* when its concentration has been lowered until a non-toxic level is reached, seems not yet to have been studied. An answer to this question would be of considerable theoretical interest, since it might shed light on the further question whether their anaerobic faculties are a primitive or a secondarily acquired character. There is no doubt that the great majority of animals which lead a predominately anaerobic life in nature are descended from aerobic ancestors. This is indicated, to use endoparasites as an example, by the fact that such worms as *Ascaris* do consume oxygen when it is available and that in the absence of oxygen their eggs develop only to a certain stage (which varies somewhat from species to species), requiring oxygen for full development. It is frequently assumed that the first life forms appearing on this earth were anaerobic and that the ability to utilize oxygen was acquired only secondarily. The most convincing arguments hinting at such a sequence have been brought forward by geochemists who assume a more or less complete absence of molecular oxygen in the earth's atmosphere preceding the origin of life.

Chapter—11
Influence of Size, Motility, Starvation, and Age on Metabolic Rate

There is a variety of nematodes and like many other groups of animals they vary tremendously in size. The smallest nematode studied by Overgaard Nielsen (1949), *Monhystera vulgoris,* weighed 0.2 μg, and a female of *Dioctophyme renale,* weighed by von Brand (1957), 43.3 g; that is some 200 million times more than the former. This range then is potentially available to study the question of the influence of size on metabolic activities. Actually, *Dioctophyme* has not been used yet for studies relevant to this subject. The largest nematode for which data are available is *Ascaris lumbricoides* which is still some 10 million times heavier than *Monhystera*.

As Kleiber (1947) pointed out everybody agrees that the metabolic rate of homeotherms is directly correlated to body size is metabolic rate is defined as heat production per specimen per unit time. In other words, nobody doubts that in one hour a mouse uses less oxygen than an elephant. There is further agreement that quite generally an inverse correlation exists between body size and metabolic rate per unit weight per unit time. Again in other words, one kg of mice would use considerably more oxygen per hour than one kg of elephant. As a first approximation then the relation between body size and metabolic rate can be formulated as "somewhere in the range between rate per animal and rate per unit weight". The first attempt to establish a closer relation between metabolic rate and body size can be found in a paper published as far back as 1839 by Sarrus and Rameaux. According to Kleiber (1947) "these authors apparently take it for granted that it is nature's aim to make the rate of heat production of large and small animals in proportion to their respective surface areas or the 2/3 power of their body weights."

This view can be considered as the basis on which the so-called *"surface law"* was developed, this is, the assumption that the metabolic rate of animals differing in size remains constant when referred to unit surface. This was firmly established, in so far as warm blooded animals are concerned, through Rubner's work (1883). His series of dogs ranging in weight from 3.12 to 31.2 kg has become a classic. Their heat production when expressed per kg per hour increased progressively with decreasing size from 1.48 Cal. to 3.67 Cal., while it varied irregularly only between 1036 and 1212 Cal., when calculated per m^2 per 24 hours.

Rubner's dog series is an excellent example for an intraspecific comprising of small and large specimens. When such comparisons are made, it is justified to calculate the actual surface area according to Meeh's formula:

$$S = K \times W^{2/3}$$

in which S represent the surface in square decimeters and W the body weight in kg. K is a constant which is valid only for animals having the same general form and differing only in size. Therefore when interspecific comparisons are to be made, the constant K has to be separately determined for each different species. This has been done repeatedly for a variety of homeothermic and a few poikilothermic animals, often with precise methods. The usefulness of these measurements for interspecific comparisons has been seriously doubted, with apparent justification. Kleiber (1947) points out that uncertainties exist as to the real meaning of "surface area." He raises the point that one ought to know, for example, whether the true surface of a rabbit includes the surface area of its large ears.

Such difficulties are avoided by accepting the procedure first suggested by Krogh (1916) of basing calculation of metabolic rates on a definite power function of the body weight. This of course does not eliminate all difficulties, since at once the vexing question arises what power function should be used.

A variety of such power functions have been recommended: most commonly 2/3 or 3/4, but none has found universal acceptance. Before elaborating this point, it seems necessary to describe the relation between size and metabolism in cold-blooded animals.

Fundamentally they are the same as those existing in homeotherms. During intraspecific comparisons and also during interspecific comparisons of organisms showing the same general organization it is quite generally observed that smaller specimens have a higher rate of metabolism per unit weight than larger ones. In detail, however, the relations seem to be more varied than in warm-blooded animals. In some cold-blooded vertebrates, the clam *Anodonta,* or certain aquatic snails the "surface law" as characterized by the 2/3 power of body weight satisfactorily explains the relationship between size and metabolic rate. In other cases, however, it does not. In certain insects or *Helix* species the increase in metabolism with increasing size is proportional to weight rather than surface. In a third group, finally, such as some aquatic snails an intermediate relation is found.

A useful method for establishing the relation between size and metabolic rate is to plot weight per specimen against oxygen uptake per specimen per unit time on a double logarithmic scale, that is to analyze the data according to the allometric procedure recommended in recent years by an increasing number of investigators. The exponential equation is.

$$Y = a.X^b$$

In this Y represents metabolic rate. X body size, a an integration constant showing the value of Y when X=1. The exponent b is the allometric constant indicative of the proportion of increase in Y with respect to X. In a logarithmic plot of Y against X, *b* is the slope of the line representing the experimental data. Therefore calculation of the tangent of the angle formed by this line with the abscissa yields the power to which the body weight must be raised to describe the relationship between body weight and metabolic rate. If this value is 1.00 the increase in metabolism parallels the increase in weight, if it is 0.67 it follows the 2/3 power of body weight.

Zeuthen (1953) applied this type of analysis to the organisms for which he could find suitable data. They range in size from bacteria to warm-blooded animals weighing several hundred kilograms. He found on interspecific comparison that organisms differing greatly in size show marked differences in their weigh/metabolism relation. One group is represented by the bacteria and the protozoa where the decline in metabolic rate per unit weight with increasing size is especially sharp. As soon as the small metazoa are reached on his graph, this rapid decline ceases and the slope of the curve indicates *a b* value of about 0.9. For larger poikilothermic invertebrates he finds 0.8 characteristic. For homoetherms he accepts the similar value of 0.75, the actual metabolic levels for the latter being significantly higher than for poikilotherms. His results are then presented in the form of bisecting lines, but he emphasized that it might also be possible to interpret the data as a very long extended S-shaped curve.

A remarkable point revealed by Zeuthen's (1953) analysis is that is intraspecific comparisons of various species three phase curves result, all resembling the curve described above for interspecific comparison. He emphasizes that cessation of growth in several species "is preceded and perhaps effected by a period during which there is a very slight increase or even a decrease in the over-all metabolism, at the same time as there is continued growth of the body." Zeuthen (1953) is inclined to see evolutionary trends in these relations and to consider the similarity of the curves resulting from inter and intraspecific comparisons as more than a coincidence. He explains the trend of the intraspecific curve as indicating ontogenetic recapitulation of the phylogenetic evolution of organisms.

Zeuthen's (1953) presentation of data rests on logarithmic graphs covering tremendous size ranges. This type of presentation tends to minimize deviations from the curves. Even through their trend is rather convincing, the previously summarized differences in weight/metabolism relationships in various groups of animals still remain. It is possible that they are related to intimate connections between metabolism and growth. This has been emphasized by von Bertalanffy (1951) who established three growth types which he considers as characteristic for the three types of weight/ metabolism relation that he acknowledges: (1) Parallelism

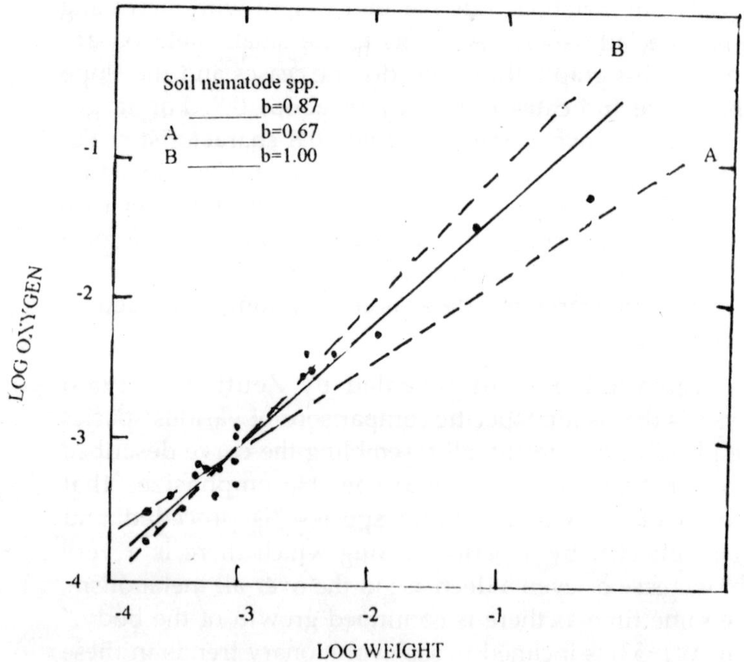

Fig. 11.1 : Allometric plot of oxygen consumption/size relationships of soil nematodes.

of increase in metabolism and surface, (2) parallelism of increase in metabolism and weight, and (3) intermediate cases with close parallelism neither between metabolism and surface nor weight.

The relations between size and metabolism have been discussed several times for nematodes. Those data which have been published in sufficient detail to permit plotting according to the allometric procedure are shown in Fig. 11.1, 11.2. The former deal with very small roundworms, which, with the exception of the larger one, the marine *Pontanema vulgare,* are all soil inhabitants. The size range is 0.0002 mg to 0.40 mg. The worms shown in Fig.11.2 are all parasites of warm-blooded hosts and vary in weight from 0.5 mg to 4780 mg. The curves shown, both being of course interspecific comparisons, prove rather definitely that nematodes follow the same general trend as the whole range of animal studied by Zeuthen (1953), that is, the *b* value is relatively high for the small species and definitely lower for the larger ones. Insofar as the latter are concerned, the line best fitting the plotted points deviates but little from that

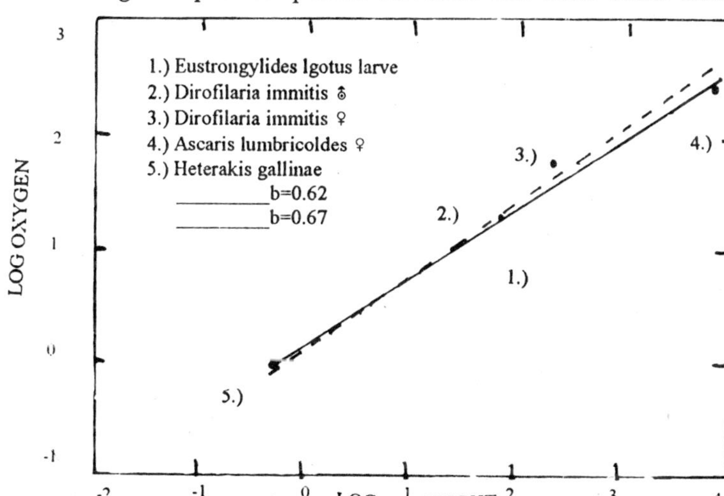

Fig. 11.2 : Allometric plot of oxygen consumption/size relationships of some parasitic nematodes of homeothermic vertebrates.

characterizing *a b* value of 0.67. In this size range then the size/metabolism relation seems to be quite close to the classical $W^{2/3}$ relation. However, it must be realized that this assumption is based on but a few data; it may become necessary to revise it once data for more species becomes available.

It is regrettable that so far nothing is known about the respiratory rate of *Dioctophyme renale*. With such data, it would be possible to see whether or not the scope of the curve for nematodes changes again when the giants are reached, a fact observed by Zeuthen (1953) for several groups of animals.

One further point deserves mention. A comparison of Figs. 11.1 and 11.2 shows that the level of metabolism is higher for the larger than the smaller worms. This is simply a consequence of the fact that the metabolic rate of the larger worms was determined at body temperature (37° to 41°C) because they parasitize homeotherms, while the small free-living nematodes were studied at 16°C.

In addition to the above data, relevant figures are contained in the paper by Rogers (1948) and Lazarus (1950). Both investigators calculated the oxygen uptake per unit surface area of a variety of parasitic nematodes, but they do not present the actual data dealing with the weight of their specimens or the amount of oxygen consumed per specimen per unit time. Their data are therefore not sufficiently detailed to allow allometric plots. Both investigators reach the conclusion that the variations between species were smaller when the respiratory rate was calculated per unit surface than when calculated on any other basis, but a rather pronounced variability still remained even after calculation based on surface area.

Larger nematode species should be quite suitable for interspecific studies, but so far only one author, Krueger, seems to have given attention to the problem insofar as parasitic nematodes are concerned. He reported first that large specimens of *Ascaris lumbricoides* consumed less oxygen per unit weight than small ones. His experimental results did not fit one smooth curve: instead he found a break at a body weight of about 1.4 gm. Below this weight, the increase in oxygen consumption per gm tissue was considerably more rapid than above it. For animals weighing between approximately 1.4 and 10 gm, he developed the following purely empirical formula:

$$O_2 \text{ Consumption } (mm^2/30 \text{ min})$$
$$=20 \text{ (Weight—1.4)+110}$$

In a second paper Krueger (1940) reinvestigated the problem with *Ascaris* specimens weighing between 0.3 and 10 gm. He found, on the average, a fairly good proportionality between metabolism and surface when the latter is calculated according to Meeh's formula with a K value of 13.96. However, he does not think that the actual surface of the worms plays the decisive role because he found rather considerable variations in respiratory rate between specimens having identical surface areas. But, as von Bertalanffy (1951) points out these variations could be due to a variety of secondary factors.

The only other relevant data available pertain to the marine nematode *Pontonema vulgare*. Because of the small size of the worms, especially the juvenile ones, the individual weights are quite uncertain. For the single determination a large number of specimens had to be used which were not always of uniform size. However, it does appear certain that the respiration of juveniles is higher than that of adults. It is highest in small juveniles and seems to decrease rather sharply as they increase in size. Once mature, however, the respiratory rate seems to be fairly independent of weight and sex.

It remains to discuss briefly why the metabolic rate is in most cases more closely related to the surface than to body weight. The theoretical foundations of the surface law have received a great deal of attention, especially insofar as homeotherms are involved. Kleiber (1947) classifies these theories into five major groups as follows:

"The metabolic rate of animals must be in proportion to their body surface,

(1) "Because the rate of heat transfer between animal and environment is proportional to the body surface area".

(2) Because the intensity of flow of nutrients, in particular oxidizable material and oxygen, is a function of the sum of internal surfaces which in turn is proportional to the body surface".

(3) "Because the rate of supply of oxidizable material and oxygen to the tissues is a function of the mean intensity of the blood current, which is proportional to the square area of the blood vessels, which in turn is proportional to the area of body surface".

(4) "Because the composition of animals is a function of their body size. The composition may be meant either anatomically; the larger the animal the lower is the ratio of the mass of metabolically active organs to the mass of metabolically inert organs; or the composition may be meant chemically: the larger the animal the lower its percentage" of 'active protoplasma'.

(5) "Because the cells of the body have an inherent requirement of oxygen consumption per unit weight, which is smaller the larger the animal'.

Kleiber (1947) rejects theories 2, 4, and 5 completely, but considers theories 1 and 3 basically sound. He integrates both these theories into one which he formulates as follows: "In natural selection, those animals prove to be better fit whose rate of oxygen consumption is regulated so as to permit the more efficient temperature regulation as well as the more efficient transport of oxygen and nutrients". Logically it would seem that an acceptance of this theory would lead to rejection of the "surface law" for poikilothermic animals, unless one would postulate two quite different explanations for one and the same phenomenon. Indeed some authors (*e.g.*, Terroine and Delpech 1931) deny the applicability of the "surface law" to poikilothermic animals. However, the data summarized on the preceding pages indicate rather clearly that in poikilothermic vertebrates and in invertebrates a weight/metabolism relationship exists which in no way differs fundamentally from that encountered in homeotherms. A comparative physiologist therefore will be reluctant to concede that Kleiber's theory really goes to the fundamentals of the matter.

Zeuthen (1953) states that the relation observed between size and metabolism indicates "that the sum of enzymes, metabolites, and perhaps inhibitors which govern metabolites activity grow less than the totality of the protoplasma". In this connection it is of interest to note that according to Rosenthal and Drabkin (1943) the cytochrome concentration of kidney cortex decreases with increasing body weight. Succinoxidase and malic dehydrogenase concentration assayed in liver, heart, kidney, and brain of mouse, rat, and cow similarly indicated decreased concentrations with increasing size. One of the possible consequences of decreased concentration of metabolic enzymes in large animals might be a decrease in their metabolic rate. Such an assumption, however, need not necessarily be true. Theoretically, cells with different amount of metabolic enzymes could still show the same metabolic rate, since a "safety factor" in the amounts of vital enzymes can be assumed and a great reserve in unutilized enzymatic potentiality must be present when the basal metabolic rate is determined in many animals, as illustrated by comparison of the metabolic rates of an insect during flight and during rest. Therefore, direct determinations in isolated it issues are necessary to decide the issue.

Unfortunately the results of such experiments have been somewhat conflicting. Kleiber (1941) and Weymouth et al. (1942) found the respiratory rate of tissue slices of homeotherms inversely correlated to body size and Weymouth et al. (1944) arrived at a similar result for tissue slices of the kelp crab, *Pugettia producta,* when tissues takes from specimens differing greatly in size were compared. On the other hand, a reliable investigator like Krebs (1950) found only a small decrease in tissue slice respiration in small, as compared with large warm-blooded animals.

A new viewpoint was injected recently into the discussion by Smith (1956). He isolated and studied liver mitochondria of rat, rabbit, sheep, and steer. He found an inverse correlation between number of mitochondria per gm tissue and size, but the individual mitochondria of the large species were larger than those isolated from the small ones. The product of total mitochondrial number times mass per gm tissue increased with the 0.77 power of body weight. He calculated that the aggregate mitochondrial surface showed approximately the same relation to the size of the animal, as that existing between size and oxygen consumption. In his words "it appears probable that the relative amounts of these elements in any given tissue will prove the controlling factor in determining the regression of oxygen utilization on total body size of the species". It certainly would be desirable to extend such investigation to other tissue besides liver and to other species, especially poikilothermic vertebrates and also invertebrates. If the above relation holds up in these cases, the analysis will certainly have progressed one big step forward, but inevitably the question will then arise why the mitochondrial number decreases with increasing body size.

One of the factors determining the metabolic level and often changing it dramatically, is motility, or in more general terms, muscular activity. It is well known that during heavy muscular work the energy requirements of man are raised several times over those characteristic of complete rest; under extreme physical work and for short periods this increase may amount to about 20 times the basal metabolic rate. But even such an increase is greatly surpassed during flight of some insects. In honey bees, the blowfly *Lucilia,* or the locust

Schistocerca increases over the resting rate of 30 to 50 times, or even more, have been observed. The increase relative to that quoted for man is probably even more than indicated by these figures, since the resting metabolism of an insect is not comparable to the basal metabolism of a mammal, but probably higher than the true basal rate. The degree of difference between working mammal and flying insect is perhaps best shown by mentioning a few absolute figures. The rate of oxygen consumption of man during the most violent exercise does not surpass 72 $mm^3/gm/min$, while in flying *Lucilia* specimens values exceeding 300 $mm^3/gm/min$ have been reliably reported. Such values appear to be paralleled in vertebrates only by humming birds who average during flight 1300 $mm^3/gm/min$ and may reach values of 2500 $mm^3/gm/min$. It does not seem appropriate to go into more details on this occasion, since nematodes do not fly.

Many nematodes must be considered fundamentally as swimming animals, although in most cases they probably cannot be separated sharply from crawlers. Soil nematodes or the round-worms parasitizing such relatively "dry" habitats as the swim bladder of fish or the body cavities of mammals are hardly primarily swimmers in the true sense of the world. Nevertheless swimming is an important mode of locomotion for many species. Swimming is for man a rather strenuous exercise and it is therefore important to realize that what can be strenuous for one species places but little additional strain on another. It appears that one of the most important factors is the size of the animal. Zeuthen (1947) has demonstrated convincingly that in planktonic micro animals the work of floating and swimming generally requires less than 50 per cent of the total energy production, and in many cases probably only a small fraction of that value. While Zeuthen's (1947) deductions are rather clear cut because based on actual determinations, it may be worthwhile pointing out that Ludwig (1928) basing himself on calculation alone came to the conclusion that the ciliary locomotion of *Paramaecium* requires an almost negligible fraction of the total energy produced.

A relation similar in principle to that described for swimming, probably also holds for other types of locomotion, although not necessarily for all. I have been unable to find data on the energy expenditure during crawling that would allow analysis according to the present point of interest, but von Buddenbrock (1934) published an interesting paper dealing with the kinetic and static work done by large and small walking animals. He determined the time it takes various animals to move a distance equivalent to their body length at their normal gait. From these data he calculated the amount of work performed by various sized animals when walking up an incline of 5 per cent for 100 seconds. He came to the conclusion that a tiger weighing 300 lbs. does 135 times the work than a 5 lb. cat per unit time. A tiger, however, moves 2.25 times faster than a cat; hence in the above example in 100 seconds it lifts its body weight 2.25 times higher. If the tiger would slow down to the speed of the cat, both would perform the same amount of work per kg weight and unit time. Since the amount of work done in climbing an incline parallels the increase in body weight, theoretically, the number of calories expended in the performance of this work should increase proportionally. However, with increasing size, the basal metabolism of mammals increases only by the 2/3 or 3/4 power of the body weight, so that the percentage by which the basal metabolic rate must be increased to do an equal amount of work per unit weight, becomes larger the larger the animal. Von Buddenbrock (1934) himself points out the similar shift in relation of muscular activity and heat loss of mammals of various sizes.

It is possible that the positive correlation between size and increase in metabolic rate explains two observations on nematodes which at first glance appear contradictory. Slater (1925) reported that *Ascaris* specimens died rapidly in the absence of oxygen when sustained motility was induced by electrical stimulation. This probably indicates that motility in this case increased the energy demands to such an extent that the anaerobic metabolism could not be raised sufficiently over the resting rate to meet them. *Ascaris* is a large worm; the second relevant observation, which appeared puzzling to the author reporting it, deals with a small one. Santmyer (1956) found the same degree of motility in specimens of *Panagrellus redivivus* that had been freshly isolated from the medium as in those that had been starved for 24 or 48 hours, although the former consumed exactly twice as much oxygen per unit weight per unit time than the latter. Santmyer (1956) states: "This difference in the energy requirements for motion in fresh and starved worms is not easily explained". He seems then tacitly to assume that a very sizable portion of the total oxygen consumption must be related to muscular activity. However it must be realized that *Panagrellus* is extremely small, the individual worm weighing but 1.21 μg. If therefore the described correlation between size and energy requirement for muscular activity holds in this case, the reduced rate of oxygen consumption of the starved worms could be explained on the basis of a reduction of the basal metabolic rate induced by starvation rather than to different requirements for motility by fed and starved specimens. As Zeuthen's (1947) results indicate motility puts very little stress on organisms of the *Panagrellus* size range. The inability of Santmyer (1956) to find differences in motility between fed and starved worms simply seems to indicate that the increase in oxygen consumption over the resting metabolism necessary to allow motility was so small in comparison to the resting rate as to remain within the limits of error of the determinations. While this explanation would seem to follow logically from the data available, the desirability of further analysis must be stressed, because the influence of the starvation factor as assumed here is at variance with what has been reported by Overgaard Nielsen (1949) for four species of free-living nematodes. He found no difference in the rate of oxygen consumption during 3 days starvation as compared to the pre-starvation rate. On the other hand, however, it has been well established that some parasitic nematodes reduce their metabolic rate materially during starvation. On the fourth day of starvation the larval *Eustrongylides ignotus* reached a level of oxygen consumption amounting to 50 per cent of the prestarvation rate, the much smaller *Trichinella spiralis* larva reached the same level after only 22 hour's starvation. It is interesting that in these two cases the starvation level was exactly the same as that found by Santmyer (1956), that is, 50 per cent of the prestarvation level. Similarly in all three instances this level, once reached, was maintained for the balance of the experiments. To what extent the *Trichinella* larvae moved, is not known, but the large *Eustrongylides* larvae are always rather sluggish, perhaps, strengthening the probability that the reduction in metabolic rate due to starvation depends primarily on a reduction of the basal rate.

One factor which is difficult to control in experiments with nematodes is age. This is regrettable since it is known that age does often have an influence on the metabolic level. It is thus well known that the basal metabolic rate of man decreases somewhat with increasing age and Weinbach (1956) has shown that the level of oxidative phosphorylation is decreased in the albumen gland of aging snails, an observation later extended to and confirmed for the liver of aged rats. Overgaard Nielsen (1949) briefly discussed the point for the marine nematode *Pontonema vulgare*. As already mentioned in another connection, he found higher rates of oxygen consumption in juvenile than in sexually mature specimens. However, whether this was due to the age factor, to the fact that juveniles are smaller than adults, or to an altered metabolism due to sexual maturity cannot be decided at present. Overgaard Nielsen (1949) himself is inclined to attribute the decline in rate shown by sexually mature adults to "some sort of biochemical maturation process" because he found the metabolic rate of adults varying in size rather constant. Rogers (1948) found a pronounced and rapid decrease in the rate of oxygen consumption directly correlated to the age of third stage, *Nippostongylus muris* larvae which were studied at 1, 4, and 12 days of age. A comparable decrease has also been reported for free-living larvae of the same species aging in non-nutrient media. Here, then, it is difficult to differentiate between the influence of age and starvation.

There can be no doubt that many nematode species, especially parasitic ones, would be especially well suited for the investigation of the relation between age and metabolism. Some species stay alive for exceedingly long periods as larvae. Viable larvae of *Trichinella spiralis* have been found 11 years after infection of hogs and even after 31 years after infection of humans. In contrast their life span is brief after sexual maturity. The males of *Trichinella* live only a few days after copulation and the females only a little linger. Another long lived larva is that of *Eustrongylides ignotus*. It has been kept alive for years in *vitro*. As concerns adult parasitic nematodes, it has been reported that some human hookworms live at least 6 to 7 years in the human intestine, although the great majority of specimens die during that time and adult *Loa loa* can live for at least 13 years in humans. These examples should suffice to show that there are available a great variety of forms differing considerably in life span and therefore singularly well fitted for a variety of experimental approaches to the general problem of metabolism and aging.

Chapter—12

Influence of pH, Ions, and Osmotic Pressure on Life Processes

A factor remaining quite constant in one habitat may be variable in others. A good example is the hydrogen ion concentration. With a few exceptions, it shows little fluctuation in oceans, while it is much more variable in certain freshwater habitats.

In the upper layers of the high seas the pH values reported vary only between 8.0 and 8.3. They depend somewhat on the water temperature and pressure. With rising temperature the pH of sea water falls slightly and a similar decrease can be observed with increasing depth. The degree to which the pH remains constant depends in sea water just as in any other solution on its buffering capacity; that is, on its capacity to equalize the withdrawal or the addition of H^+ or OH^- ions. The buffering capacity is, essentially, due to the presence of partially dissociated acids and bases, and especially to the presence of their almost totally dissociated salts. In sea water, and in many freshwater bodies, by far the most important buffering mechanism is the carbon dioxide system. The water contains bicarbonate and carbonate ions, undissociated molecules of CO_2 and carbonic acid. All these substances are in equilibrium with each other, and with the hydrogen ions as follows:

$$CO_2 + H_2O \rightleftharpoons H_2CO_3$$
$$\updownarrow$$
$$H^+ + HCO_3^-$$
$$\updownarrow$$
$$H^+ + HCO_3^-$$

The basic constituents, carbonate and bicarbonate (some borate is also present and belongs in the same class) are in excess of the acid constituents; in other words, an alkali reserve exists. Its concentration in most sea water does not exceed 0.0026 N. That is, it is very low and in fact much lower than in even weak laboratory buffer solutions. In view of the low buffer concentration of oceanic waters, it is legitimate to raise the question of the reasons for their constant pH. The answer is to be found in the tremendous size of the ocean basins. The absolute amounts of buffering solutions are, despite their low concentration, correspondingly enormous. It is, for example, estimated that the total carbon dioxide of the ocean water amounts to 25 to 27 times that of the entire gaseous atmosphere. A reflection of the low buffer concentration found in sea water is the fact that locally deviations from the typical pH value occur. For instance, a pH of 7.6 has been found at an intermediate depth off the West coast of South Africa—possibly an indication for pronounced biological activity in this particular layer. For further information Sverdrup, Johnson, and Fleming (1942) and Harvey (1945) should be consulted.

The CO_2/bicarbonate system is the most important buffer system in fresh water, but pH changes in one and the same water body and constant differences in pH values between different bodies of water are much more pronounced than in the oceans even though the buffer capacity of most fresh water is actually greater than that of sea water. The relatively frequent occurrence of local and transient pH changes in the former can be related to several factors distinguishing fresh-water and oceanic water basins. The former are usually relatively small and the actual overall amount of buffer substances is limited. Locally originating pH changes therefore involve a much greater percentage of the entire water mass and cannot be equalized as readily by exchange with neighbouring unaltered water as in the sea. Furthermore, in the small water basins the factors responsible for initiation of pH shifts are commonly more potent than in the immense expanses of the ocean. Of special importance are biological factors. Both plant and animal life are usually much more concentrated in fresh water than in the sea and due to the interplay of assimilation and respiration rather pronounced and often cyclical changes in CO_2 content of the fresh-water habitats may occur and be reflected in pH changes. Also, the amount of decaying organic material, either of local origin or swept in, is often greater in small than in large water accumulations.

In respect to permanent pH differences, it may be mentioned that in water basins containing clear water (brooks, streams, lakes) pH values lying between 6 and 8 are usually found, although a more alkaline reaction with pH up to 10 is encountered not too rarely. On the other side of the scale lie the so-called dystrophic brown waters. Their pH is generally on the acid side and in certain swamps pH values as low as 3.2 have ben observed. Such low values cannot be due only to CO_2 (which cannot be responsible for a pH lower than 3.9), but must be related to the presence of stronger acids which may have originated from the decomposition of organic material. The extreme acidity of certain volcanic lakes (pH 1.7) is due to the presence of strong mineral acids.

There exists a distinct difference between the faunas of relatively strongly acid water (pH below 5, that is, essentially, dystrophic waters such as moors) and those whose pH is above approximately 5. This distinction has been emphasized by Harnisch (1951) who points out that the strongly acid waters are shunned by many animals, such as certain species of rotatoria, most oligochaetes, the hirudinea, malacostraca, ostracoda, molluscs, fish, and amphibia. The fauna of such waters is then characterized largely in a negative way, by the complete absence of certain forms. Species which do live in strongly acid waters are not limited to them, but are found as well in waters with higher pH. There appear to be but few animal species that actually prefer to live in strongly acid surroundings. Harnisch (1951) mentions in this connection the rotatoria *Polyarthra platyptera* var *minor,* and *Anurea serrulata,* the crustacea *Streblocerus serricaudatus,* and *Acantholeberis curvirostris,* and, with reservations, some protozoa.

It seems on the whole that the great majority of aquatic invertebrates tolerates fairly large pH changes rather well, perhaps better than many fishes. It seems justified to assume that the pH of the water will be the limiting factor for the distribution of aquatic invertebrates only when a given locality experiences extremes of acidity or alkalinity. Exceptions, however, occur. One of the most instructive cases is that of *Spirostomum ambiguum.* Putter (1904) observed that this ciliate died rather rapidly when kept in the laboratory in shallow dishes. He concluded that this indicated sensitivity towards oxygen which of course would diffuse rapidly into a shallow layer of water and he assumed that *Spirostomum* leads an anaerobic life in nature. A reinvestigation by Saunders (1924) however showed that this reasoning was faulty because the pH factor had been neglected. Saunders (1924) agrees that the ciliate dies when kept in shallow dishes, but it dies because under these conditions the medium loses carbon dioxide and the pH rises to about 8.0 which is lethal to the animals. The limits of survival lie between pH 6 and 7.8. If kept in dilute buffer solutions the ciliates can live permanently in shallow waters. Still later work has then shown that *Spirostomum,* in contrast to Putter's vie, does not tolerate lack of oxygen any too well.

The pH relationships of aquatic nematodes have not yet been studied. It would be of interest to determine at least the tolerated pH limits. It seems reasonable to assume that such forms as the vinegar eel should show adaptations to rather extreme conditions. In artificial culture *Turbatrixaceti* appears to tolerate at least a pH range of 3.5 to 9.0. Growth is best in media with an initial pH of 4.5 but drops sharply at pH 3.75.

The pH of soils varies considerably from region to region. It may be mentioned as an example that in a recent survey of Italian soils supporting tobacco cultivation 79 per cent of the soil samples had a pH 7.3 (44 per cent pH between 7.9 and 8.5). In 11 per cent the pH values varied between 6.8 and 7.2 and in 10 percent between 5.6 and 6.7, while values below 5.5 were found only in 0.2 per cent of the samples. The extremes recorded in this survey are surpassed in other localities. Isbell (1958) for instance described a soil with pH values ranging from 8.1 to 9.3; in the lower layers of the same locality the pH values rose even to 10.1 and 10.2. To cite an extreme on the other side of the scale, the moss *Dicranella cerviculata* has been found on soils having a pH of 2.8.

Although many plants tolerate a fairly wide range of soil pH's, they often show a preference (through their abundance in certain localities or by better growth) for a fairly narrow pH range. Small (1954) distinguishes 6 types of plants in respect to their pH relationships; these range from acidiphilous plants (thriving only below pH 4.8) to alkaliphilous plants (found usually on soils with a pH above 7.0 and up to a pH of 8.4). It is evident that possible variations in the occurrence of plant-parasitizing nematodes could be doe to an indirect influence via the host plant rather than to a direct pH influence on some stage in the life cycle of the worms that comes in direct contact with the surrounding soil. To decide between alternative possibilities would require a laboratory analysis which does not seem to have been done as yet.

To what extent the pH of the soil may be responsible for or may at least contribute to variations in the non-parasitic soil fauna is not known. In so far as soil nematodes are concerned. Overgaard Nielsen (1949) has made a careful survey of the nematode populations of various types of soil. Although he did not carry out pH determinations, considerable differences in this respect undoubtedly existed between several of his chosen localities, since the soils studied ranged from sandy ones to bog soil. On the whole, it appears that differences in the nematode fauna of various soil types were more of a quantitative than a qualitative nature. Only few species were reported as specific for certain soils and their distribution seemed correlated with the moisture content of the soil in question (extremely dry on the one hand, permanently water-logged on the other). More recent studies have also shown that certain nematodes, *e.g. Tylenchorhynchus martini* thrive best at certain moisture levels of their surroundings. Anaerobic bacteria may be, in Johnston's view, at least partially responsible for the death of this nematode in water-saturated soils, since he observed that a species of *Clostridium* produced a toxic substance rapidly fatal to the worms. It is entirely conceivable that the pH of the surroundings may have a profound influence in such cases. It has been shown at least for a larval parasitic nematode (*Eustrongylides ignotus*) that filtrates of *Bacillus subtilis* cultures were highly toxic to the worms only at pH values above 8.0 while acidified culture filtrates were practically innocuous. It was shown in this case that the toxic substance was ammonia and that only the ammonia ion was toxic to the larvae, while ammonium salts were harmless. It must be emphasized that the pH of the medium itself was not a factor in the toxicity; solutions adjusted with NaOH to the same pH as the toxic culture filtrates were non-lethal.

The last large group of habitats with pH relations of significance to the present discussion, are the parasitic habitats. Practically all plant and animal tissues are potential sites for parasites. It will not be feasible to go into much detail, but a general survey may be helpful for a possible future analysis of the pH relations of plant-parasitizing nematodes *in situ* which is heretofore lacking.

A fairly detailed survey of the pH relations of plants has recently been presented by Small (1954). He points out that the pH of plant tissues is usually acid. It ranges, with few exceptions, between pH 4.0 and 6.2 for vegetative tissues, while flowers show a wider range with variations between pH 2 and 8. The pH of natural cell saps varies between 1.4 and 7.4. Values below 4.0 are more common among gymnosperms than angiosperms. The extremely low values sometimes encountered in cell saps are unquestionably due to the presence of organic acids, especially oxalic, citric and malic acid. A vacuolar pH of below 1.0 has recently been reported for the alga *Desmarestia;* in this case the unusually low value is attributed to the presence of sulfuric acid in the vacuolar fluid. From the standpoint of nematode physiology it may be relevant that neighbouring tissues of one and the same plant may have different pH values. Thus, according to Small (1954) the sap of epidermal cells of potatoes or sunflowers usually has a pH of 4.0 to 5.2 while that of the parenchyma normally varies only between 5.6 and 5.9. Whether tissue changes induced by parasites may change the pH of plants has apparently not yet been studied. It would be not entirely surprising if such changes were to occur since the buffer capacity of plants, though distinctly stronger than those of sea or fresh waters, is not overly well developed. Besides bicarbonate, phosphate and organic compounds such as malate and citrate supply the most commonly encountered buffer systems.

Parasites living in animals may find themselves in habitats with very stable pH. One of the best examples are the blood parasites. The pH of human blood normally fluctuates within quite narrow limits around a value of 7.4. The alkali reserve of the blood is equivalent to approximately a .047 M sodium carbonate solution, that is, blood is quite well buffered due to the presence of bicarbonate, phosphate, and protein. Conditions so extreme as to result in larger pH shifts of the blood are not well tolerated. It is well known that severe acidosis first induces come and that death may ensue when the blood pH falls to such moderately acid pH as 6.95. On the other side, alkalosis so severe as to raise the blood pH above 7.8 may also be fatal. It is then not surprising that parasites adapted to live in surroundings with such normally stable pH also may show a fairly narrow optimal pH range. The respiration of *Plasmodium gallinaceum,* for instance, stays constant only between pH 7.6 and 8.0 and that of *Trypanosoma lewisi* between pH 6.7 and 7.8. Another blood parasite, *Schistosoma mansoni,* is more tolerant, and keeps its oxygen consumption unaltered over the pH range 6.8 to 8.9.

It probable that tissue and cellular parasites also live, in most instance, in surroundings with fairly stable pH. According to the determinations of Frunder (1951) the pH of tissues such as muscle measured *in situ* is close to that of blood with a pII of 7.3 only in absolute rest. Any disturbance, even such normal physiological ones as activity or a drop in environmental temperature, leads to a more or less pronounced acidosis of the tissues. He found that under the influence of the above physiological stimuli the tissue pH may vary between 6.0 and 7.3. Fundamentally all such changes have the same basis, namely the local oxygen supply becomes to a lesser or higher degree suboptimal in respect to the demand, leading to the production of acids, chiefly lactic. Such physiological tissue acidoses usually last only a short time, up to a few hours. In contrast, pathological acidoses may last for longer periods and be considerably more severe. Frunder (1951) studied the *in situ* pH relationships of muscles experimentally infected with single species of bacteria. Briefly and incompletely summarized, he found little acidosis during chronic infections, but pronounced pH changes in acute ones. In cases of tissue necrosis and abscess formation the pH could be as low as 5.3. After a study of other pathological acidoses, such as those due to inflammation induced by chemicals (turpentine and others) he reached the general conclusion that pathological acidoses, in contrast to the physiological ones, are due to decreased respiration accompanied by an increased glycolysis of the injured tissues. It certainly would appear worth-while to study the corresponding relations in the case of abscesses due to animal parasites, such as *Entamoeba histolytica* liver abscesses, or the abscesses formed subcutaneously after injection of certain trichomonads. One might find a clue to the often peculiar distribution of the parasites within an abscess.

Insofar as cellular parasites are concerned, their pH relations have not been studied in detail. Even more than in the case of tissue parasites it would seem probable that parasites occupying a large part of a cell (*e.g.* malaria parasites, coocidia, *Endotrypanum,* and others) may exert considerable influence on the pH of their immediate environment through the excretion of metabolities and mechanical disturbance of the cellular organization which may well result in functional changes of the host's cellular metabolism. One of the reasons why this phase of the host/parasite relation seems not to have been studied, may be the difficulty of determining accurately the intracellular pH of single cells. The results of calorimetric procedures are often uncertain in their interpretation and may lead to considerable errors unless applied with great caution. According to the data available in the literature the pH of single cells may vary from about 5.6 to slightly alkaline values. It would be an oversimplification to assume the same pH for all parts of a single cell; it has been shown on the contrary that the pH values of the nucleus often differ from those of the protoplasm.

It may be pertinent to point out that even tissue parasites may be capable of influencing the pH of their environment, as can be inferred from the following *in vitro* experiments. Von Branch and Simpson, (1944) observed that the pH of nutritive solutions in which axenic larval *Eustrongylides ignotus* were kept, was shifted towards the acid side when initially either slightly alkaline or slightly acid, but towards the alkaline side if initially, strongly acid. This capacity of exerting a certain regulatory influence on the environmental pH is probably widespread among nematodes. Prior to the experiments quoted above, but less convincing because not carried out with axenic worms, are the relevant observations of Whitlock, Link, and Leasure (1939) on *Strongylus vulgaris* and those of Gettier (1942) on *Camallanus scabrae.*

While blood- and other tissue parasites definitely, and cellular parasites probably, live under fairly stable pH conditions, this is less true for parasites of the intestinal tract. A few data taken from Leasure, Link, and Whitlock's (1939) paper dealing with the pH values found in the digestive tract of a long series of horses, may serve as an example of the variabilities encountered. The pH range encountered in the stomach was 2.01 to 8.60, in the ileum 5.88 to 8.03, and in the rectum 5.39 to 7.13 to mention only three of the seven sections investigated. A survey of many further data of the pH values determined in the intestinal tract of a great variety of vertebrates can be found in Mennega (1938) and von Brand and Jahn (1942).

There are many factors which will influence the intestinal pH and will bring about the above variations. Mention can be made of the amount and the quality of the food ingested, the time elapsed since the last meal, the type of intestinal flora and so no. In so far as the quality of the food is concerned once especially relevant example may suffice. The administration of large amounts of lactose lowers the intestinal pH markedly and this may be one of the reasons, though certainly not the only one, why a lactose or milk diet is often detrimental to parasites.

Because the pH of various sections of the vertebrate intestinal tract shows such great variations it is not surprising that the question has been discussed whether the pH factor may perhaps play a role in the distribution of various parasite species within this habitat. In so far as nematodes specifically are concerned, this possibility has been investigated by Davey (1938). According to his findings, *Ostertagia ctrcumcincta* can tolerate pH variations ranging from 3.2 to 9.0 and this may well be one of the reasons why this species invades the abomasum of sheep (pH 3.15 to 5.25) but not the stomach of dogs (pH 2 or below) or horse (pH 1.1 to 6.8). However, the observations of Davey do not explain why this species is limited to the stomach of sheep and does not settle in the lower parts of the intestinal tract. The possibility of a pH preference range, in contrast to a tolerated range has no been studied and would probably be very difficult, if not impossible to study, because of the limitations of the material. Davey (1938) reported also that *Trichostrongyulus colubrifrmis* and *T. vitrus* which live in the usually acid duodenum readily tolerated a pH of 5.6. Nematode species found in the less acid middle and lower small intestine (*Nematodirus filicollis, N. spathiger, Cooperia oncophora, C. curticei,* and *Strongyloides papillosus*) died at pH values ranging from 3.9 to 4.6. It seems possible then that the acidity characteristic of the duodenum (when the pylorus opens and the strongly acid abomasal contents reach duodenum) may be the factor, or may be perhaps at least a contributing factor in preventing the above species from parasitizing the duodenum.

In addition to being lethal beyond a certain range the hydrogen ion concentration way play a role in the establishment of nematodes in particular habitats. Larva nematodes invading a final host generally undergo molts during their further development. Crofton (1947) has shown that the second ecdysis of *Trichostrongylus retortaeformis,* which transforms the infective larva into the first parasitic larva, occurs almost exclusively in the pH range 4 to 5 which is also the approximate range encountered in the stomach of rabbits. In HCI solutions with pH below 4 the larvae either died or changes other than those leading to ecdysis occurred on the sheaths; and in solutions with a pH above 5 only very few larvae molted.

Experiments to test the reactions of organisms to pH changes have been mentioned repeatedly on the preceding pages. In order to maintain a reasonably stable pH during such experiments, the addition of buffers is usually required. It is evident that the addition of buffers to the medium not only stabilizes the pH, but also changes the osmotic pressure to some extent and may introduce new ions as well as increase the concentration of some which are normally present. One should be aware of the possibility that these factors may influence organisms in ways totally unrelated to pH regulation. It has, for instance, been shown that certain freshwater crustacea (*Diaptomus gracilis* and *Daphnia* is, in solutions containing primary potassium *pulex*) are rapidly killed in Sorensen buffers, that phosphate and secondary sodium phosphate. The toxicity of the solutions was traced to the potassium; the animals survived well in corresponding buffers prepared with sodium salts alone.

It is of course a well known fact that both cations and anions may exert specific influences unrelated to their contribution towards the physicochemical constitution of the medium. It is thus usually assumed that the stimulus exerted by carbonic acid on the respiratory center of vertebrates is rather specifically due to the CO_2 itself and not to the hydrogen ions. Other examples are the anesthetic action of the Mg^{++}ion or the effect of the Ca^{++}ion on protoplasmic activity. Only one other aspect of the general problem of ionic biological influence may be mentioned, because some of the relevant work has been done with a nematode; it deals with the impact of various ions on the rate of respiration. Von Brand (1943), investigating the larval *Eustrongylides ignotus,* tested single salts in concentrations isotonic to a 1 per cent NaCl solution. All cations were used in the form of their chlorides and all anions in the form of their sodium salts. All salts, with the exception of $NaNO_2$ and to some extent KCI, were well tolerated and stimulated the oxygen consumption according to the following series:

Cations: $Na^+ < Mg^{++} \leq Ca^{++}, NH^+_3 < K^+$

Anions: $Cl^- \leq SO_4^{--} < NO_2^-, NO_3^- \cdot PO_4^{---}$

It must be stressed that the above procedure of testing single salts is useful only with organisms as resistant as *Eustrongylides ignotus*. The great majority of animals, and especially isolated organs or tissues, do not function properly if their ionic environment consists of only one salt.

From the standpoint of nematode physiology, as indeed for comparative physiology in general, the concept of ion antagonism is very important. It has been known for decades that solutions of pure sodium chloride, even when isotonic, were highly toxic to vertebrate tissues. Isolated frog hearts perfused with such a solution soon cease contracting. Ringer was the first to show that the addition of small amounts of potassium and calcium to the perfusion fluid was highly beneficial and allowed the heart to contract rhythmically for a long time. Solutions developed on basis of this observation are known as balanced salt solutions and are frequently designated by the name of their originator (thus Ringer's Tyrode's, or Locke's solution, to name a few of those most commonly used). Such solutions are essential for the successful study of many problems involving parasitic nematodes, as the following examples will show. The examples involve animal parasites since plant parasites, to my knowledge, have not yet been studied in this respect.

Stoll (1940) compared the influence of Tyrode's, Ringer's and physiological NaCl solutions, and of tapwater and distilled water on the *in vitro* ecdysis at the end of the first parasitic stage of *Haemonchus contortus* and on the survival of the worms. He found the first two of the above solutions, especially if slightly diluted, far superior to the others, thus giving clear indication for the functioning for the ion antagonism mechanism. This latter was still more clearly brought out in the study of Fenwick (1939) who demonstrated convincingly that the *in vitro* life span of *Ascaris* larvae was materially prolonged when small amounts of either K^+, Ca^{++} or Mg^{++} were added to a sodium-chloride solution. The best solution, both in regard to total concentration and ionic composition contained 0.137 M NaCl, 0.0027 M KCl, 0.0018 M $CaCl_2$ and 0.001 M $MgCl_2$. Another well documented case is that of *Litomosoides carinii*. Bueding (1949), after the study of the influence of the ionic environment on metabolism and motility of the worm, developed his basic filarial medium, containing 0.137 M NaCl, 0.0027 M KCl, 0.0003 M $CaCl_2$, 0.001 M $MgCl_2$ and 0.06 M sodium phosphate buffer of pH 7.6.

Reference has been made repeatedly to the osmotic concentration of solutions used for the maintenance of nematodes. It is probable that in nature this factor plays a considerable role in determining the distribution of nematodes. In this connection it is of interest to note that in free-living nematodes a sharp dividing line exists between marine species on the one hand and freshwater and soil species on the other. Overgaard Nielsen (1949) pointed out that all investigators agree that both groups have practically no species in common.

The experimental analysis of the osmotic relationships of plant-parasitic nematodes has not progressed very far. Machmer (1958) emphasized that their tolerance towards soil salinities in unknown. He observed that the Citrus nematode, *Tylenchulus semipenetrans* was recovered from salt-treated plots in numbers 3 to 20 times as great as those found in control plots. With the root knot nematode, *Meloidogyne incognita acrita* a similar, though less pronounced beneficial effect of salt treatment of soil was observed.

Nematodes seem to be, generally speaking, fairly resistant to variations in environmental osmotic concentrations. To give only a few examples: Davey (1938) found that *Ostertagia circumcincta* could be maintained in Ringer's solutions prepared to vary in NaCl concentration from 0.4 to 1.3 per cent. Von Brand (1942) reported an unaltered respiratory rate of larval *Eustrongylides ignotus* during immersion for several hours in solutions varying from distilled water to 1.5 per cent NaCl, while 2 and 4 per cent NaCl solutions led to a progressive decline in oxygen consumption. Von Brand and Simpson (1942) observed that the same worm could be maintained *in vitro* for several months at 37.5°C in media containing either 0.5 or 1.0 per cent NaCl as well as organic components.

The osmotic relations of some nematodes have been studied experimentally by immersion into hypo or hypertonic solutions and observation on concomitant weight changes or some other life activity, such as pulsations of the excretory ampulla. These experiments have revealed that, contrary to the view frequently expressed in the literature, the cuticle of nematodes is permeable to water. If *Parasaris equorum,* for instance, is immersed into a hypotonic solution, the freezing point of its coelomic fluid becomes depressed and the weight of the worm increases, while in the hypertonic solutions opposite effects take place.

Although the nematode cuticle is permeable to water, it may not be equally so in all species. Weinstein (1952) found an inverse relation between pulsation rate of the excretory ampulla and the solute concentration of media in which filariform larvae of *Ancylostoma caninum* and *Nippostrongylus muris* were kept, but the pulsation rate was about 10 time as fast in the case of *Nippostrongylus* than that of *Ancylostoma,* indicating possibly a more rapid water influx into the former worm. However, Weinstein (1952) points out that an alternative mechanism may also be operative, an active method of osmotic regulation. Such a possibility was first discussed for nematodes by Panikkar and Sproston (1941) and has become quite probable through the observations of Stephynson (1942). He reported that the swelling experienced by a soil nematode (*Rhabditisterrestris*) during immersion in distilled water was reduced after about 15 minutes and conversely that the body shrunken by immersion in hypertonic saline eventually began to re-expand.

It is of course well known that invertebrates vary greatly in their ability to regulate the concentration of their body fluids. Many marine invertebrates have only poorly developed osmoregulatory faculties; they swell up and die rapidly in fresh water. Fresh-water invertebrates, on the contrary, must evidently quite generally be able to avoid excessive losses of osmotically active materials. Finally certain animals are known to live part of their life in the sea, and part in fresh waters. The marine crab *Eriocheir siensis,* for instance, ventures far into fresh-water streams. In sea water its blood is slightly hypotonic, in fresh water it is strongly hypertonic. *Eriocheir* in fresh water loses large amounts of ions, probably largely through the skin. Its ability to tolerate fresh water so well rests on the well developed ability of its gills to absorb ions. This should not be visualized as a passive influx, but as a selective active absorption. Certain ions are absorbed, others are rejected, and apparently different mechanisms exist for cation and anion absorption.

Although osmotic and ionic regulation are in some ways closely allied, they are not identical. The ionic composition of the body fluids of many marine animals is fundamentally similar to that of sea water, but usually the concentration of certain ions differs too much from that of sea water to allow the assumption of a simple passive equilibrium between the two. It is probable that ionic regulation occurs also in nematodes. Hobson, Stephenson, and Eden (1952) reported that the coelomic fluid of *Ascaris* is hypo-osmotic in respect to its environment. Although the electrolyte composition of the two is on the whole very similar, the ionic composition of the body fluid remains far more constant than that of the intestinal contents of the host. When the worms were exposed to solutions of varying ionic composition, the Ca^{++} and Mg^{++} concentrations of the body fluid were in no way affected by those of the environment, while the Cl^- concentration was.

Chapter—13

Influence of Temperature on Life Processes

Temperature is one of the most important environmental factors influencing the metabolic level of organisms. In the animal kingdom, a distinction must be made between warm-blooded (homoeothermic) and cold-blooded (poikiloghermic) species since they react differently to changes in environmental temperature. In warm-blooded animals the rate of metabolism rises if the ambient temperature sinks below that of a thermoneutral environment, the level of which varies somewhat from species to species. The increased metabolic level is due mostly to increased muscular activity induced by the low temperature (increased muscle tone, shivering) which produces sufficient heat to maintain the body temperature at is normal level. If, on the other hand, the environmental temperature rises appreciably above that of thermo-neutral surrounding again the metabolic rate rises. This is due essentially to a sharp increase in glandular (sweating) and cardiorespiratory activity, that is, an increase in functions that play a considerable role in the physical (in contrast to the chemical) heat regulatory mechanisms.

In cold-blooded animals, on the contrary, the rate of metabolism in general follows the changes of the surrounding temperature, until in the upper, and sometimes in the lower temperature range as well, a temperature detrimental to the animal is reached. At this point of the upper temperature range, a sudden drop in metabolic rate manifests itself, but at the corresponding point in the lower temperature range such a sudden drop cannot be expected. Due to the fact that most invertebrates can live at low temperatures, their metabolic rate becomes so low at the lowest tolerated temperature that any further possible drop due to cellular damage may not be noticeable. In such cases, the best indication of damage inflicted by cold in injury that becomes evident only when the animal is brought back to a normal environmental temperature.

Although the difference between a typical warm-blooded and a typical cold-blooded animal is pronounced, certain, transitional cases occur which to some extent obscure the dividing line. Bats, for instance, during sleep (not only during hibernation, but the year round) show hardly a trace of heat regulation, in contrast to their periods of wakefulness, and the heat regulation of such animals as *Echidna* or *Bradypus* is only weakly developed. Imperfect thermoregulation is found in certain lemurs and it is well known that thermoregulation is imperfact in very young specimens of many homoeothermic species. The mammals capable of true hibernation (certain insectivores and rodents, but not such animals as bears or badgers) must be mentioned in this connection. During the cold period of the year they essentially lose the ability of regulating their body temperature at the high level characteristic for the warmer time of the year. A certain power of regulation is nevertheless maintained, as indicated by the fact that their body temperature never falls below a certain minimal value (varying from species to species and usually lying above 0°C) regardless of how low the external temperature may fall.

On the other hand, cases are known in which cold-blooded animals can reach and often maintain a temperature higher than the ambient one. Fish as a rule have a body temperature close to that of the surrounding water, bu. fish belonging to the family Scombridae (*e.g., Thynnus, Pelamydes,* or *Bonites*) have been observed with temperature 5 to 10° higher than that of the sea. It is assumed, at least for the tuna, that the thick muscle layers prevent a rapid dissipation of the heat produced by these active tissues. A different mechanism must be assumed in several terrestrial animals, such as lizards which may reach a body temperature of 40°C by sitting quietly in the sun when the surrounding temperature is 9-10°C. Since air, in contrast to water, is not a good heat conductor, the heat absorbed by the lizard from the sun is lost so rapidly as it is absorbed.

A quite similar mechanism is also operative in many beetles and other insects, especially when their outer integument is dark, and is responsible for the fact that sun-exposed inanimate objects are warmer than the air. It would nevertheless be an over simplification to assume for the above invertebrates a behaviour exactly paralleling the inanimate objects. Strelnikov (1934) has shown that, under identical conditions, the temperature of a living specimen of the beetle *Sternodes caspius* rose by 14°, while that of a dead one rose by only 9°. He points out that heat increases all vital processes in the living animal, that is, the heat produced by metabolic processes has, in the case of the living beetle, to be added to that gained by absorption from the sun.

A different way of raising body temperature is operative during flight preparation of such insects as moths, butterflies, or bees. They must reach a certain minimum temperature before they are capable of performing the vigorous muscular movements required for flight regardless of the air temperature. They reach this point by performing a series of deep and rapid respiratory movements accompanied by vibrations of the wing muscles, thus producing heat faster than it is dissipated.

Finally there exists among insects a mechanism which could be called collective thermoregulation. It is especially well developed in bees, but is, with some variations, found also in other social insects, such as hornets and certain ants.

During the winter the bees keep their temperature above a minimum of about 8°C by forming a cluster which becomes tighter the cooler the surroundings, and in which heat is generated by muscular movements, the temperature in the center of the cluster varying between about 15 and 30°C. A constant exchange in position relative to the interior of the cluster occurs between the relatively cold "outside" and the relatively warm "inside" bees, insuring that the former do not freeze to death. During the summer, on the other hand, overheating (especially detrimental to the brood) is avoided by fanning, and probably by water transport. The heat regulation of the social insects, like many other of their life activities, has many fascinating aspects, especially in respect to the varying importance of chemical and physical regulation found in different species, but such details are beyond the scope of this account; an interesting presentation of this problem has been given by Himmer (1932).

It is not likely that any kind of thermoregulation occurs in nematodes, but their temperature relationships are nevertheless of interest because they tolerate substantial temperature variations both in nature and under experimental conditions. Species of roundworms have been found in the Arctic and, at the other extreme, they have been reported from hotsprings with temperatures up to 51°C, at which temperature *Dorylaimus granuliferus* was found. While this is not the highest temperature at which living organisms are encountered, it is nevertheless far above the temperature tolerated for any length of time by almost any invertebrate.

The question of heat tolerance is of course of importance to plant nematologists, since heat-killing of plant-parasitizing nematodes is a well established control measure. It should be realized that the determination of the thermal death point of an organism is largely a matter of experimental conditions, since not only the temperature proper, but also the length of exposure is of utmost importance. In other words, in order to give an adequate description of heat death, death rate curves at various temperatures should be established. This, for instance, was done by Santmyer (1955) in a comparative way for the free-living nematodes *Panagrellus redivivis* and the plant-parasite *Meloidogyne incognita acrita*. Neither species could survive exposure to 46ºC for one hour, but at temperatures between 42 and 45ºC the free-living species was significantly more heat-resistant than the parasite.

Santmyer (1955) is apparently inclined to ascribe heat death of the nematodes in question to enzyme inhibition, pointing out that most animal enzymes have an optimum around 40ºC and are rapidly inactivated above 50ºC. This,

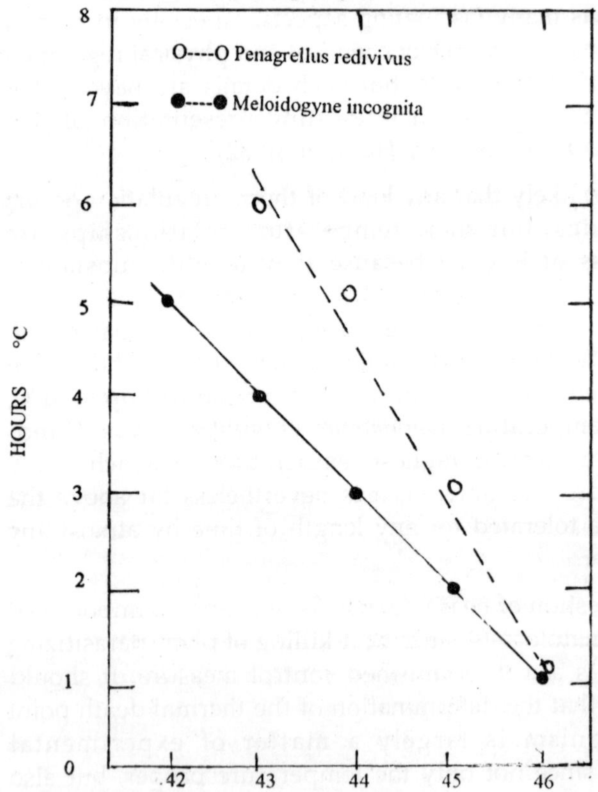

Fig. 13.1 : Thermal death time of two nematodes species according to Santmyer. (1955).

certainly, is a possibility which probably has more to recommend itself than the old view that protein coagulation is the main factor involved in heat dealth. It must, however, be realized that neither hypothesis can explain all cases of heat death. It is thus well known that different arctic animals die from heat at temperatures that are far below the inactivating temperature of the common enzymes or the coagulation point of proteins. The clam *Pecten groenlandicus*, for instance, dies when exposed to a temperature of 10°C. It is possible that in such instances an effect on the cellular lipids is involved, although the nature of the supposed effect is not clear. It is, as Helibrum (1938) points out, " a remarkable fact that animals and plants which live at lower temperature have fats with lower melting points than those accustomed to live at higher temperatures" and "the differences in sensitivity of various organisms toward heat was correlated with the differences in melting points of their lipids." It would then seem quite possible that heat death, when occasioned by none too extreme temperatures (that is excluding frankly protein-denaturing temperatures) may not have the same basic physiological mechanism in all cases.

Heat resistance is often dependent upon the hydration of the tissues. Organisms capable of approximately complete dehydration are much more heat resistant than when hydrated. Overgaard Nielsen (1949) has thus shown that dedicated specimens of the nematode *Plectus granulosus* could withstand 66°C for at least 4 days and Hinton (1951) demonstrated that dried chironomid larvae (*Polypedilum vanderplanki*) survived readily a 14-hour exposure to 60°C and recovered, though only temporarily, from a 21-minute exposure at 126-130°C or a 5-minute exposure at 199-210°C. In water, the same larvae did not survive temperatures higher than 45°C.

In regard to cold resistance, many invertebrates survive the coldest periods of the year even in high latitudes without apparent injury. Poikilothermic terrestrial animals frequently experience extremely low temperatures. Animals living in fresh water or in the sea are not exposed to similar extremes, but members of the intertidal fauna will often encounter very low temperatures and be subject to rather frequent and violent temperature changes, resulting from their being exposed to air at low tide and covered by water at high tide. Similarly, parasites which develop in a cold-blooded host and are transferred during the cold season to a warm-blooded final host, must be capable of withstanding a sudden temperature change of considerable magnitude. The degree of coldhardiness of many invertebrates has been tested experimentally. To give only one example at this point, it has been shown that the free-living nematode *Anguillula silusiae* can revive after having been subjected to a temperature as low as—192°C although only a fraction of one per cent of the exposed specimens survived.

The questions must then be raised as to (1) the factors which are responsible for coldhardiness and (2) the mechanism of cold injury.

In studying the literature relating to the first question, it becomes evident that one very important point in determining the degree of resistance to cold is the state of hydration of the tissues. Resistance generally increases when a certain degree of dehydration is achieved prior to exposure to the cold temperature. Gehenio and Luyet (1951) have shown that *Turbatrix aceti,* to choose a nematode as an example, withstands ultra rapid freezing better when slightly dehydrated than when taken directly from the fluid medium. In the latter case less than one per cent were alive one hour after being frozen at—77°C, while 90 per cent were alive when the specimens had been kept for 16 hours at 95 per cent relative humidity prior to freezing. The ultra rapid freezing method employed by these authors is the so-called vitrification method by which freezing is achieved so rapidly that no intracellular ice crystals are formed, but the water assumes a vitreous state.

The relationship between cold resistance and desiccation can be tested in many cases only with specimens that have at most been only moderately desiccated, because many animals are damaged by high degrees of water loss. Many nematodes belong in this class, but several species of plant-parasitizing nematodes and especially the moss nematodes do withstand water losses which amount to practically completely desiccation. It has been shown that in this state nematodes and other members of the moss fauna, such as tardigrades and rotifers, survive for more than 20 months' exposure to—190°C and can resist even—272°C. The bulb and stem nematode *Ditylenchus dipsaci* survived—80°C for at least 28 days after lyophilization and storage in vacuo. Becquerel (1950) exposed members of the moss fauna first dehydrated in a high vacuum to temperatures lower than 0.05° Kelvin and found them able to withstand this treatment. In such extreme cases there is justification in raising the question whether a state of suspended animation in the true sense of the word was induced, that is, whether life processes actually halted temporarily. It must be realized in this connection that even the high degree of desiccation characteristic for dried moss fauna apparently does not as such induce complete stoppage of metabolic processes. Labunetz (1950) observed that the reserve substances of desiccated tardigrades decreased when the animals were kept at 15°C. Pigon and Weglarska (1953) found the respiratory rate of a dehydrated *Macrobiotus hufelandi* 600 times slower than that of an active one (20-22°C) and Hinton (1951) found that a dehydrated chironomid larva (*Polypedilum vanderplaki*) consumed per hour 0.0023 μl. of oxygen, as compared to 0.3933 μl. for an active larva (35.5°C).

It is, on the other hand, difficult to see how cellular enzymatic activity could persist near absolute zero. It is of course true that upon prolonged exposure to such abnormally low temperature, an increased percentage of the animals dies as the time goes by. This, however, does not necessarily indicate a continuation of metabolic processes during the period of storage at low temperatures. The failure to revive could in such instances be due to progressive disorganization of the cellular colloids or some similar nonenzymatic process which would not be indicative of either persistence of stoppage of life. To solve the problem of whether desiccated moss nematodes survive in a state of true suspended animation will require the application of the most refined techniques capable of demonstrating minute traces of metabolic endproducts, *e.g.*, carbon dioxide. It would be a worthwhile undertaking since the philosophical implications of the question surpass purely biological interest. It may be mentioned that such profound students of the problems as Becquerel (1950) and Schmidt (1955) are convinced that true anabiosis occurs in the above cases.

An important factor in determining the degree of cold resistance is the medium in which the organisms are exposed to low temperatures. For instance, several species of parasitic protozoa and larval parasitic nematodes remained viable for shorter or longer periods during cold storage after having been frozen in what can be termed more or less normal media. But it has been found subsequently that parasitic protozoa in general survive much better in a medium containing several per cent of glycerol while parasitic nematodes apparently have not been studied in this respect. The favourable influence of glycerol on survival of animal cells seems, however, to be quite general, having been reported for such diverse tissues as ovarian tissue, erythrocytes, or spermatozoa. Glycerol solutions of the strength employed are hypertonic. In view of the preceding discussion it would appear tempting to explain their favourable influence by assuming the induction of a slight tissue dehydration. It is, however, questionable whether such a simple assumption suffices. Luyet and Keane (1953) have described curious weight changes of tissues in glycerol solutions which are not entirely compatible with the above assumption. Furthermore, it has been reported that glycerol, in contrast to sugars, is much less effective in protecting plant tissues than animal tissues against frost damage. Whatever the final explanation of the protective action of glycerol will be, it is probable that it is not confined to artificial laboratory systems. In this regard it is remarkable that relatively high glycerol concentrations have been reported recently from the blood of cold-resistant insects.

In the case of one frost-hardy insect at least, *Bracon cephi,* according to Salt (1958) the glycerol present is responsible for abnormally low super-cooling points (—47°C) and freezing points (—7 to —17°C). Super-cooling (*i.e.,* body fluids remaining liquid at temperatures below their freezing points) frequently has been assumed to be an important factor in allowing survival at low temperatures. Ice crystal-formation is one of the widely discussed and accepted mechanisms involved in cell damage. This assumes that the formation of inter-cellular ice crystals is damaging to cells. The facts that ultrarapid freezing (vitrification) and rapid heating after freezing favour survival of many organisms and frequently avoid the damage seem after slow freezing or thawing, are in good accord with the above view and it appears certain that many cases of cold-damage can be explained on the basis of intra-cellular ice crystal formation. However, a number of recent observations cast some doubt on the universal applicability of this theory. It has thus been shown that several species of insects definitely survive the formation of ice within their tissues and the same holds true for several species of intertidal molluscs, although the ice crystal-formation in these cases may have been limited to intracellular spaces. For instance, an Alaskan chironomid larva under natural conditions survives—40°C for months, although it becomes brittle at—10°C in the laboratory. Under experimental conditions this animal survived repeated freezing and thawing and it could be shown that at—15°C 90 per cent of all the body water was frozen, but even at —35°C a little unfrozen body fluid was still present. Similarly, the intertidal molluscs *Modiolus modiolus* and *Littorina littorea* were not damaged by temperatures at which 71 and 76 per cent of their body waters respectively were frozen. It is clear than such cases are not only indicative of great resistance towards ice crystal-formation, but to other changes as well. When so much water is frozen, evidently the concentration of inorganic and organic components of the unfrozen fluids will rise sharply and reach a level that many animals in the unfrozen state could not survive, even if belonging to species fairly resistant to desiccation. It is obvious that a full understanding of both cold-resistance and frost-damage will require much additional study.

It must also be realized that cold damage and frost damage are not always identical. Many tropical poikilothermic animals are damaged by temperatures above the freezing point of their tissues. It is well known that many snails of the temperate zone withstand the rigors of winter without damage and the cold resistance of intertidal snails has already been mentioned. In contrast, a two-hour exposure of the tropical snail *Australorbis glabratus* to 0.3°C resulted in almost 50 per cent of the specimens being damaged so much that they died in a subsequent recovery period at 30°C. Similarly, arctic crustacea are active and appear entirely healthy down to the freezing point of the water, while comparable tropical forms die at temperatures varying between 5 and 10°C.

Why tropical invertebrates should be susceptible to cold damage at temperatures above the freezing point its unknown. The observation, however, proves that the formation of ice crystals, whether intra- or extra-cellularly, is not in all cases responsible for cold damage. Metabolic adaptations to cold exist in arctic invertebrates; it may well be that they lie at the root of the different degrees of cold resistance existing between them and the tropical forms. Scholander *et. al.* (1953) have thus demonstrated that at low temperatures the metabolic rate of arctic aquatic animals remains at a higher level than in corresponding tropical species. On the other hand, however, arctic insects and spiders showed at most traces, if any, of such adaptation. The problem is far from solved.

The temperature interval between damaging high and damaging low temperatures can be used to study temperature influence without interference of temperature-induced abnormalities. Theoretically, any life process can be studied in this respect, and many have been, such as motility, heartbeat, embryonic development, metabolism, to name only a few. There is no great difference in response to changes in environmental temperature between biological processes which have the same fundamental nature, that is, which rest on chemical processes. They increase in rate between 2 and 3 and occasionally up to 5 times with a temperature increase of 10°C, just as chemical processes studied in a test tube do. On the other hand, physical processes, such as photochemical reactions, frequently increase by only 10 per cent or even less for a 10°C rise. Such processes are rare in animals, if they occur at all without being connected with some chemical reaction.

The present discussion will be limited to the chemical processes and will deal essentially with temperature influence on the respiratory metabolism, a phase of animal physiology

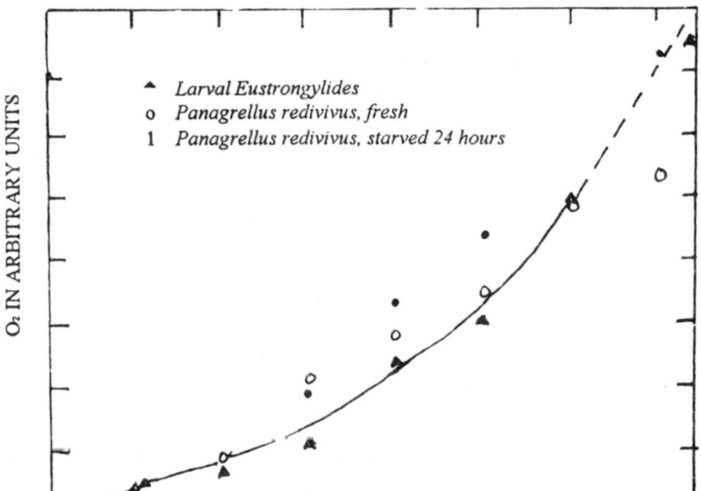

Fig. 13.2 : Oxygen consumption of two nematode species projected on Krogh's normal curve. *Eustrongylides* data from von Brand (1943), *Panagrellus* data from Santmyer (1956).

especially suited for discussion of the relevant relationships. Others examples could have been used, *e.g.*, heartbeat, ciliary movement, and others, but they would show essentially the same temperature relationship as the example selected here. In order to study temperature influence on respiration uncomplicated by other factors standard, conditions should be applied. In the case of invertebrates this means ideally, the use of the same specimen, or group of specimens at all temperatures studied, elimination of digestion and absorption of food, and exclusion of motility, either by using narcotized animals or by selecting immotile stages, such as insect pupae. Control of motility is essential when temperature may induce vigorous movements, such as flight, but in many cases of aquatic and semi-aquatic animals it is unnecessary because, as will be explained in another section in greater detail, their resting and moving metabolism rates differ little. It is even possible to argue that in such cases it may be preferable to avoid possible complications due to narcosis.

Because it allows rough differentiation between chemical and physical processes, the temperature coefficient Q_{10} has frequently been calculated for the respiration of many animals. It is not necessary to determine the respective rates at temperatures actually 10° apart; Q_{10} can be calculated from determinations done at any 2 temperatures by the formula:

$$Q_{10} = \left(\frac{K_2}{K_1}\right)^{\frac{10}{t_2 - t_1}}$$

Q_{10} in most cases studied is not a constant value over a wide range of temperatures: generally it has a tendency to decline with increasing temperature. A special situation seems to prevail below freezing. In plant material and in the case of an undetermined *Chironomus* larva, Scholander *et. al.* (1953) found Q_{10}'s of 2 to 5 above freezing. Below freezing an abrupt drop in respiration occurred and the Q_{10} reached abnormally high values of 20 to 50. The above authors are of the opinion that the low respiratory rate below freezing may be related to 3 factors: (1) the Q_{10} per se, (2) dehydration, and (3) a diffusion barrier set up by the ice crystals formed. They also justifiably point to the significance of their findings as a rational explanation for the very long survival of frozen organisms. It certainly would be of great interest to conduct similar experiments with dehydrated and cooled nematodes.

Because the Q_{10} usually does not stay constant, when a wider temperature range is studied, its value in establishing the temperature relationships frequently has been doubted and other ways of describing them have been sought. A "normal curve," applicable to such diversely organized animals as mosquitoes, fish, curarized frogs, and dogs has been established by Krogh (1916) for the temperature range of near 0°C to about 30°C. Experience accumulated over the intervening years has shown this curve to be very useful in describing not only the temperature relationships of the respiration of many organisms, but also those of several other life activities, such as carbohydrate metabolism. But experience has also shown that the response of the respiration of several invertebrates to temperature changes deviates rather markedly from the "normal curve."

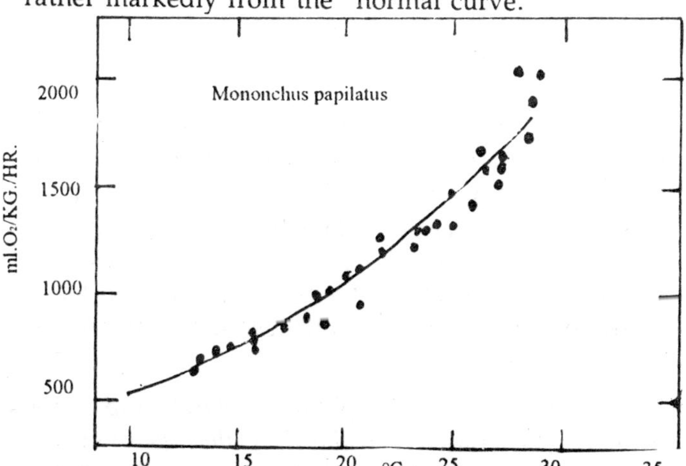

Fig. 13.3 : Oxygen consumption of *Mononchus papillatus* projected on Krogh's normal curve.

Insofar as nematodes are concerned, Krogh's normal curve expresses very well the temperature relationship of the oxygen consumption of the parasitic *Eustrongylides ignotus* larva and of the adult free-living *Mononchus papillatus*. The respiration of *Panagretlus redivivus* showed greater deviations from the curve either in material freshly taken from the culture medium, or starved for 24 hours prior to the determinations.

While Krogh's normal curve has apparently been applied only to biological processes, another expression of temperature relationships is used not only by biologists, but very frequently by physico-chemists as well. This is Arrhenius' equation:

$$K_2 = K_1 \times \frac{\mu}{e^2}\left(\frac{1}{T_1} - \frac{1}{T_2}\right)$$

The symbol μ in this equation represents the energy of activation, or the temperature characteristic of the chemical reaction involved. It is calculated easily from the following transformation of the above equation:

$$\mu = \frac{2(T_1 \times T_2)}{T_2 - T_1} (\text{In } K_2 - \text{In } K_1)$$

In applying Arrhenius' equation, the biologist usually is interested in two things: (1) whether the process under study is adequately expressed by the formula and (2) if such proves to be the case, what are the μ values found. The simplest way to test whether a series of data fits the equation is to plot the log of the velocities (k values) against the reciprocal of the absolute temperature. If the points group themselves along a straight line, the equation describes the process satisfactorily.

An example of such a straight line relationship is the respiration of the snail *Australorbis glabratus* over the temperature range 0.3 to 37°C, which is a range somewhat wider than that readily tolerated by the snail. In many other

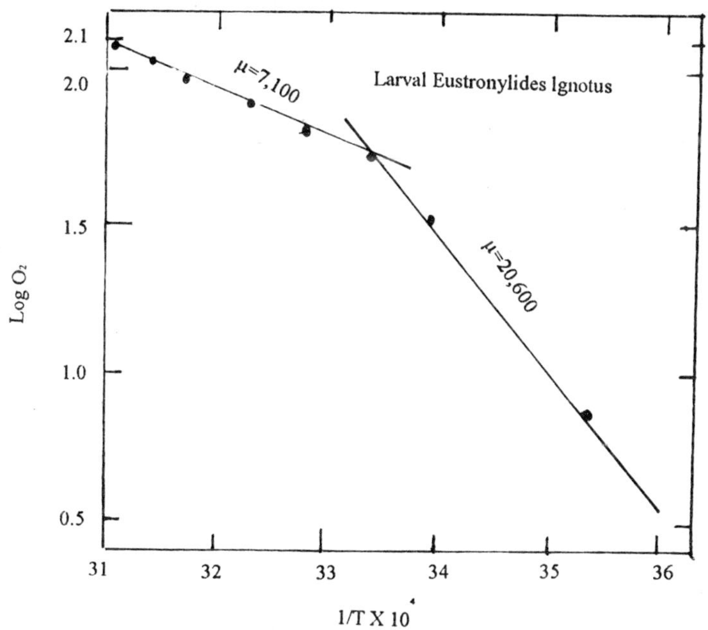

Fig. 13.4 : Oxygen consumption of *Eustrongylides* expressed according to Arrhenius formula.

cases, however, a single line is not sufficient to represent the temperature relations adequately. Frequently two, and occasionally even more, interesting lines result from application of Arrhenius' equation. Examples relating to nematodes are shown in figure 8.4 and 8.5.

The occurrence of intersecting lines, and therefore different μ values at different temperature ranges has been known for many years and has received special attention from Crozier and his students. It is evident that the end result of a series of reactions is determined when a biological or physiological process, such as respiration, is measured. In each series of reactions the slowest one will determine the speed of the over-all process and the slowest, place-setting reaction therefore deserves the often-applied term "master reaction." If the temperature characteristics of the various members of a catenary reaction were different, as is probable in many instances, a different member of the series

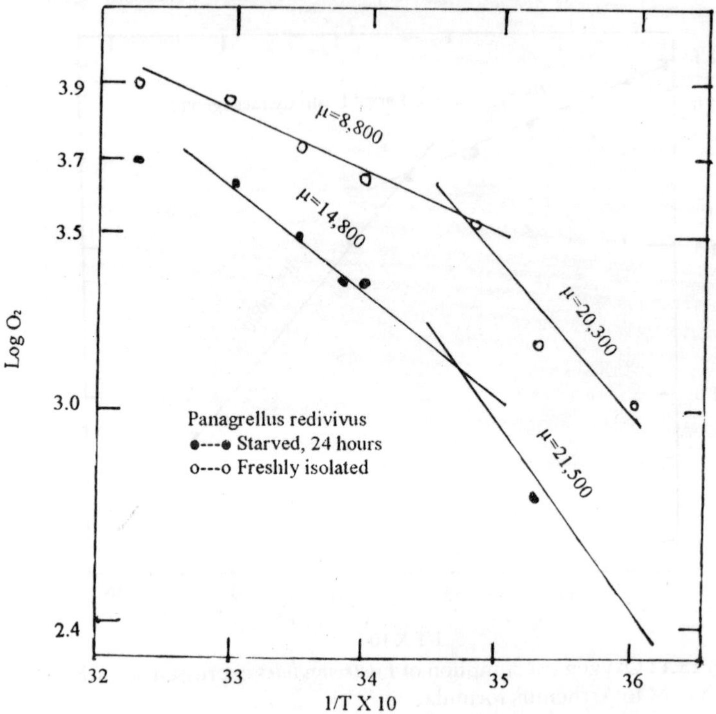

Fig. 13.5 : Oxygen consumption of *Panagrellus* expressed according to Arrhenius formula.

could be expected to assume the role of master reaction in different temperature ranges. The change-over from one pace-setting reaction to another would then be responsible for an abrupt change in the μ value of the over-all reaction.

Physical chemists have been able to associate definite temperature characteristics with certain chemical reactions, for example, 16,500 calories per gram molecule seems to be characteristic for iron-catalyzed oxidation, or 20,000 calories for hydrogen ion catalysis. Although it would be tempting to identify the values found for biological processes with some definite reaction, such an attempt is, at the present state of our knowledge, bound to fail. It is true, as Crozier (1925, 1926) points out, that the magnitude of μ values, derived from a great variety of biological processes, does not change continuously over the actual range of observed values and cannot be expressed by a single smooth curve. On the contrary, the frequency of μ values is discontinuous and shows between 8,000 and 32,000 calories at least 9 frequency peaks. Similarly, the break in the curves occurs much more frequently at some temperatures than at others, being found especially frequently at 15°C. It does seem probable that such observations imply a specific nature of the temperature characteristic which, however, cannot be characterized at the present time.

The situation is extremely complicated, as a final example may show. Krueger (1940) studied the dependency of respiration on temperature using *Ascaris lumbricoides* as the experimental animal. He found different temperature relationships for the oxygen uptake and the carbon dioxide production. The μ values for the former process varied between, 5,000 and 15,000 with peaks at 7,500, 10,000 12,000, and 13,500, while those for the carbon dioxide production varied between 9,000 and 25,000 with peaks at 10,000, 15,000, 18,000, and 21,000. It is possible that in this particular case the persistence of carbon dioxide producing anaerobic processes in the presence of oxygen lies at the root of this dichotomy, but in view of the exceptional variability of Krueger's values, a reinvestigation of *Ascaris* would seem desirable.

Chapter—14
Nematode Infections

The preceding chapters have dealt with parasites consisting of but a single cell or of groups of independent cells. We turn now to a consideration of those small multicellular or metazoan animals known as "nematodes" or "nemas," or as Shakespeare had King Richard II say, "Let's talk. . . of worms." Some of these animals are well known for the diseases they cause in man; *e.g.*, *Wuchereria bancrofti* (Cobbold), transmitted by mosquitoes, is the cause of human filariasis, and *Necator americanus* (Stiles) is the cause of hookworm infection. Plants are also attacked, the well-known sugar-beet nematode being an example, and such infestations are of considerable economic and agricultural importance. Many, however, parasitize and destroy insects, and it is these with which we shall be concerned here.

The phylum Nemathelminthes (roundworms) includes three classes: Nematoda, Nematomorpha, and Acanthocephala. The members of these classes have the following characteristics in common: (1) the body is elongate, unsegmented, and usually threadlike; (2) it is enclosed in a hardened cuticula; (3) it has no appendages, cilia, vascular system, or special respiratory organs.

Those Nemathelminthes for which insects serve as primary hosts are included in the classes Nematoda and Nematomorpha. In their general external appearance, all the nematodes are very much alike. Rarely is there any distinct variation in the diameter of the individual nema, although the body tapers toward one or both ends. The anterior end is slightly more blunt than is the posterior end. The smaller living nematodes are more or less transparent, but the larger species are usually opaque. The movement of most Nematoda is fairly characteristic. In a clear fluid medium, the worms undergo rather vigorous coiling and twisting to the right and to the left without making much progress. If solid particles are present in the fluid, however, the organism changes its movement to one of serpentine character, winding in and out of the particles.

Of the class Nematomorpha, members of the subclass Gordiacea (hairworms) are parasites of insects. They differ from the Nematoda in being larger, as a rule, and more uniformly cylindrical, with bluntly rounded ends, and having a faintly colored exterior.

Acanthocephala are known as thorn-headed worms. They are highly specialized for parasitic life and have no free-living stage and no trace of an alimentary tract at any stage of their development. Some of them spend the larval stage within an insect host and the adult stage in mammals.

In general, there are three main stages in the developmental cycle of nematodes: eggs, larvae (including four growth stages), and adults. The eggs are extremely minute—being microscopic in size. The developing embryos are usually discharged as eggs, but in a few species they hatch within the uterus of the female worm and are brought forth viviparously. The embryo may require weeks or months for its growth and may wait within its shell for indefinite periods before it is passively introduced into a new host. Upon hatching, the larvae have the main morphological characteristics of adult nematodes except that the reproductive system and the secondary sexual characters are lacking. Usually the young larvae spend a short period as free-living organisms, frequently in an aquatic environment of water or mud. The larvae of nearly all nematodes undergo four molts before becoming adults. These molts may take place in the egg, during the worm's free existence, in the tissues of an intermediate host, or in the definitive host. Most nematodes are bisexual, the females producing fertile eggs after copulation. Males are frequently less common than females, though usually both sexes are plentiful. The males reach maturity a little sooner and do not live quite so long as do females; this sometimes gives the impression that only females are present.

The cultivation of nematodes on artificial or synthetic media has been attained in a few instances.

Insect Hosts Affected. Within any order of insects, the number of known specific hosts to any particular group of organisms frequently depends simply on the degree to which the insects in any one order have been studied from this viewpoint. Therefore, merely citing the number of nematodes described from the various orders of insects would not necessarily indicate the true range of parasitism as it occurs in nature.

In 1928 van Zwaluwenburg presented a list showing the number of insect species involved as hosts to nematodes according to the various insect orders. Out of a total of 759 insect species (in 16 orders) with which nematodes were then known to be associated, the majority (268) were in the order Lepidoptera. As van Zwaluwenburg suggests, this is because of the intensive work of Schultz (1900), who gave special attention to this order. Next in number of host species is the order Coleoptera with 172 species, then Orthoptera with 116 species, and Diptera with 96 species. Lesser and decreasing numbers were of the order Hymenoptera, Isoptera, Hemiptera, Odonata, Siphonaptera, and others. The total number of insect hosts of nematodes known today can only be estimated. La Rivers (1949) has listed 97 insect hosts (some new and some repetitions) which have been reported in the literature up to 1946 and since Zwaluwenburg's paper appeared.

As to the number of species of nematodes associated with insects, only speculative estimates can be suggested.

The number of species already reported from this habitat appears to be in excess of 1,000 species.

Relations to Host. The biological relations between nematodes and insects vary all the way from those of mere fortuitous association to those of strict and destructive parasitism. Van Zwaluwenburg (1928) grouped the entomophilic roundworms into five classifications: primary parasitism, secondary parasitism, internal mechanical association, external mechanical association, and commensalism. In the last-named category were placed those nematodes which live within the burrows or nests of beetles, termites, and ants, and which feed upon the frass and debris that accumulates in the nests. Some authors (*e.g.*, Filipjev and Stekhoven, 1941) simply separate the insect nematodes into two large groups: those living within the alimentary tracts of insects, and those inhabiting the body cavity of insects. Since there are so many variations and gradations in the types and degrees of "parasitic" associations between insects and nematodes, it is difficult to draw a true line between them.

Perhaps one of the most convenient groupings to follow is that used by Christie (1941), who divided the nematodes associated with invertebrates into three groups: (1) Those nematodes which live in the alimentary tract of the invertebrate and which are not included in the next group. In most cases the life cycles are simple. (2) Those nematodes which are more or less closely related to free-living species and which often have a combination of saprophagous and parasitic habits. They may live and reproduce in the cadaver of the host, which may or may not have been killed by the parasites. Others of this group may pass through one or more free-living generations which alternate with one or more parasitic generations. Christie designates these organisms as novitious parasites and semiparasites. (3) Those nematodes which parasitize the body cavity or the tissues of their host. These worms are highly specialized, obligately parasitic, and at the most spend only a transitory period in the alimentary tract of the invertebrate. Of these three groups the insect pathologist is perhaps most interested in the last, but the other two have members that are also important from a pathological standpoint as well as from the standpoint of their role in biological control.

The remainder of this chapter will be devoted to a brief consideration of a few important members of the three groups just designated.

Nematode Parasites of the Insect Gut

As it does for many microorganisms, the alimentary tract of an insect affords ideal living conditions for certain nematodes. A great majority of those nematodes which have this habitat are grouped in Filipjev's (1934) category Oxyurata. According to Christie (1941), nematodes belonging to the families Thelastomatidae and Rhigonematidae and to the subfamily Ransomnematinea are parasites of the alimentary tract of animals, and there is an occasional species of the family Diplogasteridae that has acquired this mode of life. Christie also tells us that the literature contains descriptions of between 60 and 70 species of thelastomatids parasitic in the gut of insects and myriapods. Most of these descriptions are of a taxonomic character only, and very little information is available as to the biological relationships existing between them and their hosts. So it is with the other groups. In general, very little apparent harm is caused the host by the parasites, which have simple life cycles and rarely invade vital tissues of the insect. In most cases, the eggs of the nematode do not hatch in the gut of the insect but pass out with the feces. Once outside the host they undergo further development until they reach an infective stage. Infection of a fresh host occurs when these infective eggs are ingested by the insect along with its food.

The biology of this group of nematodes has been only meagerly studied, but enough information has been gathered to indicate that the oxyurids inhabit the guts of those insects which have well-developed digestive systems. Such digestive systems afford almost perfect digestion of the food, combined with a slow passage of the food through the gut, a comparatively long stasis of the fecal pellets in the rectum, and a rich bacterial flora (Filipjevac and Stekhoven, 1941). Accordingly, these nematodes are rare in the caterpillars of Lepidoptera and in species of Locustidae—groups in which the food passes rapidly through the alimentary tract and is eliminated only partly digested. Insects that do harbor these nematodes include species of Blattidae, Scarabaeidae, Passalidae, Hydrophilidae, and others. Many of the oxyurid nematodes live in the rectum of the insect, ingesting the rectal contents and absorbing mostly those food substances which are of little value to the host.

Examples. Since very little harm is caused the host and practically no pathology results from the infestation, a detailed discussion of these nematodes will not be necessary here. Two species, however, may be cited as examples of the group.

Cephalobium microbivorum Cobb, a diplogastrid, lives in the gut of the black field cricket, *Gryllus assimilis* (Fabr.). An individual insect may harbor up to 30 or more nemas, the eggs of which pass out of the host with the feces. The insect apparently acquires the "infection" by way of the mouth. Infected crickets have been collected in Virginia and in Kansas.

Leidynema appendiculatum (Leidy), a thelastomatid, is a parasite of the cockroaches *Blatta orientalis* (Linn.) and *Periplaneta americana* (Linn.) in Kansas and probably elsewhere. The egg of this nematode passes out with the feces of the insect and undergoes a short period of development, forming a small tadpolelike larva. Still within the egg the larva, at first motile, becomes inactive and forms the infective stage in about 3 to 7 days. The egg completes its life cycle when it is ingested by the insect and hatches in the hind portion of the midgut where the worm matures. No evidence of harm to the host insect has been obtained.

At this point the fact might be mentioned that some nematodes (*Spirurata* and *Filariata*) are parasites of vertebrates but have insects as intermediate hosts. In certain cases the insect serves as a transmitting agent for the larval nematodes. The insects, depending on the species, may acquire the nemas by the sucking of infectious blood or by feeding upon or contacting the droppings or refuse of animals. The nematologist can usually quite readily differentiate these nematodes from those which parasitize or are specifically associated with insects.

Nematodes Semiparasitic in Insects

Neoaplectana glaseri Steiner in the Japanese Beetle

In May, 1929, while digging for grubs of the Japanese beetle, *Popillia japonica* Newm., near Haddonfield, New Jersey, Glaser and Fox (1930) found a number of dead fully grown larvae, which upon dissection were observed to be infected with nematodes. Later the same season parasitized pupae and adults were collected. Specimens of the nematode were sent to Steiner (1929), who named it *Neoaplectana glaseri*, commenting that it is probable that the worm has for an original host some native insect or insects ad has only recently adapted itself to the Japanese beetle. Steiner placed the nematode in the family Oxyuridae, but it has since been placed in the family Steinernematidae by Christie (and Anguillulidae by Filipjev). Subsequent to its discovery, recoveries of the parasite from

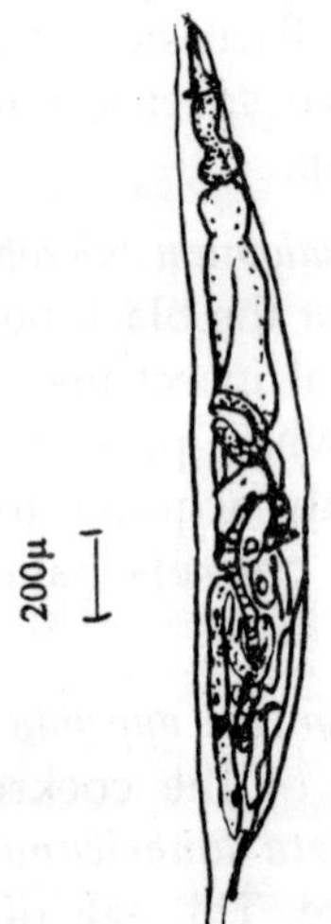

Fig. 14.1 : *Leidynema appendiculatum* (Leidy), a parasite of cockroaches. Adult, female.

the original Haddonfield area have been made regularly. Outside of this limited locality no isolations of the nematode have been accomplished, even though a diligent search has been made in numerous other parts of New Jersey and in Pennsylvania. In addition to the Japanese beetle, several other species of insects are susceptible to infection by *Neoaplectana glaseri*. We are informed by Glaser, McCoy, and Girth (1940) that parasitism of the larvae of *Anomala orientalis* Wth occurs with a facility at least equal to that observed in the Japanese-beetle grubs. Other susceptible insects include the larvae of the following: *Autoserica castanea* (Arrow), *Macrodactylus subspinosus* (Fabr.), *Ochrosidia villosa* (Burn.), *Cotalpa* sp., *Cotinis nitida* (Linn.), *Xylorycles salyrus* (Fabr.), several species of *Phyllophaga;* the larvae of the white-fringed beetle, *Pantomorus leucoloma* (Boh.); and the larva of the European corn borer, *Pyrausta nubilalis* (Hbn.). Some attempts to infect silkworms and army-worms have been unsuccessful.

Symptoms. Normal healthy grubs of the Japanese beetle are white in colour, fairly active, firm to the touch, and have a good appetite at summer temperatures. When infected with *Neoaplectana glaseri,* the grubs have a diminished appetite, become flaccid, and are less active. They become colored and may be somewhat mottled in appearance; at first the coloration is spotty, but later, just before and after death, it is more uniform. The most characteristic colour is a rusty or ocherous brown; but dark-brown, light-brown, and dirty-white colourations also occur. Grubs that have turned black in colour are usually not parasitized by Nematodes.

To determine with certainty if a larva is parasitized by *Neoaplectana,* a microscopic examination is necessary. The body contents of parasitized grub are usually swarming with nemas. As many as 2,400 infective-stage nematodes have been recovered from one Japanese-beetle grub—usually the number is about 1,500. Occasionally nemas of the genus *Diplogaster* are encountered. In dying and recently dead larvae and pupae, nearly all stages of *Neoaplectana* exist. In insects that have been dead for a long time, the dark second-stage nemas predominate.

Experimentally the average length of time from infection to death is approximately 11 days, and infection usually results in death. Occasionally a grub becomes parasitized very lightly—perhaps only one nematode originally becomes successfully established—but the host may eventually be killed.

Life History. The infective second-stage forms are acquired by the Japanese-beetle larva through the mouth. After entering the insect the nematodes soon develop into mature males and females and copulate. The female is ovoviviparous, the eggs hatching within the uterus. After remaining in the uterus a while, the young nematode larvae pass out through the vulva one at a time. The average female produces abut 15 young, which are discharged into the alimentary tract of the host. Under optimum conditions each generation requires from 5 to 7 days for completion. As the new generation develops within the grub, the parasites may become so numerous that the insect dies and the nematodes invade the entire body. The worms usually pass through one or two more generations in the cadavers of the host, consuming nearly all the contents and leaving only a sac formed by the integument and the head capsule, and filled with a thin fluid swarming with larval parasites. Since further nematode development continues in the dead host, the nematode is considered a saprozoic or semiparasitic organism rather than a true parasite. Glaser, McCoy, and Girth (1940) have suggested that *N. glaseri* is a species in transition from the free-living saprozoic mode of existence to a parasitic one. As many as three generations may occur within one host, two of these appearing after the host's death—a behavior not characteristic of the nematodes of vertebrates.

In the older cadavers the second-stage nemas are the predominant type. These invade the soil and remain in a free-living state until they are ingested by another grub. Sometimes exceptionally large female nemas, which may produce enormous numbers of eggs, are found in the beetle larva.

As long as conditions are favorable and there is an abundant food supply in the insect, the life cycle of the nematode is completed in three molts, the third stage being omitted (Glaser, McCoy, and Girth, 1940). When the food is exhausted and conditions for continuous development are unfavorable, growth ceases at the end of the second stage, the alimentary tract empties, the body becomes more slender, and the parasite changes into the third-stage larva. In making this change, the cuticle of the second-stage larva is retained and the enclosed third-stage larva is said to be "ensheathed" or, as Fuchs (1915), Bovien (1932), and others call it, the "dauer" stage. In the case of *N. glaseri* the sheath is not very tenacious and is lost soon after the larva assumes a free-living existence in the soil, where it may persist, in a more or less active condition, for at least 2½ years.

According to Glaser (1932), the life history of *Neoaplectana glaseri* is essentially the same when the worm is reared on an artificial medium as it is in its insect host. Ensheathing can be brought about by placing the nematodes in an environment favorable for survival but not for growth to the preadult stage. This is done by removing the nematodes from the cultures, washing, and keeping them in an isotonic salt solution until they have ensheathed.

Cultivation. It was Montaigne who reminded us that "man cannot make a worm," but, after many fruitless attempts, Glaser (1931) did succeed in cultivating *Neoaplectana glaseri* on an artificial medium. The base of this medium was veal-infusion agar, having a reaction of pH 7.4. A day before the medium is to be inoculated with nematodes, 2 milliliters of a 10 per cent dextrose solution are added to a sterile Petri dish. The veal-infusion agar is poured into the dish and mixed with the sugar solution. When cool, the surface of the medium is flooded with a heavy suspension of a living pure culture of baker's yeast. The plate is incubated at room temperature until the yeast has grown uniformly over the entire surface of the medium. This takes about 24 hours, after which the nematodes—initially a washed suspension from infected grubs—are inoculated into the medium. The entire culture is then incubated at room temperature (24 to 27°C.).

The nematodes feed upon the living yeast cells until after about 2 weeks the yeast is depleted and the culture must be transferred to fresh medium. After about the seventh or eighth transfer on this medium the culture tends to die out, the nematodes failing to produce young. If the worms are passed through grubs several times they may regain their ability to survive seven or eight transfers on artificial media. If, however, supplementary food materials, such as desiccated cow's ovaries, are added to the medium it is possible to maintain the nematodes for many more transfers.

In 1940, Glaser described a test-tube method of growing the nematodes in yeast-free and bacteria-free cultures. Using aseptic techniques, mouse, beef, or rabbit tissue (ovary or kidney) was placed at the base of a veal-infusion-agar slant, and the evaporation from the tube was held to a minimum. In this, and in a similar liquid medium, the worms grew well, the tissue being almost entirely digested in from 18 to 24 days.

The two media just described, though excellent for maintaining stock cultures, are too expensive and do not yield large enough quantities of material desirable for field distribution. Accordingly, McCoy and Glaser (1936) developed a medium using fermented potato mash. This medium was placed in trays and inoculated with 200 to 400 thousand nematodes from veal-infusion-agar plates. After an incubation period of from 6 to 10 days at about 21°C., a yield of about 4 million nematodes per tray was obtained. Used in a similar way but more productive and convenient was a medium developed subsequent to this by McCoy and Girth (1938), in which an infused veal pulp was used as a base.

The veal-pulp medium was kept from bacterial decomposition (bacteria are easily introduced with the nematodes) by the use of sodium methyl-*p*-hydroxybenzoate, with or without the addition of dilute formaldehyde solution, added as a preservative. Lean veal was used in the preparation of the medium, and this was ground through a food chopper, mixed with twice its weight of distilled water, and held in the refrigerator for 18 to 24 hours. The infusion was then poured through a flannel cloth and the pulp was mixed with preservative and placed in culture dishes. The nematodes mere inoculated at a rate of about 600 nematodes per square centimeter of culture surface. After inoculation the cultures were incubated at a temperature of about 21°C. for approximately 7 days. The yield ranged from 9,000 to 12,000 nematodes per square centimeter of culture area.

Use as Control Measure. With the development of methods for propagating *Neoaplectana glaseri,* state (New Jersey) and Federal agencies about 1934 became interested in the possibility of using nematodes as a means of controlling the Japanese beetle in eastern United States. In cooperation with the Rockefeller Institute, these agencies conducted field experiments, the results of which indicated the potential value of nematodes as a control factor. In general, however, it appears that the widespread use of nematodes has had certain drawbacks that have made it impractical when compared with other means of control, including the use of the milky diseases of the Japanese beetle. In any case parasitization of the beetle by the worms when it does occur, cannot but be beneficial.

In 1940 Girth, McCoy, and Glaser reported the results of 2½ years of field experiments conducted in New Jersey. Seventy-three tests were performed with ensheathed nematode larvae during this period. Two methods of treatment were used: a subsurface treatment in which a certain number of nematodes were placed in a small hole under the sod of the turf being treated; and a surface treatment in which the nematodes, in an aqueous suspension, were applied from a sprinkling can or spray tank onto the surface of the soil. More satisfactory results were obtained when the latter method was used.

The percentage of total population of Japanese beetles parasitized ranged from 0.3 to 81.5 per cent, depending on soil moisture, nematode dosage, soil temperature, and density of beetle population. The optimum conditions were (1) a soil temperature (at 1.5 inch depth) of 60ºF. or higher; (2) soil moisture of 20 per cent or higher, with the soil not flooded; (3) a heavy host population; and (4) turf or other permanent cover. Nematode infestations have survived under field conditions for 8.5 years when the host population was maintained by stocking it with fresh grubs. Girth, McCoy, and Glaser found that the parasite was able to maintain itself under natural field conditions for 6.5 years, with a low host population existing for 5 years of this period. They also reported that nematode infestation in nature is spread by the flight of parasitized adult beetles and by the migration of nematodes and nematode-infected grubs through the soil.

The use of *Neoaplectana glaseri* as a control agent is one of the few instances in which nematodes have been used artificially in the control of insect pests. The role of nematodes generally in the natural destruction of noxious insects is known to be significant, and these worms probably supplement the activity of parasitic insects and predators. Such a conclusion as to the value of this type of control is certainly indicated by such reports as those from Japan, where mermithid infestations in leaf-hoppers have been reported very high (40 to 70 per cent) in certain years. Similar high infestations have ben observed in certain scolytid grubs in Europe. Other examples have been cited by Oldham (1933).

Other Semiparasitic Nematodes and Novitious Parasites

Similar to *Neoaplectana glaseri* Steiner is *Neoaplectana chresima* Steiner, originally found near Moorestown, New Jersey, in dead and dying corn earworms, *Heliothis armigera* (Hbn.). It has also been found in larvae of the Japanese beetle, *Popillia japonica* Newm., and experimental infections have been produced in five additional insect species. The biology of this nematode has been described by Glaser, McCoy, and Girth (1942), who were able to cultivate it on artificial media. With certain exceptions, its life cycle is similar to that of *N. glaseri*. In the Japanese beetle infection by *N. chresima* causes the insect to lose its appetite and become sluggish, and the grub cadavers assume a dirty dull yellow color. The larval contents become fairly viscid and have a dirty-yellow hue. Experimentally, Japanese-beetle grubs die about 4 days following exposure to soil or food contaminated with *N. chresima*. The possible use of this parasite in the control of its insect hosts has not been investigated.

Neoaplectana bibionis Bov. is a parasite of dipterous insects in Denmark. In the third stage of its development, dauer larvae are formed, which remain unchanged in the gut of the insect host until the latter eventually dies from other causes. When this happens, the nematode larvae move into the host's tissues, continuing their development until a large population is built up. When the insect cadaver is consumed, the young nematodes migrate out into the soil and become dauer larvae. *Neoaplectana affinis* Bov. also occurs in Denmark and parasitizes the same insects as does *N. bibionis*.

Diplogaster labiata Cobb occurs in the elm borer, *Saperda tridentata* Oliv., and was originally found near Manhattan, Kansas. It reproduces in the intestine of the living adult insect and may accumulate in sufficient numbers to rupture the gut and kill the beetle. Infected female hosts are usually sterile. Some species of *Diplogaster* may be associated externally with insects. Bovien (1937), for example, describes two species (*D. stercorarius* and *D. magnibucca*) that are carried by the dung beetle, *Aphodius fimetarius* (Linn.). The favorite site of attachment of these nemas is the lower side of the beetle's elytra at a point where the latter are attached to the body of the insect. Other species (*e.g.*, *D. aphodii*) may be found as endoparasites of dung beetles.

Pristionchus aerivora (Cobb) is best known in its relationship to termites, *e.g., Leucotermes lucifugus* Rossi, but it has also been reported from dead pupae of the corn earworm, *Heliothis armigera* (Hbn.), and in dead pupae of the roseleaf beetle, *Nodonota puncticollis* (Say). In the case of the termite, the nematodes are found only in the head of the insect and vary from 1 to 75 per infected insect. When the infection is a heavy one, the termite becomes sluggish and dies, after which the nemas reproduce in the carcass. The specific name of this species is derived from the habit, shared by some other nematodes, of swallowing air which may be seen to pass down the esophagus to the anterior end of the intestine where it is absorbed.

The genus *Rhabditis* contains a large number of saprozoic free-living species, species that live in terrestrial snails and worms, species that inhabit manure heaps or the galleries of bark beetles where they encyst on the exterior of the beetles, and a few species that live for a time in the bodies of certain insects. An example of the later is *Rhabiditis janeti* (Lac. Duth.), which invades the salivary glands of the ant, *Formica rufa* Linn. The host-parasite relation here has not been clarified. Fuchs (1937) has separated out from the genus *Rhabditis* the subgenus *Parasitorhabditis,* the adults of which live in the galleries of bark beetles. For example, he distinguishes numerous varieties of *Parasitorhabditis obtusa* Fuchs that are associated with different species of these beetles both internally and externally.

Other species of nematodes having semiparasitic relations to insects have been reported, but the life histories of most of these are incompletely known.

Nematode Parasites of the Body Cavity and Tissues of Insects

This group of nematodes is considered by Christie (1941) to include the oldest parasitic nematodes in the sense that their progenitors were the first to assume a parasitic mode of life and have gradually become highly adapted to this way of living. Those species which parasitize the body cavity and tissues of insects are, for the most part, included in the families Tetradnematidae, Mermithidae, and Allantonematidae. A few selected examples from each of these families will be considered briefly in the pages that follow.

Tetradonematidae

Only two species of tetradonematids associated with insects are well known: *Tetradonema plicans* Cobb and *Aproctonema entomophagum* Keilin.

The first of these species, *T. plicans,* was found as a parasite of the larval, pupal, and adult stages of the gnat *Sciara coprophila* Lint. in Kansas. Each insect harbored an average of 10 nematodes. The adult stage of the worm is passed in the body of its host, where it also lays eggs. The exact route of infection is not known with certainty, but it has been assumed that the eggs are swallowed and hatch and that the nematode larvae penetrate through the wall of the gut into the body cavity. It is possible, however, that the larvae enter the host by penetrating the body wall. Once within the body cavity, the young nematodes develop rapidly and copulate, and the females lay eggs before the insect pupates. If the host larva is infected rather early in its life, the chances are that it will succumb before pupation occurs. In such individuals the fat tissue is consumed, and the body of the insect is more transparent than is normally the case. When there is a light infection the larval insect may pupate, and the pupa may die, or metamorphosis may continue until the adult emerges. The infected adults are fairly normal in their activity and appearance, but they lack functional reproductive organs.

Aproctonema entomophagum Keilin was discovered in larvae of gnat, *Sciara pullula* Winn., in England. According to Keilin and Robinson (1933), each infected host usually harbors several female nematodes and a varying number of smaller males. After the parasites reach maturity in the body cavity of the insect, they copulate. After this the males die

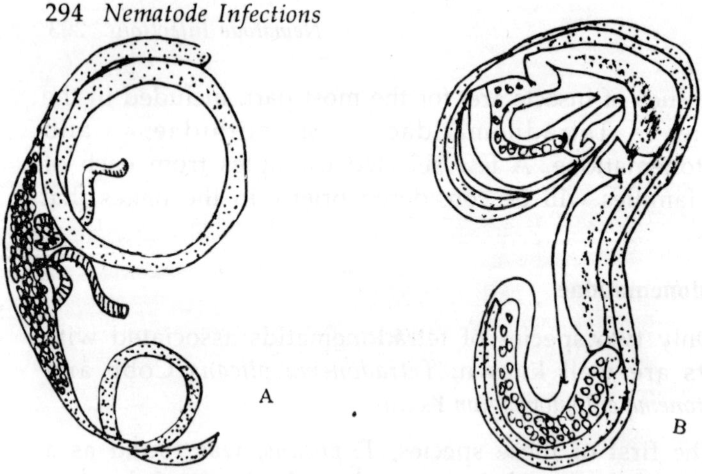

Fig. 14.2 : *A. ɪetradonema plicans* Cobb, a parasite of the gnat *Sciara coprophila* Lint. An egg-laying female with males attached. *B. Aproctonema entomophagum* Keilin, a parasite of the gnat *Sciara pullula* Winn. A spermatized female.

and the females emerge by penetrating to the outside, whereupon egg laying begins almost immediately. When egg laying ceases, the female dies. After hatching, the young nematode larvae invade a new host, apparently by penetrating the body wall. Early infection of the larval insect usually portends its death; late infection may enable the insect (at least the female) to complete its metamorphosis which, however, may be delayed. The infected adult females do not have functional reproductive organs.

Mermithidae

The mermithids constitute a noteworthy group of many inset-parasitizing species. Most of them are very slender organisms, in most cases not longer than 10 to 20 centimeters in length, with the males usually shorter than the females. Their colour is generally white because of the transparent cuticle and the presence of a large fat body; some species are faintly coloured with a yellowish, brownish, or blackish hue. The cuticle may be thick or thin. Those mermithids having thick cuticles parasitize terrestrial insects; those having thin cuticles inhabit the body of aquatic insects.

We shall not here attempt to generalize regarding the nature of the biological relationships between mermithids and their insect hosts, nor shall we try to give a general picture of the pathology resulting from the parasitization. Instead these points will be dealt with in the course of considering a few of the better known examples of the mermithid parasites.

Mermithid Infections in Grasshoppers. Grasshoppers are subject to parasitization by several species of mermithids.

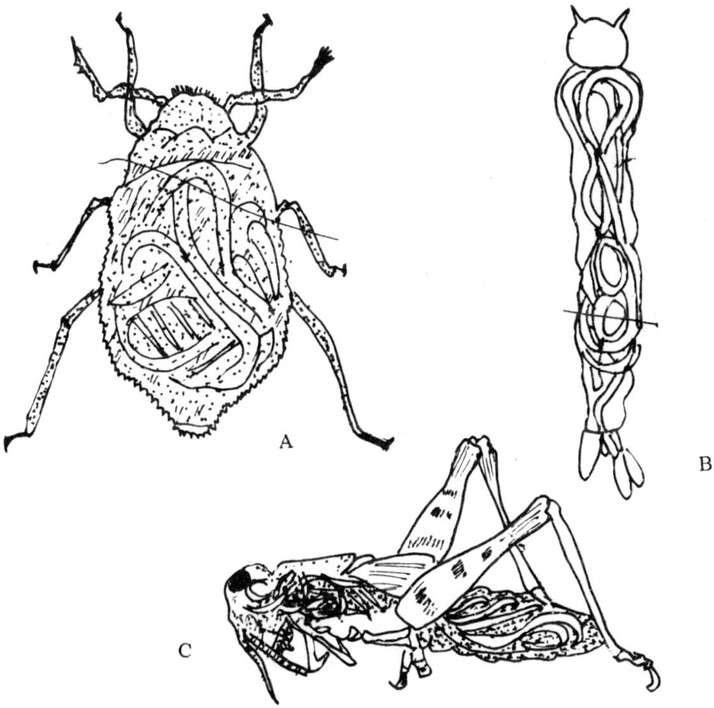

Fig. 14.3 : Examples of insects parasitized by nematodes. *A.* A root aphid (*Anoecia*) infected with an undetermined nematode. *B.* Larva of *Aëdes aegypti* (Linn.) infected with two nematodes (*Mermis* sp.), one of which is about to emerge via the anus. *C.* Grasshopper nymph (*Melanoplus*) containing one fully grown female *Agamermis decaudata* C., S., & Ch.

In the United States two species of nematodes are prominent in this regard: *Agamermis decaudata* C., S., & Ch. and *Mermis subnigrescens* Cobb. Although *A. decaudata* is sometimes found in insects other than grasshoppers, *M. subnigrescens* appears to be strictly a grasshopper parasite. Both species occur in the northeastern and north central parts of the United States. Important papers on these nematodes have been published by Christie (1936, 1937) and by Cobb (1926, 1929), and the following account is based on these reports.

Agamermis decaudata has been found in grasshoppers (both Locustidae and Tettigoniidae), crickets (Gryllidae), and occasionally in leaf-hoppers and in beetles. So far as in now known, the species of grasshoppers most commonly serving as hosts are *Melanoplus femur-rubrum* (DeG.) and *Conocephalus brevipennis* (Scudder).

The free-living stages of *A. decaudata* inhabit small cavities in the soil usually from 5 to 15 centimeters below the surface. The adults in such cavities generally consist of one female and up to eight males coiled together so as to form a "knot." After copulation, egg laying takes place and, in Virginia, continues from about the first of July until cold weather sets in during the fall. Cleavage and embryonic development occur after the eggs are laid, and the first molt takes place within the eggshell; most of the eggs laid during a summer do not hatch until the following spring. The newly hatched second-stage larvae are at once infective as they migrate to the surface of the soil and climb grass and other low vegetation when it is wet with dew or rain. Newly hatched grasshopper nymphs are sought out and their body walls penetrated by the nematode, which thus gains entrance to the body cavity of the insect. The penetration usually takes place under the edges of the pronotum, between the abdominal segments, or at other locations where the integument is thin. Entrance through the chitinous covering is effected by the use of the nematode's stylet, "probably aided by the dissolving action of a chitin solvent secreted by one or more of the most anterior esophageal glands" (Christie). As the nema enters the insect, the posterior two-fifths of its body length—which up to this point has served as a propelling and food-storage organ—breaks off at a given spot (the "node") and is left outside. Within the body cavity of the grasshopper the parasite grows rapidly and undergoes pronounced external and internal morphological changes. The intestine, which serves as a reservoir for nutrient materials, assumes an enormous size, filling all the body-cavity space not occupied by the other organs. There is usually only one parasite per host. About 1 to 1½ months later the males force their way through the body wall of their insect hosts, followed 1 or 2 months later by the females. The worms fall to the surface of the ground and enter the soil where during the winter the males and the females remain separate, each individual forming a separate "knot."

The next spring the nematode undergoes its final molt, the males seek females, and copulation takes place. About the first of July egg laying begins and continues until interrupted by cold weather. The nest spring the year-old female continues laying its eggs until, toward the close of the summer, its reserve food supply has become exhausted. Most of them do not survive a third winter in the soil.

The pathology of the infection by *Agamermis decaudata* is not very apparent from external examination of the host grasshopper. The insects sometimes have distended abdomens and may appear sluggish and incapable of sustained flight. Internally, the gonads are visibly affected, particularly in the females. The ovaries are always greatly reduced in size, and infected female grasshoppers are probably not capable of laying eggs. The host grasshopper always succumbs when the parasite emerges from it.

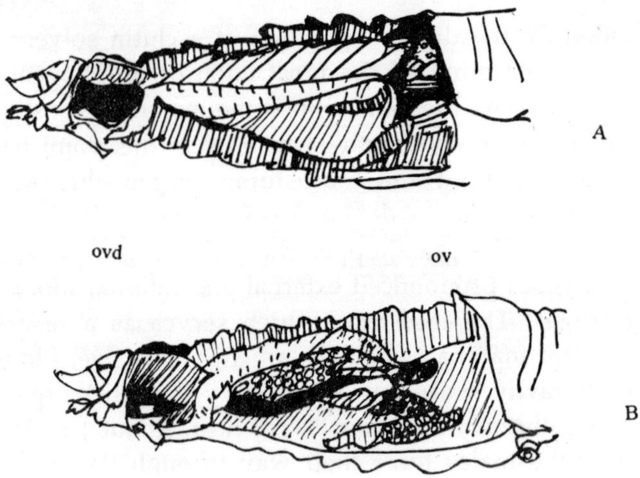

Fig. 14.4 : Effect of mermithid nematodes on their hosts. Dissections of adult female grasshoppers, *Melanoplus femur-rubrum* (DeG.), showing effect of parasitization on reproductive organs. *A*. Normal grasshopper. *B*. Grasshopper parasitized by *Agamermis decaudata* C. S. & Ch.

Mermis subnigrescens Cobb has been found infecting nine different species of grasshoppers, including both Locustidae and Tettigoniidae, and two additional species have been experimentally infected. Attempts to infect other insects, such as crickets, have been unsuccessful.

The life cycle of this nematode is similar to that of *Agamermis decaudata* C. S. & Ch. but differs in one important respect. Instead of depositing its eggs in the soil, the gravid females of *Mermis subnigrescens* migrate to the surface of the soil, climb up the vegetation, and deposit their eggs thereon. The eggs are swallowed by grasshoppers which feed on the vegetation. Upon reaching the insect's alimentary tract, the eggs hatch, and the young larvae penetrate the wall of the gut and enter the body cavity where they continue their development. Since the grasshoppers become infected while feeding, they are vulnerable throughout their entire life. As the nymphs grow older, they eat more vegetation, and the change that the insect will become infected is correspondingly greater. A hundred or more parasites may be harbored by a single grasshopper; usually, however, a host contains from 1 to 5 parasites.

Within the body cavity of the host, the male nematodes remain from 4 to 6 weeks, and the females from 8 to 10 weeks. The parasites emerge by forcing their way through the body wall of the insect, after which they enter the soil.

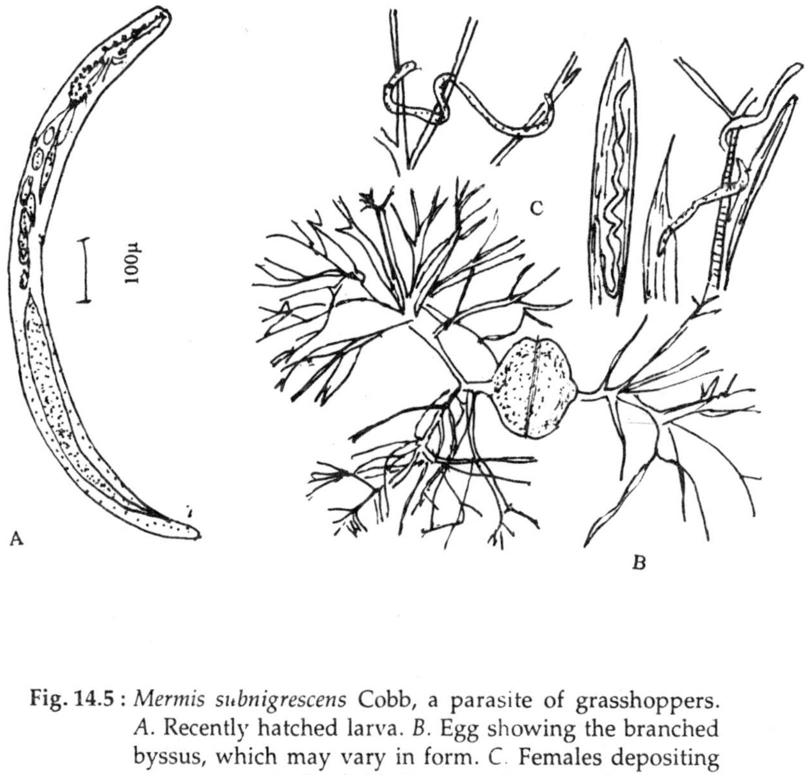

Fig. 14.5 : *Mermis subnigrescens* Cobb, a parasite of grasshoppers. *A*. Recently hatched larva. *B*. Egg showing the branched byssus, which may vary in form. *C*. Females depositing eggs on grass in the field. Not according to scale.

The emergence of *M. subnigrescens* results in the death of the host. If the insect happens to harbor parasites of different ages, all those that are too immature to escape and survive in the soil perish with the host when the older nematodes emerge.

Unlike *Agamermis decaudata,* individual *M. subnigrescens* remain separated in the soil and rarely do a female and one or more males coil up into a knot. Copulation may take place, but it is not necessary and females reared in the absence of males produce viable eggs. According of Christie (1937), most of the nematodes emerge from the host during the summer and autumn and molt the following April. By July the females begin to exhibit a brownish color due to accumulating eggs, and by September they are nearly black in appearance. Although the eggs are viable at this time, they are not laid until the following spring, usually beginning in May and continuing throughout the summer until the end of July or August. Egg laying takes place in the daylight during periods of rain or heavy dews when the vegetation is wet.

Before the parasite emerges, causing the death of the host, certain changes are brought about in the body of the grasshopper as a result of the parasitization. The development and function of the gonads, particularly the ovaries, is suppressed. This effect is more pronounced in those insects which harbor greater numbers of parasites. The growth of both male and female grasshoppers is considerably retarded, infected individuals remaining in the nymphal stages longer than uninfected insects.

Both *Agamermis decaudata* and *Mermis subnigrescens* are of economic importance in that they serve as factors in the natural control of the grasshoppers they infect. The hardiness of *M. subnigrescens* makes it particularly important as a means of control, since it is able to maintain itself in large numbers where the grasshopper population is consistently low. The latter species appears to be the more suitable of the two for colonizing, although very little investigation of such a possibility has been carried out.

Mermithid Parasites of Ants. For many years it has been known that numerous species of ants are often infected with mermithids which bring about more or less typical changes in their hosts. Sometimes the parasitized ant appears only slightly different from a normal individual—perhaps there is only a slight difference in color or the gaster is somewhat distended. On the other hand, the pathology resulting from the parasitism may be marked, modifying the external anatomy so that the infected individual is not identical to any normal caste but shows female, worker, or soldier characters in varying degrees only. Such ants are known as "intercastes."

Various types of intercastes occur, and these are sometimes designated by different names. "Mermithogynes" may be found in the genus *Lasius* where infected females resemble normal females but usually have a smaller head, shorter wings, and a partly distended gaster. Several different intercastes may occur in the genus *Pheidole*. These have been termed "mermithergates" and have a mixture of female, worker, and soldier characters. In most cases they resemble workers and soldiers. Five of these types have been recognized. Two types of intercastes have been found in *Pheidole pallidula* (Nyl.). One of these shows very little variation from normal; the other is a modified soldier type which Vandel (1930) designates as "mermithostratiotes," reserving the term "mermithergates" for modified worker types.

The life cycles of the nematodes (*Agamermis, Hexamermis, Allomermis,* and others) that parasitize ants have not been studied to any great extent. The route of infection appears to be oral in some cases but not in others; in most instances it is unknown. Typical in this regard in the uncertainty in the case of *Allomermis myrmecophila* (Baylis), which parasitizes ants (*Lasius alienus* (Förster), *L. flavus* (Fabr.), and *L. niger* (Linn.)) in England, Germany, and probably elsewhere. Crawley and Baylis (1921) believe that the ants become infected while in the larval stage, whereas Gösswald (1929, 1930) presents evidence that indicates that the eggs of the nematode are ingested. Similar uncertainty concerns the rest of the life history of this mermithid. It does appears that after the parasite has completed its development, it emerges from the insect, either through the anus or sometimes

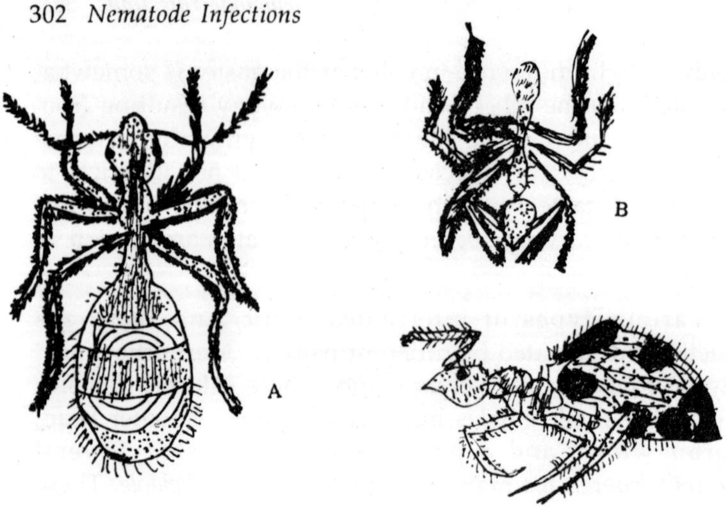

Fig. 14.6 : Examples of mermithergate ants (*Pheidole*). *A* Normal worker. *B* and *C*. Dorsal and lateral views of infected workers showing worm coiled in the abdomen.

between two of the ventral plates of the gaster. It enters the soil and lays large numbers of eggs. Males have not been seen, and the females apparently are able to produce viable eggs without copulation.

Allantonematidae

The allantonematids constitute a large group of nematodes parasitic in insects and having a life cycle distinctly different from that found in any other group of nematodes. Judged from an over-all viewpoint, they constitute an important factor in the natural control of insects since not only do they kill large numbers of insects, but many of them sterilize their hosts, thus preventing the laying of viable eggs. Consequently their aggregate destructive effect on insect populations must be considerable.

As Christie (1941) has described them, the adult gravid females occupy the body cavity of the insect, frequently in small numbers, often one per host. The eggs are either deposited here in the body cavity of the host or they hatch before deposition. In either case, the young larvae commence their development in the host insect, molting once or twice, depending on the species, and then' escape from the host. This is accomplished either by entering the alimentary tract and passing out through the anus or by entering the female reproductive system and passing out through the genital aperture. In most cases, both male and female insects are infected by the nematodes. The free-living stage, usually of short duration, is passed wherever the host insect undergoes its early development. During this period, the worms molt at least once, in most cases probably twice, and become adults.

Upon entering a new host, the female nematode increases greatly in size. When fully grown, the female is usually curved ventrally and assumes a sausagelike shape; there are, however, exceptions to this. If the female does not actually lay her eggs in the body cavity of the host, her uterus becomes distended with developing eggs and larvae, which fill a large part of her body, and some of her organs degenerate. Usually the larvae pass out through the vulva of the female into the body cavity of the insect. In some nematode species the size of the female does not increase enough to provide space for the expanding reproductive organs. In *Sphaerularia bombi* Duf., for example, the uterus is everted through the vulva, and the entire reproductive system develops outside the body of the nematode. This prolapsed uterus increases so much in size that the body proper amounts

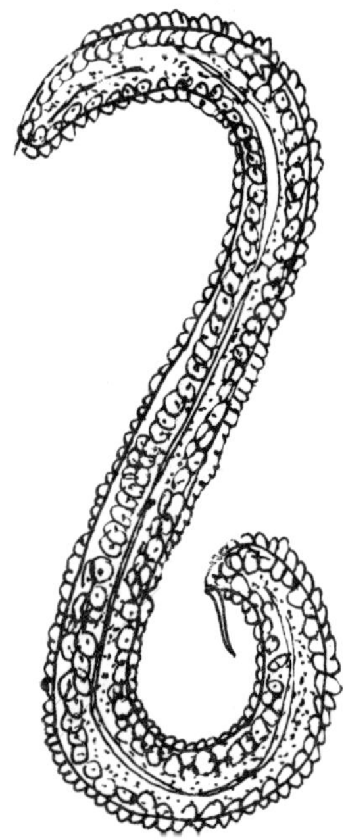

Fig. 14.7 The female reproductive system of *Sphaerularia bombi Duf.* The nematode's small body proper remains attached at one end of the prolapsed uterus.

to little more than an attached vestigial structure, apparently without function.

As Christie points out in his discussion of the group, there are several deviations from this typical life cyçle. In some cases, the males as well as the females enter the body cavity of the insect. Sometimes neither the males nor the females become parasitic, this role being taken over by the larval stages. A few species have heterogeneous life cycles, having parthenogenic and gamogenetic generations.

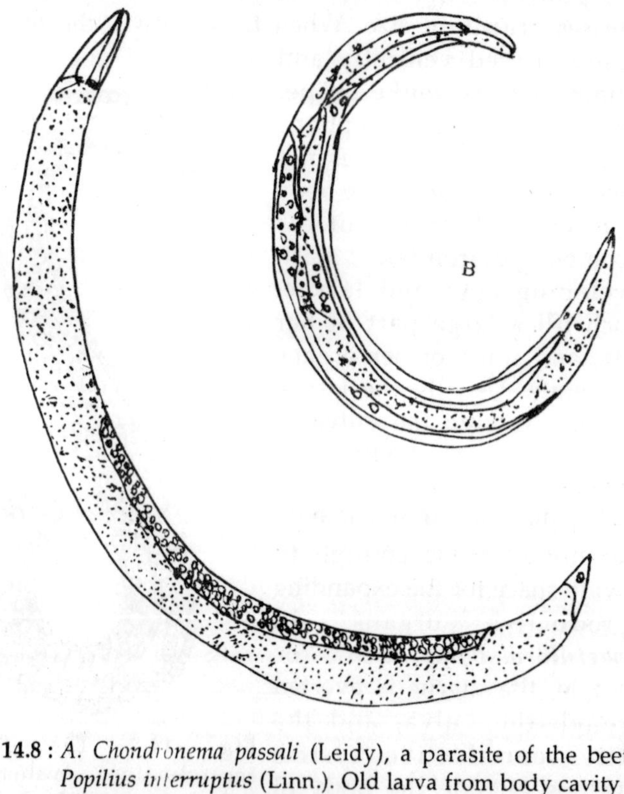

Fig. 14.8 : *A. Chondronema passali* (Leidy), a parasite of the beetle *Popilius interruptus* (Linn.). Old larva from body cavity of host. *B. Allantonema mirable* Leuck., a parasite of the pine weevil, *Hylobius abiëtis.* An adult preparasitic female after copulation.

It is difficult to select representative examples from this large group of parasitic nematodes. The best we can do, and keep within the limits of this chapter, is to mention the high points in the life histories of some of them and to point out some of the interesting biological and pathological relationships existing between them and their insect hosts.

Allantonematids Parasitizing Beetles. About 100 years ago Leidy discovered the nematode now known as *Chondronema passali* (Leidy) in the body cavity of the beetle *Popilius interruptus* (Linn.). A large percentage of any particular beetle population may be infected, and each beetle may harbor from 500 to 1,000 parasites, which may be in all stages of development except adults. Once outside their hosts, neither the males nor the females become parasitic but remain in the galleries formed by the beetle in decaying logs and stumps. The eggs hatch in the body of the female worm, and the very young larvae enter the host possibly perorally.

The body cavity of the pine weevil, *Hylobius abiëtis* (Linn.), is sometimes found parasitized by *Allantonema mirable* Leuck in Europe. Unlike most allantonematids, the fully grown female of this nematode is oval in shape, about 1.5 to 2 millimeters long and about 0.75 to 1 millimeter wide. It consists essentially of a sac filled with the uterus containing eggs and larvae. When the eggs hatch they pass into the body cavity of the host, undergo two molts, and eventually leave by penetrating the insect's alimentary tract and passing out through the anus. This escape is made in the vicinity of the place where the adult beetles lay their eggs so that the female nematode may later, after copulation, penetrate into the body cavity of the weevil larvae.

In 1921, Cobb reported finding a nematode parasite, which he named *Howardula benigna* Cobb, in the cucumber beetle, *Diabrotica vittata* (Fab.). It occurs less commonly in *D. trivittata* (Mann) and in *D. duodecimpunctata* (Fab.). The nematode larvae pass out with the eggs of the insect. Male beetles, as well as females, are infected but the fate of the nematode larvae within the male is not known; it is possible that they are transferred to the female beetle during copulation.

The bark beetle *Ips typographus* (Linn.) is parasitized by at least two species of allantonematids: *Aphelenchulus diplogaster* (Linst.), and *Parasitylenchus dispar* subsp. *typographi* Fuchs. The bark beetle *Pityogenes bidentatus* (Hbst.) has been found to harbor *Aphelenchulus tomici* Bov. which, like *A. diplogaster* (Lint.), passes out of its host through the anus and undergoes its free-living development in the frass of the beetle galleries. Several species of the nematode genus *Bradynema* also occur in beetles.

Allantonematids Parasitizing Flies. One of the best known allantonematids parasitizing flies is *Tylenchinema oscinellae* Goodey observed by Goodey (1930) as a body-cavity parasite of the frit fly, *Oscinella frit* (Linn.), in England. The fly has three generations a year and the life cycle of the nematode is, accordingly closely correlated with that of the insect—it also undergoes three generations a year. Within the body cavity of the insect the female nematode gives birth to living young. These young larvae accumulate in the hemocoele and continue their development until eventually they effect their escape from the host by penetrating the food reservoir of the fly's digestive system through the anus. After a short free-living existence, the impregnated female enters the body cavity of the frit-fly larva, probably by the direct penetration of the body wall. The pathological effects of the parasitization are not apparent on the external characters of the host. Internally, however, both the male and female gonads are sterilized. Occasionally the fly gets the upper hand and is able to build up its gonads in a normal way; in such cases the nematode fails to grow and becomes degenerate.

The small dipterous insect *Scatopse fuscipes* Meig. is known to be parasitized by the body-cavity parasite *Scatonema wülkeri,* described by Bovien in 1932. This nematode is of particular interest because its progeny can reach full maturity and even copulate inside the maternal uterus, or within the body cavity of the host. The life cycle of the parasite may be completed in a period of a few weeks while the host is still in the larval stage. The free-living stage is of short duration. The females penetrate the body wall of the fly larva at almost any point on the surface of the insect's body. Most of the host larvae contain but one parasite, some two, and a few have been found harboring three nemas. The parasitization does not appear to have any marked effect upon the adult host, and its presence does not result in the sterility of the insect. In the case of the host larvae, however, the parasites fill a good share of the body cavity, the fat body of the insect is consumed, and finally the insect succumbs to the attack of the parasites.

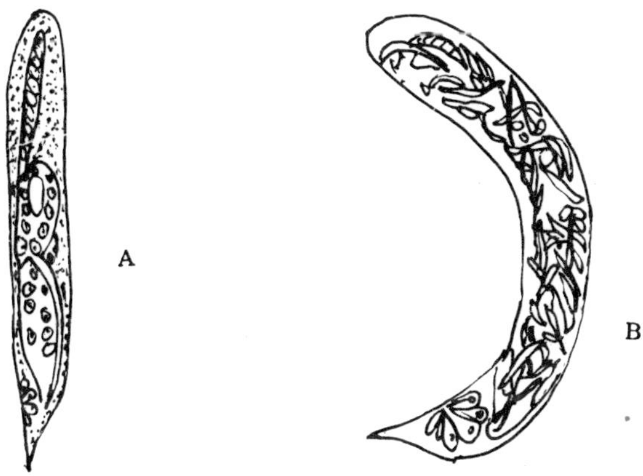

Fig.14.9 : *Scatonema wülkeri* Bov., a parasite of *Scatopse fuscipes* Meig. *A*. An adult parasitic female a few days after entering host. *B*. Fully grown gravid female.

Heterotylenchus aberrans Bov. is a parasite of the body cavity of the onion maggot, *Hylemya antiqua* (Meig.), in Denmark from where it was reported by Bovien (1937). Each parasitized fly usually contains from one to four large adult females (gamogenetic generation) together with a larger number of small adult females (parthenogenetic generation). The gamogenetic females deposit their eggs in the body cavity of the host, where they hatch and where the larvae develop into parthenogenetic females. These females deposit eggs in the body cavity of the insect, and from these hatch larvae of both sexes. When these larvae are ready to undergo their final molt, they penetrate the insect's ovaries and migrate to and accumulate in the oviducts, from which they escape through the genital aperture. The infected female flies

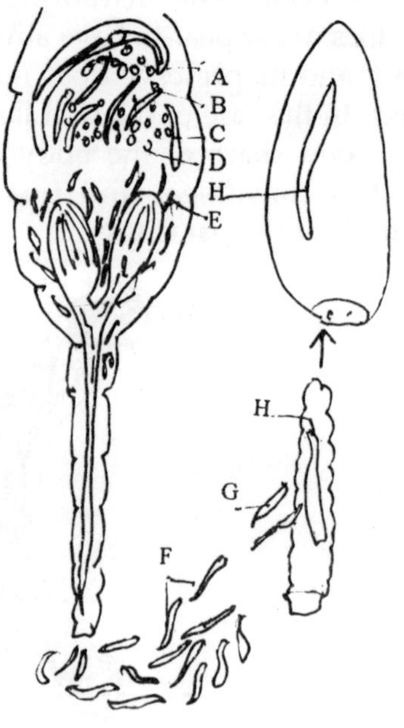

Fig. 14.10: A diagrammatic representation of the life cycle of *Heterotylenchus aberrans* Bov., a parasite of the onion fly, *Hylemya antiqua* (Meig.). The adult parasitic female of the gamogenetic generation (*A*) lays eggs (*B*), which develop into females of the parthenogenetic generation (C). These females lay eggs (*D*), and the resulting larvae (*E*) enter the reproductive organs of the female fly and pass out through the genital aperture. Outside the host these larvae develop into adults of the gamogenetic generation (*F*) and copulate, whereupon males die and impregnated females (G) enter fly larvae. While the fly matures and pupates, the female grows (*H*) to reach full stature (*A*) and lays eggs (*B*).

are rendered sterile because their ovaries fail to develop; the male flies, although infected, probably retain their fertility. The fat tissues of the body are also greatly reduced. How the nematodes escape from the male insects, if they do, has not been determined. In their free-living existence, the nematodes reach maturity and copulate, after which the males die and the females enter the insect, presumably by penetrating the body wall.

Another species of nematode parasitic in a dipterous insect is *Tripius gibbosus* (Leuck.) found in *Cecidomyia pini* (DeG.) in Germany. During the parasitic development of the female, the uterus is gradually everted through the vulva. The uterus continues to develop on the outside of the nematode until it forms an oval structure attached to the smaller body proper.

Fergusobia curriei (Cur.) occurs in gall flies of the genus *Gergusonina* in Australia. This nematode is also found in the galls produced by the insect and may be considered a true plant parasite. Some authorities regard its relationship to the insect as one of symbiosis (mutualism) rather than parasitism. This appears to be a very unusual case of combined parasitism and mutualism, depending upon whether the worm is in the plant or in the insect host.

Allantonematids Parasitizing Other Insects. One of the most interesting and remarkable of nematode parasites of insects is *Sphaerularia bombi* Duf., which parasitizes several species of *Bombus* (bumblebees), as well as *Vespa rufa* Linn. and *Vespa vulgaris* Linn., in Europe and in North America. Each host harbors one, or only a few nemas. *S. bombi* has a fairly typical allantonematid life cycle: Eggs are deposited in the body cavity of the host, and the larvae, after a period of development, pass out of the insect by way of the anus and enter the soil. Here they reach maturity, and copulate; the males die, and the females enter their new hosts, which are usually queen bees. Only the queen bees are parasitized because they hibernate in the soil and the nematodes gain access to them while the insects are penetrating the ground in the autumn. The ovaries of the infected queens are retarded in development and produce few or no eggs, and such queens probably never establish colonies.

After entering its insect host, the female of *S. bombi* undergoes little or no increase in size, but the uterus gradually is everted through the vulva. The everted uterus carries with it the other reproductive organs and the modified intestine or "fat body," and completes its development on the outside of the body proper. The prolapsed uterus increases to an enormous size while the original body of the nematode remains a small functionless structure that may become detached. The pathological changes in *Sphaerularia*-infected bees, particularly in the ovaries, have ben described by Palm (1948), who considers the parasite to be a genuinely pathogenic worm.

The thrips *Aptinothrips rufus* (Gmelin) is subject to attack by *"Tylenchus" aptini* Sharga. This nematode deposits its eggs in the body cavity of the host, and the larvae eventually leave by way of the alimentary tract and anus. Other insects serving as primary hosts to allantonematid nematodes include aphids and earwigs.

Nematomorpha

The Nematomorpha are in many ways similar to the Nematoda except that they are usually larger, more uniformly cylindrical, have bluntly rounded ends, and have a faintly coloured cuticle. There are other structural and physiological differences, but in many respects their life cycle is similar to that of mermithids. Because of their superficial resemblance to thick hairs they are sometimes known as "hairworms." Insects, diplopods, molluscs, and crustaceans constitute their favorite hosts.

Gordiid larvae gain entrance into their insect hosts by being ingested along with the insect's food or water, and through ingestion by the insect of the encysted worm After being ingested by the insect the cyst bursts and the larva escapes, perforates the intestinal wall of the host, and reaches the body cavity. If the host is large enough, the larva may accomplish its development in this situation. If, however, the insect is a small one, it may serve only as an intermediate host, and the worm will become an adult after the intermediate host has been ingested by the final host. After leaving their host, the male and female worms copulate and oviposition soon follows. The adult worms may live for 6 months or more in fresh water or in mud.

Those insects which serve as intermediate hosts of Nematomorpha are frequently small aquatic insects (*e.g.*, Chironomidae, Ephemeridae, Trichoptera). It these insects are not swallowed by another insect, the worms may remain in them for a year or more until the parasites finally disintegrate and become absorbed by the host's tissue. On the other hand, when the intermediate hosts are ingested by a carnivorous or by an omnivorous insect, the worms complete their development in the latter, or final, hosts. In the case of some of the worms that complete their development in terrestrial insects, the intermediate hosts may be tadpoles, young frogs that have just metamorphosed, and also snails. Filipjev and Stekhoven (1941) list the following groups of insects as hosts of Nematomorpha: Coleoptera (Silphidae, Tenebrionidae, Dytiscidae), Orthoptera (Tettigoniidae, Locustidae, Mantidae, Blattidae), Odonta, and Trichoptera.

The effects of Nematomorpha on their hosts are usually not great. Sometimes the fat body of the insect is decreased, and there may be other minor anatomical or histological alterations. Some insects (*e.g.*, Carabidae) may have their fat body reduced as a result of the parasitization, which may, in addition, lead to a degeneration of the reproductive system and parasitic castration of the host. The intestinal tract may also be arrested in its function. Some insects die when the worms escape from them; others may survive this process and continue their normal existence.

The Nematomorpha are widely distributed and are found in both marine and fresh-water environments, and, as far as insects are concerned, in both terrestrial and water-inhabiting species. Most of the forms in which we are interested here belong to the two families Gordiidae and Chordodidae. The single important genus of the first family is *Gordius;* the latter family includes the genus *Chordodes* and 7 or 8 other genera.

Platyhelminthes

A few species of flatworms are known to be associated with insects, but in most of these cases the insect serves merely as an intermediate host. Examples of such worms and their intermediate hosts include *Dipylidium caninum* (Linn.), the dog tapeworm (in fleas), *Hymenolepis nana* (v. Sieb). (in fleas and probably certain beetles), *H. diminuta* (Rud.) (in certain lepidopterans, earwigs, fleas, beetles, and cockroaches), *Raillietina echinobothrida* (Meg.) (in ants), *R. cesticillus* (Molid) (in certain beetles), *Moniezia benedeni* (Mon.) (in certain mites), and *Choanotaenia infundibulum* (Bloch) (in the housefly and dung beetles).

An interesting case of flatworm parasitization of an insect has been reported by Subramaniam and Naidu (1944) in India. These workers observed a minute plerocercoid (*Sparaganum*) in the fat tissue of a sandfly (Psychodidae). The worm showed branching and segmentation and had a small club-shaped scolex. The nature of the relationship between the worm and its host has not been elucidated.

One of the few cases of an association between trematodes and insects is that in which *Prosthogonimus pellucidus* (v. Linst.) of fowl and ducks is found as cercaria in nymphs of dragonflies, especially *Libellula quadri-maculata* Linn. The birds become infected when they ingest the dragonflies.

Chapter—15

Host-Parasite Interaction in Plants

Plant parasitism by nematodes is confined to three orders: Dorylaimida (class Adenophorea), Tylenchida and Aphelenchida (class Secernentea). There is vast difference in the nature of parasitism by genera and species of these orders. Parasitism is established when the nematode pierces the cell wall and starts ingesting cell contents. The injection of saliva during feeding may only help in ingestion of food or it may additionally induce changes in the host. The differences between nematode groups in their parasitism are seen during this process.

1. Origin and Evolution of Plant Parasitism

Plant parasitism in nematodes is multiple in origin and may have evolved from free-living, stylet-bearing soil nematodes. There is no evidence to suggest any major role of adaptation in the evolution of parasitism. Although local adaptation may possibly be responsible for some differences as, for example, the local races of a nematode breaking resistance of a resistant host variety, there is little possibility of a major pest evolving through adaptation. This can be attributed to a low number of generations per year through which the more specialized parasite passes. Two main trends in the development of plant parasitism among nematodes are considered. First, there is a trend from ectoparasitism to migratory endoparasitism to sedentary endoparasitism.

With the development of sedentary endoparasitism, which is restricted to the females, there is also morphological adaptation leading to a high rate or reproduction. Second, there is a trend from indiscriminate feeding on the epidermis and cortex with accompanying cell necrosis and tissue damage to descrete feeding sites with specific plant responses and a suppression of necrosis. Exceptions are genera like *Ditylenchus* and *Aphelenchoides* which are both ecto- and endoparasites. Some ectoparasites become semiendoparasites but the host response remains primitive, that is, necrosis occurs.

It appears that the plant parasites belonging to the order Dorylaimida, especially the Trichodoridae (*Trichodorus, Paratrichodorus*), are most primitive forms followed by other Dorylaimids (*Longidorus, Xiphinema*). The Tylenchids are more highly evolved and in this group the sedentary parasites of roots and the cereal seed gall nematode, *Anguina*, can be considered as most highly evolved parasites. They establish good synchronization between their life cycle and that of the host and often have an elaborate system of deriving nutrients from the host.

2. Proof of Pathogenicity

To establish that a particular nematode causes a particular disease, the Koch's postulates must be proved. The four postulates for proof of pathogenicity are:

(i) Constant association of the suspected organism with this disease.

(ii) Isolation association of the suspected organism with the disease.

(iii) Reproduction of the disease by inoculation with a pure culture of the organism.

(iv) Reisolation of the organism from the inoculated diseased host, and identification of it with the original inoculant.

The Koch's postulates were originally developed for diseases caused by bacteria which were easy to isolate and grow on synthetic media. Literal application of these postulates to nematode disease is difficult for two main reasons. One reason is that plant parasitic nematodes generally cannot be grown on artificial media. The second reason is that in many cases the nematode forms a complex with other plant pathogens (fungi, bacteria and viruses). Since the second postulate cannot be proved the third naturally becomes inoperative. However, since the second law simply stresses the necessity of obtaining aseptic monospecific inoculum, methods have been developed to separate the nematodes from infested plants, free them from contaminants, grow a host in aseptic conditions and inoculate it with the nematode. The inoculation should result in the specific disease symptoms unless association of other organism is essential for the symptoms.

3. Mechanisms Involved in Parasitism

The phytonematodes use their stylet to puncture the epidermis and the cell walls and for displacing the cells while moving intercellulary. During feeding and causing mechanical damage with their stylet, they secrete hydrolytic enzymes from their esophageal glands which dissolve cell walls or act as digestive enzymes. Most common enzymes in nematodes are cellulase, protease, and amylase. Enzymes detected in nematode secretions and homogenates and listed below:

Pectinmethylesterase:	*Ditylenchus triformis, D. myceliophagus, D. dipsaci* (alfalfa race)
Non-specific esterases:	*Meloidogyne hapla, M. javanica*:
Alkaline phosphatase:	*Meloidogyne* species.
Acid phosphatase:	*D. triformis, Meloidogyne*, spp., *Tylenchulus* semipenetrans.
Amylase:	*Aphelenchoides sacchari, T. semipenetrans, D. dipsaci, D.*

	destructor, D. triformis, Heterodera schachtii, Globodera rostochiensis, Meloidogyne, javanica.
Cellulase	*Aphelenchoides avenae, A. sacchari, D. dipsaci, D. myceliophagus, D. triformis, Radopholus similis, Helicotylenchus nannus, Pratylenchus penetrans, P. zeae, Heterodera trifolii, H. schachtii, M. arenaria, M incognita, T. semipenetrans.*
Polygalacturonase	*A. avenae, P. zeae, D. dipsaci, D. myceliophagus, D. destructor.*
Invertase	*A. sacchari, M. arenaria, R. similis, P. penetrans, P. redivivus.*
B-glucosidase:	*Aphelenchoides fragariae, P. penetrans, Globodera rostochiensis.*
B-galactosidase:	*Globodera rostochiensis*
Pectinase:	*D. dipsaci, D. destructor, R. similis, P. penetrans, M. arenaria, M. hapla.*
Proteinase:	*G. rostochiensis, H. schachtii, D. dipsaci, D. destructor, D. allii, D. triformis, A. ritzemabosi, Pratylenchus redivivus.*
Chitinase:	*D. dipsaci, D. myceliophagus, D. destructor.*

The enzymes not only dissolve cell walls and help in ingestion and digestion of food, they include metabolic changes in the host also. Enzyme activities in nematodes are influenced by parasitic habits. In migratory endoparasite *Pratylenchus penetrans* the cellulase activity is seven times more than in the sedentary endoparasite *Heterodera trifolii*.

The root parasitic, nematodes, whether primitive or highly evolved in the mode of parasitism, establish a modified feeding site by disrupting or stimulating normal root cell function to satisfy their nutritional demands for growth and reproduction. Two main strategies to meet these demands are the creation of large volumes of cytoplasm near the feeding site and the modification of this cytoplasm for the efficient withdrawal of nutrient. The establishment of specific feeding sites is closely linked with feeding strategies.

4. Parasitism of the Dorylaimida

The phytonematodes of this group are migratory root ectoparasites which either browse along the root surface (*Trichodorus, Paratrichodorus*), or feed for longer periods at specific sites from deeper root tissues (*Longidorus, Xiphinema*). Apart from causing direct damage to roots, these nematodes are of great economic importance as vectors of several plant viruses. Species of *Trichodorus* and *Paratrichodorus* transmit Tobraviruses or NETU (nematode transmitted tubular viruses) whereas *Longidorus* and *Xiphinema* transmit the Nepoviruses (nematode transmitted polyhedral viruses).

(i) *Trichodorid Nematodes*: These are also known as stubby root nematodes. The members of Trichodoridae usually aggregate near the tips of rapidly growing roots where they often feed gregariously on epidermal cells and root hair, occasionally on subepidermal cells. When root growth is retarded due to massive destruction of the epidermal layer, the parasitism is transferred to the apical region of the root. Roots that stop growing lose their attraction. The nematode then moves away in search of new feeding sites, usually emerging in lateral rots which when heavily attacked give rise to the characteristic stubby root symptoms.

In all Trichodorid nematodes the stylet is a ventrally curved mural tooth with a solid tip. Obviously, this is only to ensure piercing of cell wall, not for food ingestion. The nutrients are removed from the cell through a self-made feeding tube composed of hardened saliva. This feeding tube is left in the host cell after each feeding cycle which takes only a few minutes. After achieving perforation of cell wall with rapid stylet thrusts, the nematode first inserts only the tip of the stylet repeatedly through the hole during salivation. The saliva hardens to form the feeding the tube along the stylet. Saliva is injected into the cytoplasm also and a massive accumulation of cytoplasm occurs at the perforation site. The host nucleus is drawn to this aggregation. It swells and soon becomes optically empty. When sufficient cytoplasm has accumulated at the perforation site, the nematode starts making deeper thrusts of stylet (with feeding tube) for ingestion of the accumulated cytoplasm. Probably the tonoplast is also ruptured and cytoplasm and cell sap mix together. This liquefaction of cell contents facilities ingestion through the feeding tube. Cell death soon becomes evident by the coagulation of the remaining cytoplasm by action of enzymes.

(ii) *Longidorid Nematodes*: These nematodes are relatively of large size and possess a hollow axial stylet, about 100 to 300 microns long. Thus, they can feed deep within the plant root. The two parasitic members, *Xiphinema* (dagger nematodes) and *Longidorus*, differ in their parasitic specificity.

Some species of *Xiphinema* may feed for many hours or even days from cells with the differentiated vascular cylinder while others like *X. index* and *X. diversicaudatum* prefer root tips as their feeding site. Terminal galls on root tips are seen in the attack of *X. index*. These galls appear to be attractive for the nematode and provide nutrients for egg production. The galls contain much enlarge multinucleate cells which arise by synchronous mitosis without cytokinesis. These cells are metabolically highly active and seem to be indispensable for reproduction of the nematode. The non-host plants react by necrosis of the cells. The modification of cells or the necrotic reaction are triggered by enzymes in the saliva of the nematode.

Species of *Longidorus* attack only root tips which show terminal galls as in the attack of *Xiphinema*. In contrast to the latter, which in most cases, feed for several minutes at a time from several layers within the root tip galls. *Longidorus* species insert their stylet nearly fully protected into the root tips before they start salivation and feeding at the same spot for several hours. Hypertrophy of root tip cells in the initial stage of gall formation is followed by hyperplasia and then by secondary hypertrophy at the actual feeding site. However, the root tip modification induced by these species is not uniform for the whole genus. Some species induce formation of multinucleate cells in root tips similar to those in the attack of *Xiphinema*.

5. Parasitism of the Tylenchida

The Tylenchids are more advanced than Dorylaimids. They show great diversity in parasitic behaviour and adaptation which culminate in sedentary endoparasitism among the root parasites (*Meloidogyne, Globodera, Heterodera, Tylenchulus*) as well as in some parasites of aerial parts (*Anguina*).

(A) Root Parasites

The types of parasitism and related modifications amongst the root parasitic Tylenchida are given below:

(i) *Migratory Ectoparasitism*: The genera *Tylenchus, Tylenchorhyncus, Paratylenchus, Hemicycliophora* and *Belonolaimus* are important members of this group. These nematodes remain vermiform throughout their life. *Tylenchus* species and some species of *Tylenchorhynchus*, with their rather short stylet feed from surface cells for a few minutes or longer. *Pratylenchus* species feed for many hours or even days at the same spot on epidermal cells or root hair without inflicting any apparent damage. Some are equipped with long stylet for food withdrawal from deeper tissues along the root (*Belonolaimus*) or from root tips (*Hemicycliophora*). The root tips are transformed into terminal galls.

(ii) *Sedentary Ectoparasitism*: Some genera of the family Paratylenchidae (family Criconematidae, subfamily Paratylenchinae) such as Cacopaurus and *Glacilacus* show sedentary ectoparasitism. These nematodes possess long stylets and feed from deeper tissues of the root. The adult females become slightly swollen and remain attached (sessile) to the feeding site.

(iii) *Migratory Ect-endoparasitism*: The nematodes remain vermiform. Not only the stylet but the entire head region of the body is inserted into the cortical cells from which nutrients are removed (species of *Macroposthonia, Rotylenchus, Seutellonema*).

(iv) *Migratory Endoparasitism*: This type of parasitism is common in members of the family Pratylenchidae (or subfamily Pratylenchinae of the family Hoplolaimidae) such as in *Pratylenchus* (lesion nematode), *Radopholus* (burrowing nematode), and *Hirschmanniella* (rice nematode). The nematodes remain vermiform and may invade or leave the roots at any stage of their development. They usually feed cortical cells which are then destroyed by intracellular migration of the nematodes.

In the attack of the root lesion nematode, *Pratylenchus*, species, the consistent host response is necrosis which is both due to mechanical damage to the cells as well chemical damage during feeding. In peach roots attacked by *P. penetrans* it has been shown that the enzyme β-glucosidase secreted by the nematode in its saliva hydrolyses amygdalin in the root cells and releases phytotoxic compounds benzaldehyde and hydrogen cyanide. These compounds cause necrosis of tissues.

The burrowing nematode, *Radopholus*, penetrates the epidermis and moves through the rot cortex by apparent dissolution of the cells which are ingested by the nematode. The nematode, thus, forms tunnels and cavities in the cortex and stele. A large number of them can be seen accumulated in the phloem-cambium ring which is often destroyed by necrosis. The females lay eggs in the cavities.

(v) *Sedentary Endoparasitism*: The nematodes showing this type of parasitism are considered highly evolved root parasites. They invade the root, partially or completely, at a definite developmental stage, for example, as second stage juveniles in *Meloidogyne, Heterodera* and *Globodera* or as young females as in *Rotylenchulus reniformis*. The invading nematodes migrate to specific sites where they induce the formation of highly specialized nurse cell systems. These cells provide a continuous source of nutrients until the parasite dies. Female nematodes become saccate, remain completely inside the host tissue or their posterior portion may partly protrude on the outside, and usually produce several hundred eggs. Males may never feed as in *Tylenchulus semipenetrans* and *Rotylenchulus reniformis* or feed at an early juvenile stage as in root knot and cyst nematodes.

In *Meloidogyne,* the second stage juveniles enter roots and move quickly through the region or cell elongation and establish themselves to a permanent feeding site usually with the head adjacent to or protruding into the vascular cylinder. The body of the nematode lies entirely within the cortex in various positions but usually parallel to the root axis and close to the stele with the anterior end pointed away from the root tip. After sometime they become sedentary and feeding is limited to a few cells that constitute the nurse cell system. A similar pattern is seen in *Heterodera* and *Globodera* but while the root knot nematodes induce gall formation the cyst nematodes do not produce galls although they also feed from a system of nurse cells. In *Tylenchulus semipenetrans,* the second stage juveniles penetrate epidermal and hypodermal cells and feed on the hypodermis and other layers of cortical parenchyma. They also establish a nurse cell system in the cortex and the developing young females remain attached to this site. In *Rotylenchulus reniformis*, young females penetrate the roots some distance away from the root tip. They move inter and intracellularly to the feeding site which may be cortex as in cowpea, phloem as in cotton and castor roots, or the pericycle as in some dicotyledonous crops. Whatever is the feeding site the nematodes induce a nurse cell system for their nutrition.

In the above five kinds of root parasitism specialized nurse cell systems are found only in the sedentary endoparasites. In this group also the different genera vary in their histopathological effects. These are described later. Parasitism by some other phytonematodes not included in the above five types in given below:

(B) Leaf, Bulb, and Stem Parasites

(i) *Parasitism by Anguina*: The genus includes species that are fully mobile ectoparasites for a part of their life and then become endoparasites and lose mobility. They induce formation of organization galls and exhibit good integration of their life cycle with that of their host plants, thus showing characters of most highly evolved nematode parasites of aerial parts of plants. The galls produced in leaves, steams and inflorescences are highly organized structures geared primarily to provide resistance to desiccation. There is no evidence to suggest that the gall tissue provide nutrients to the nematode. The galls are small, pronounced, localized swellings arising from the leaves, on the stem, or in the inflorescence. Internally they contain several layers of cells.

(ii) *Parasitism by Ditylenchus*: The stem nematodes inhibit the intercellular spaces of parenchymatous and cortical tissues of leaves and stem where the enzymes injected by them cause hypertrophy and hyperplasia of cells resulting in swellings. These swellings are spongy and soft. The middle lamella is dissolved (digested) by pectic enzymes of the nematode and cells become spherical. Large air spaces are seen in the swellings. The swellings and giant cells induced by *D. dipsaci* are similar to galls formed by the root knot nematode. Dissolution of middle lamella is associated with reproduction of the nematode. In resistant varieties necrosis develops around the invading nematode and it dies or fails to mature. There is no cell separation and hypertrophy.

(iii) *Parasitism by Aphelenchoides:* The leaf and bud nematodes, *Aphelenchoides* spp., feed either internally or externally upon the leaves of their hosts. All stages of their development are seen in leaf tissues. They are fully mobile in their adult stage. In the infection of chrysanthemums by *A. ritzemabosi* the first reaction of the leaf is the production of patches of an opaque brown substance within the cells near the nematodes. The browning then becomes more extensive and the number of chloroplasts in the cells is reduced. Eventually this browning extends to the epidermis; few chloroplasts remain and cells in the mesophyll are ruptured producing large cavities. During necrosis the mesophyll and palisade tissues are equally affected but only in the final stage of decay do the vascular bundles and collenchyma break down. In mature leaves there is no hypertrophy or hyperplasia but in young leaves such reactions probably occur. The browning is attributed to chlorogenic and isochlorogenic acid.

6. Development of Nurse Cell Systems of Sedentary Root Endoparasites

This aspect of pathogenesis has been reviewed by many workers. The nurse cell systems of the sedentary endoparasites have, as a common feature, a high degree of metabolic activity. There is pronounced increase in cytoplasmic density in ribosomes, polyribosomes and cell organelles. Nuclei are modified to increase the capacity for nuclear-cytoplasmic exchanges.

Once introduced the nurse cell systems continue to grow and are probably maintained by the nutrient sink situation imposed by the continuous nutrient demand of the developing and reproducing parasite. Metabolic activity is most pronounced when the females start laying eggs. At this stage the nurse cells are packed with cytoplasm. When egg laying is complete, there are signs of disintegration of the system.

Walls of the nurse cells, during active metabolic condition, especially those adjacent to the vascular elements (xylem) exhibit remarkable ingrowth leading to amplification of plasmalemma. These are "transfer cells" meant for short distance transport of solutes to the area of nutrien sink. When ingrowths are not present, solutes move through plasmodesmata, especially in the pit fields.

The nurse cell systems are nematode specific. These have been classified into four categories:

(a) Groups of uninucleate, not enlarged, descrete parenchyma cells located in the cortex or the vascular cylinder near the mouth of the parasite. These are seen in the infection of *Tylenchulus semipenetrans*.

(b) The nurse cell is a single, uninucleate giant cell. It is usually initiated in the pericycle and expands as it matures into the vascular cylinder which may become greatly distorted. Transport of solutes may occur through cell wall ingrowth but more commonly through plasmodesmata in the numerous pit fields.

(c) Probably the most common type of nurse cell system develops by the expansion of several cells at the feeding site. Wall expansion is accompanied by partial wall dissolution, mainly at pit fields where the wall is thinnest. This causes mixing of contents of several cells and a multinucleate syncytium is established. Mitosis is not stimulated in the syncytia. The nematode feeds from cell which draws nutrients from neighbouring cells. The syncytia induced *Rotylenchulus reniformis* are confined to the pericycle. Beyond the syncytial cells proximity to the nematode, other pericycle cells are metabolically stimulated but they remain discrete and uninucleate. Such syncytial nurse cell system is also seen in *Naccobus, Globodera, Heterodera* and *Atalodera* but the method of solute transfer differs. Extensive transfer cells with wall ingrowths around the xylem elements are common in *Globodera* and *Heterodera*. In others solute transfer occurs through plasmodesmata in pit fields adjacent to xylem vessels.

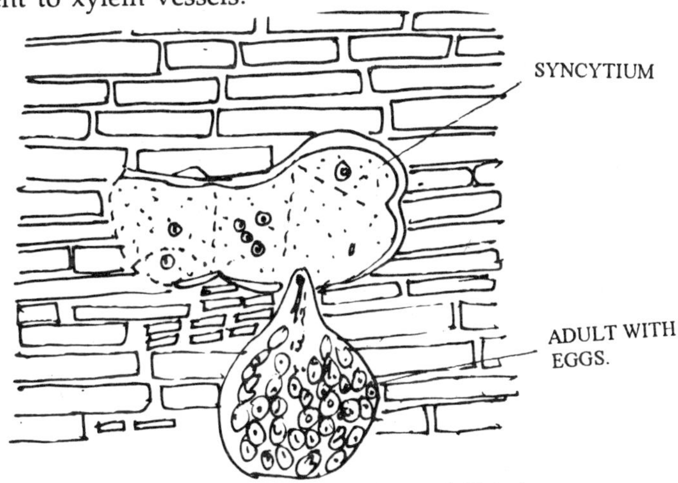

Fig. 15.1 : Induction of syncytium by *Heterodera* and *Globodera*.

(d) The fourth category comprises nurse cell systems that consist of a variable number of discrete multinucleate giant cells which develop within the differentiating vascular cylinder by the expansion of individual cells accompanied by synchronous mitosis without cytokinesis. Wall ingrowths develop adjacent to vascular elements for nutrient transfer. Such nurse cells are known to be induced and maintained by only *Meloidogyne* species which also induce copious gall formation around the infection site hence called root knot.

The nurse cell system not only profoundly affect morphology and physiology of the root, they also increase the susceptibility of the plant to other soil-borne fungal and bacterial pathogens.

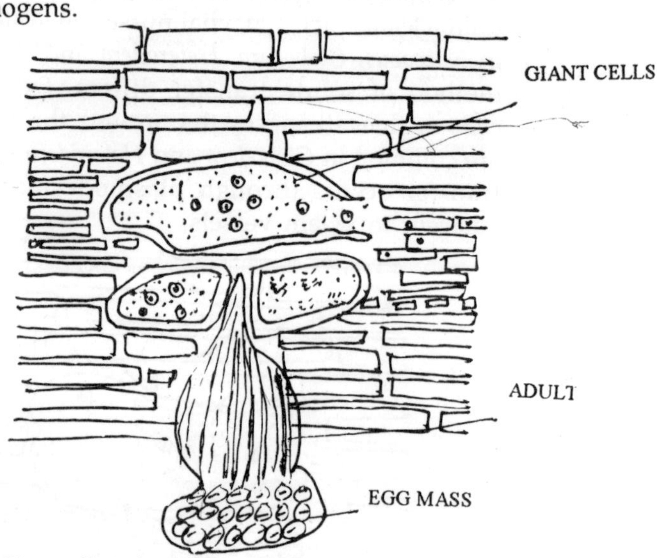

Fig. 15.2 : Giant cells induced by *Meloidogyne* in root tissues

7. Some Biochemical Aspects of Parasitism by Sedentary Root Endoparasites

Numerous reports have been published on biological changes associated with infection by sedentary root endoparasites. Growth regulators both in the affected tissue and in the parasite, amino acids, proteins, nucleic acids, and phenolic compounds have been studied.

As has been stated earlier the nurse cells are metabolically very high active. In syncytia there is intense RNA and DNA synthesis. Walls of the giant cells contain cellulose and pectin but no lignin, suberin or ninhydrin positive substances. The protoplasm contains carbohydrates, fat, RNA and large amounts of proteins. The gall tissues contain increased amounts of cellulose, sugars, keto, acids, free amino acids, nucleic acids, phosphorus, and nitrogen as compared to healthy tissue but starch is absent.

The mechanism of gall formation in *Meloidogyne* infections has not been thoroughly investigated. Probably growth regulators play an important role. Changes caused by some nematode species in cells of a susceptible host are similar to those caused by exogenous indole acetic acid, that is, hypertrophy, hyperplasia, adventitious roots, nuclear division without cell division, and the break down of the cell wall. Probably, the mechanism of auxin action depends upon the induction of RNA, synthesis. Indole compounds detected in galls caused by *M. javanica, M. hapla* and *M. incognita* are: indole-3-acetic acid (IAA), indole-3-acetic acid ethylester (IAE), indole-3-acetonitrile (IAN) and indole-3-butryic acid (IBA). Two possible source of these compounds in the roots galls have been suggested. First, nematodes inject through their saliva the enzymes glycosidase and protease into the host cells and release free auxins from the complexes in the host. The proteases breaking down proteins may also release tryptophan, the IAA precursor, and also amino acids such as phenylalanine, alanine, histidine, and serine, which promote auxin synthesis.

The second possibility is that the nematode itself releases auxins during feeding. Perhaps, indole compounds are formed in nematodes as end products of metabolism and are excreted by endoparasites into the plant tissue or by ectoparasites into the root region. IAA, IAN, IAE, and IBA have been detected in larvae and egg masses of *Meloidogyne* species. IAA in larvae of *Heterodera schachtii* and indole-3 acetic acid methylester (IAM) in larvae of *D. dipsaci* have also been reported.

8. Host Specificity

Phytonematodes are always specialized parasites, that is, they can reproduce only when they feed on certain kinds of plants. The range of such plants may be very large for such nematodes as *D. dipsaci* and *Meloidogyne incognita* or very restricted as in *G. rostochiensis* and *H. avenae*. While *H. schachtii* has several hundred known hosts, *H. glycines*, a closely related species on soybean, has less than 20 hosts.

The hosts of nematode species are often botanically related but many nematodes such as *Meloidogyne* species have hosts in several hundred plant species belonging to widely separated genera and families. The factors which determine whether a plant will be a suitable host for a nematode or not are largely unknown. Generally, the structural make up of the plant does not affect nematodes. There is evidence that some plants are not hosts or are even immune to attack of nematodes because they contain or produce toxins that inhibit nematode development. Others possibly lack some chemical essential for the nutrition of the parasite. The fact that nematodes can penetrate even non-hosts or resistant hosts but fail to develop and reproduce indicates that host specificity is determined to some factor(s) within the plant. However, in potato cyst nematode it appears that specificity is influenced by host chemicals external to the plant since in *Globodera rostochiensis* only the hatching factor from the host roots can stimulate larval emergence from cysts. On the other hand, root diffusates from even non-host wild beet plants stimulate larval emergence in *H. schachtii*. Root exudates or emanations that often attract nematodes to roots are non-specific.

9. Resistance of Susceptibility

In the successful host-parasite relationship enters the host, feeds, metabolizes, grows and produces. These steps are accompanied by alterations in host morphology and physiology that manifest in disease symptoms, damage or injury, and loss. When the host does not restrict these processes and the parasite is free to carry on its life activities in or on the host, the latter is susceptible. If the host puts restrictions to any of these activities is shows signs of resistance. When vital activities of the parasite such as growth and reproduction are checked the host is considered resistant, and if all steps mentioned above are prevented the plant is immune. Thus, from susceptibility to immunity there can be many grades of resistance.

Resistance of plants to nematodes has been variously interpreted as "lack of ability or failure of juveniles to enter the host", or as "any characteristic of plant or any interaction between host and parasite which retards or prevents the occurrence of parasitic relationship between the nematode and its host. Veech defines resistance as the absence or inhibition of disease development upon challenge by a pathogenic agent. In other words, anything that prevents, retards, or restricts disease development contributes to host resistance. Feeding and the resultant growth and reproduction of the nematode parasite are the main criteria to decide resistance and susceptibility. Based on the degree of plant resistance one can group the plants into different categories according to ability of the nematode parasite to reproduce on them. Taylor had given the following convenient and practical classification:

(i) Susceptible: plants in which nematode reproduction is normal.

(ii) Slightly resistant: plants in which reproduction is about 25 to 50 per cent of that in susceptible plants.

(iii) Moderately resistant: plants in which reproduction is 10 to 25 per cent of that in susceptible plants.

(iv) Very resistant: reproduction is one to 10 per cent of the susceptible plants.

(v) Highly resistant: plants in which reproduction is less than one per cent of that in the susceptible plants.

(vi) Immune: plants in which no reproduction of the nematode occurs.

In slightly or moderately resistant plants the life cycle of the nematode may be longer than in susceptible plants; while the resistant or highly resistant plants only a small proportion of the nematodes may complete a life cycle or produce eggs. Nematodes may enter an immune plant and pass through several stages of development but they do not produce fertile eggs.

Crop plants may also be *tolerant* to nematode invasion. Tolerance is not related to resistance in genetic terms. The plants are able to make satisfactory growth and produce a fair yield even though they come under susceptible category. In the root-knot disease such plants may form only small galls or the roots system is so copious that it compensates for the root damage caused by the nematode.

I. Mechanism of Resistance: Mechanisms of resistance to nematodes are complex and involve both the host and the nematode parasite. The mechanisms may operate from outside the host, on surface of the host, or within the host. There is no role of physical and chemical barriers in the host in resistance. The plants may contain or produce toxins that kill the nematodes. These have been sometimes called the pre-existing-nematode compounds which provide resistance outside or within the host. *Asparagus officinalis* var. Mary Washington is resistant to the ectoparasite *Trichodorus christiei*. On young plants the nematodes are able to feed but populations decline at the same rate as in follows soil. In the presence of older plants in which fleshy storage roots have formed, the population decline is more rapid than in fallow soil. Obviously, the nematotoxic principle develops in the fleshy roots and is released in the soil. Other host plant roots in the vicinity of asparagus roots are also protected. The roots contain a water soluble glycoside with a low molecular weight aglycone which has been found to kill nematodes. In addition to *T. christiei*, the compound is toxic to many other nematodes such as *Pratylenchus penetrans, Tylenchorhynchus claytoni, Helicotylenchus nannus, Paratylencus projectus, Belonolaimus gracilis* and some species of *Meloidogyne*.

The effect of the toxic compound is related to the function of acetylcholinesterase in the nematode nervous system which is neutralized.

A similar role of French and African marigold (*Tagetes patula* and *T. erecta*) has been proved against several nematode species. Root knot nematodes fail to induce giant cell formation in marigold roots. When grown between tomato and potato, marigold has been shown in protect these crops against root knot. Used as a cover crop, marigold is reported to give a high degree of control of *Pratylenchus penetrans* and *Tylenchorhynchus dubius* but other nematodes such as *Paratylenchus* and *Criconemoides* are not effected. Two nematicidal compounds have been isolated from marigold roots. These are alphaterthienyl and a derivative of bithienyl. The resistance and nematicidal effect of *Eragrostis curvula* against root knot nematodes is due to the presence of high level of pyrocatechol in roots. There are many graminaceous plants including millets that disinfest the soil. However, the concentration of the toxic compounds from their roots at which they can produce the nematicidal effect can be reached only after a prolonged stand of the plant (three to four months).

It was thought that since many nematodes are attracted to host roots along a gradient of root exudates or emanations, resistance may be linked to this phenomenon of attraction. However, it is now conclusively proved that in resistant as well as in susceptible plans attraction of an penetration by juveniles occurs almost equally well. It is after penetration that a resistant plant challenges the parasite. Similarly, there are views of some workers emphasizing that a resistant plant may lack a substance required by the nematode for its development of the substance may be short supply. There is no evidence to prove that some components of the nematode diet are lacking in resistant plants. However, indirectly the resistance is believed to the linked to feeding of the nematode.

To grow and reproduce, the nematode must have access to uninterrupted supply of food. Like in most of the advanced parasites the nematodes induce changes in the host plant that enable the nematode to ensure uninterrupted supply of host cell cytoplasm. In resistant plants this part of host reaction is absent or suppressed resulting in interruption of food supply to the nematode which fails to develop and mature. The changes in the host induced by the nematodes are nurse cell systems (sycytium, giant cells) and galls. Resistance is provided to the plant through failure of these systems. There has been extensive work in recent years to find out the mechanisms that suppress or inactivate the nurse cells and galls.

Hypersensitive host reaction has been considered since long as possible mechanism of resistance within the host. The importance of this reaction in resistance has been emphasized in studies of root knot and cyst nematodes. The hypersensitive reaction (necroses) prevents development of syncytium or giant cells or renders them non-functional, thereby interrupting food supply to the nematode.

Resistance to root infection by root knot nematode usually manifests itself at an earlier stage than in cyst nematodes. The development of *Meloidogyne* species is often blocked by a rapid hypersensitive reaction which may be associated with the accumulation of nematostatic phytoalexins at the right time and place. Phenolics are also known to stimulate indoleacetic acid-oxidase activity thereby inducing necrotic reaction. The hypersensitive reaction resulting in formation of a protective necrosis instead of the normal giant cells at the feeding site was first studied by Paulson and Webster. In addition, the reaction may be linked to cyanide-insensitive respiration. In resistant tomato infested with *Meloidogyne incognita* the cyanide sensitive and cyanide insensitive respiration both increase but in susceptible plants there is no alteration in respiration. The necrosis in cells around the feeding site prevents the nematode to develop into adult females. The juveniles die at an early stage of development. The necrotic response of the plant is genetically controlled.

In roots resistant to cyst nematode, the syncytium is usually induced but disintegrates a few days later so that males with limited food requirement develop. Although hypersensitive type of reaction occurs around the body of the juveniles it does not appear to inhibit initiation of feeding sites by the nematode and may not be part of resistance mechanism at this stage. Changes occurring after the initiation of syncytia can be directly implicated in resistance mechanisms. There are two types of responses after syncytium initiation. The major gene mediated response results in isolation of the developing syncytium by necrosis of the surrounding cells and rapid degeneration of the syncytium while in polygene mediated response the development of transfer cells (wall ingrowths) in inhibited and there is no necrosis. In this case also the syncytium degenerates because in both types of responses flow of nutrients to the syncytium is stopped. The developing nematodes are deprived of food supply resulting in either death or preponderance of males.

II. *Roles of Phenolics in Resistance*: Phenolic compounds are the best known factors involved in susceptible-resistant responses of plants. There is a definite correlation between the degree of plant resistance and the phenolics present in the plant tissue. Most phenols occur in plants in the bound form as glycoside (polyphenols) of low physiological and chemical activity. Activation requires their decomposition to free phenols. Nematodes are able to do this by secreting β-glycosidase into the host cells. Four kinds of host responses invasion in which phenolics are known to play the main role were suggested by Giebel:

(a) Browning and slow formation of wide necroses in plants susceptible to migratory ectoparasites.

(b) Quick browning and formation of non-expandable necroses in plants resistant to migratory nematodes.

(c) IAA-oxidase inhibition thus favouring auxin accumulation and giant cell and gall formation in plants susceptible to sedentary endoparasites.

(d) IAA-oxidase stimulation favouring auxin decomposition and formation of necroses in plant resistant to sedentary endoparasites.

The presence of chlorogenic acid (a phenolic) is thought to the cause of browning and resistant reaction of chrysanthemums to *A. ritzemabosi* and of root knot resistant tobacco and tomato cultivars to *M. incognita*. Artificial introduction of *β*-glycosidase causes necrosis in potato resistant to *Globodera restochiensis* and formation of syncytium in susceptible potato.

10. Biological Races in Nematodes

A biological race can be defined as a segment of nematode species differing from the rest of the species in some physiological characteristic such as pathogenicity. In practice, biological race or biotype is usually distinguished by its ability to attack a certain plant usually considered highly resistant or immune to the same nematode species.

Many biological races of nematodes have been reported. Some of the most important are those found in root knot nematodes, *Meloidogyne* species, bulb and stem nematode, *Ditylenchus dipsaci*, and the burrowing nematode, *Radopholus similis*. It is however, believed that many races occur in other nematode species such as *H. avenae, H. schachtii, Ditylenchus radicicola*, etc.

Two biotypes were identified in the golden nematode of potato (*Globodera rostochiensis*). The original race was called biotype. A and the resistance breaking race was called biotype B. The latter is now described as a separate species, *G. pallida* In *Melaidagyne,* four races of *M. incognita,* two of *M. arenaria*, and one each of *M. javanica* and *M. hapla* have been identified Seinhorst distinguished 11 biologic races in *D. dipsaci* on the basis of their pathogenicity on nine plant species. In *R. similis* two races are known. One parasitises banana but not citrus (the banana race) and the other parasitises banana as well as citrus (the citrus race).

The biologic races are important in breeding for resistance to nematodes and in deciding proper crop sequence in rotations for nematode control. They are also useful to nematologists who rely to some extent on the pathological responses of the plant to identify the nematodes. Race identification is based on pathogenicity of populations on different hosts.

11. Symptoms of Nematode Infection

A successful host-parasite relationship results in appearance of symptoms on the host. Symptoms of nematode infestation are mostly non-specific. Poor growth of the plant stunting, patchiness of the crop stand, and discoloured foliage are such abnormalities that can be attributed to many other causes such as attack of fungi, bacteria, and viruses as well as soil factors including nutritional disorders and unsuitable soil physical conditions. While some nematodes are primary pathogens, others, by feeding on the host plant, may allow other pathogens or secondary invaders which cause the damage and visible symptoms. In many diseases, there is clear association between nematodes and other pathogens.

(A) Above ground symptoms

(i) *Reduced growth:* Stunting and slow growth are common symptoms of nematode attack. Such plants appear in patches due to non-random distribution of nematodes in the field or introduction of nematode infested seed in nematode-free soil. Stunted growth is common in the attack of *Pratylenchus penetrans* and *P. pratensis* in different crops, *Globodera rostochiensis* in potato and *H. glycines* in soybean. In fruit trees stunting is accompanied by decline and die-back such as in the infestation of citrus by *R. similis* and *T. semipenetrans*. Generally, stunted growth occurs when nematode populations are high. Sometimes, low populations enhance plant growth.

(ii) *Discolouration of foliage:* Leaf chlorosis often accompanies stunted growth of crops and die-back in fruit trees. Yellowing in indicative of nutritional deficiency caused by root destruction and immobility of nutrients from root to shoot. The discolouration may range from light yellow to deep red, purple, or even black. Sometimes foliar symptoms are very specific. *Aphelenchoides besseyi* causes pale yellow or white tips or white tips or rice leaves.

(iii) *Distorted and abnormal growth:* Distortion of aerial parts is caused by the stem and leaf feeding nematodes such as *Ditylenchus. Aphelenchoides,* and Anguina species. These distortions include swellings on leaves and stems, twisting of leaves and replacement of grain with cockles.

(B) Below ground symptoms

(i) *Reduced root system*: Depletion of the root system is the most common below ground symptom of nematode attack. *Trichodorus* spp. and *Tylenchorhynchus claytoni* inhibit root growth without producing other recognizable symptoms. Short stubby and malformed roots are symptoms of attack of *Trichodorus, Belonolaimus gracilis, Xiphinema diversicaudatum,* etc. The roots may show lesions as in the attack of *Pratylenchus* and *Radopholus*.

(ii) *Root proliferation*: Some species of nematodes do not cause general decay of roots. The injury induces the plant to grow more roots, in clusters, especially behind the damaged portion. *H. avenae, G.*

rostochiensis, *H. glycines* and *H. schachtii* on cereals, potato, soybean, and sugar-beet, respectively, as well as *Trichodorus christiei*, *Belonolaimus gracilis* and *Meloidogyne* spp. produce such symptoms.

(iii) *Root galls and cysts*: Galling of roots is the most characteristic symptom of attack of root knot nematodes. However, many others nematodes such as *Xiphinema* and *G. rostochiensis* also cause swellings on underground parts of their hosts. In the attack of cyst nematodes, the presence of white to brown cysts projecting on the root surface is a characteristic symptom.

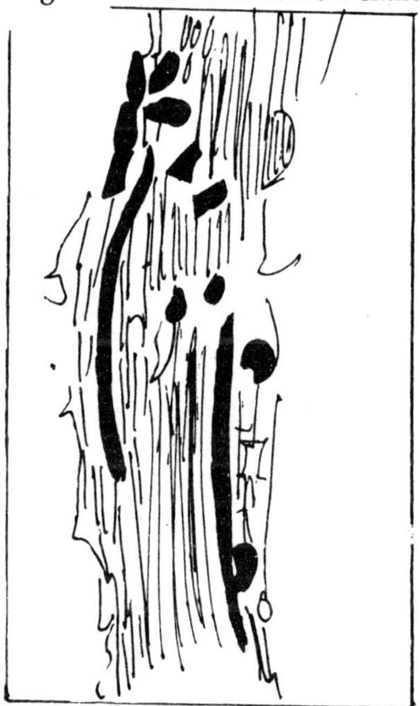

Fig. 15.3 : Feeding of *Pratyleacher*.

12. Relationship of Nematodes with Other Plant Pathogens

In the preceding pages it was mentioned that nematodes are often involved in disease complexes in which nematode is the primary pathogen and some fungus or bacterium is the secondary pathogen. Both can cause injury and disease independently but when in association there is synergistic effect and damage is much more along with alteration in symptoms and host-parasite relationship. Apart from synergistic effect, nematodes generally provide avenues for entry to those pathogens which normally cannot penetrate the intact host. In some cases there is obligatory relationship also. This aspect of host-parasite relationship has been reviewed by many nematologists.

(A) Nematode-Fungus Associations

The possibility of involvement of nematodes in aggravation of a fungal disease was hinted as early as in 1892 when it was found that cotton wilt (*Fusarium oxysporum* f.sp *vasinfectum*) was more severe in the presence of the root knot nematodes. Since then considerable attention has been paid to disease complexes involving phytonematodes and fungal parasites of plants.

(a) Association with root pathogens

(i) *Sedentary Entoparasites*: Root knot nematodes are the most extensively studied nematode parasites that influence fungal root diseases. They are known to predispose the host plants to attack of *Fusarium, Pythium, Phytophthora, Rhizoctonia, Sclerotium* and *Verticillium*. These fungi are independent root pathogens and can infect the host without the aid of nematodes but when the nematodes are present the effects are aggravated.

Of the numerous references describing effects of root knot nematodes on soil-borne fungal parasites of roots, the common in the reference to vascular wilts caused by *Fusarium* spp., especially on varieties resistant to the fungus. The incidence of wilt of cotton (*F. oxysporum* f.sp. *pisi* race 1), sweet potato (*F. oxysporum* f.sp *batatas*) and many others have been found to increase when *Meloidogyne* spp., especially *M. incognita* have invaded the roots. The increase in wilt incidence is related to the number of nematodes infesting the plant before invasion of the fungus, race of the nematode, and resistance of the host to the fungus, the effect being more pronounced in resistant varieties. It would appear that wounds caused by the nematode are directly responsible for increase in wilt incidence but the interactions are much more complicated and less understood. Propagule counts of *Fusarium* in the rhizosphere of tomato infested by root knot nematode increase and actinomycete counts decrease suggesting changes in root exudates. Root galls of tobacco are much more rapidly attacked by *Fusarium* than healthy roots.

Breakdown of resistance or inconsistency in the pattern of inheritance of resistance to wilt are most important effects of root knot nematode invasion. Changes in host physiology due to activities of the nematode are given as a reason for resistance breakdown. The fact that an association with nematodes for sometime is essential for an increased effect of the fungus suggests a change in host physiology. Resistance to vascular wilt is often related to formation of gel and tyloses (occlusion bodies) in the xylem with which the sedentary endoparasitic nematodes are also closely associated. It is possible that activities of the nematodes upset the mechanism by which the host forms these bodies. *Verticillium albo-artum* and *V. dahliae* also cause vascular wilt but in their cause root knot nematode is not as important as other nematode species. Increased susceptibility of cotton to this fungus in the presence of root knot nematode has, however, been reported.

Black shank, a destructive stem and root disease of tobacco caused by the fungus *Phytophthora nicotianae* var. *nicotianae* is greatly enhanced by earlier invasion of *M. incognita*. The fungus needs no assistance to infect susceptible tobacco cultivars. However, where the interaction between the fungus and the nematode occurs resistant varieties suffer heavy losses. There is extensive colonization of the hypertrophied and hyperplastic regions of the root galls and the fungus even colonizes the anterior and posterior regions of the female nematode body as well as the egg masses.

Interaction between *Meloidogyne* and *Rhizoctonia solani* has been reported on cotton, soybean, peas, tobacco, tomato, and okra. The root knot nematode predisposes the plants to severe damage from certain fungi normally harmless to them. *R. solani* and *Pythium ultimum* are only minor pathogens of tobacco, especially after the plants have grown for some time and have attained tissue maturity. However, if plants, even those approaching maturity, have been previously invaded (about four weeks earlier) by *M. incognita* the two fungi are capable of invading roots and causing rapid necroses. No damage or decay occurs if roots are inoculated with either of the two fungi in the absence of the nematode. Little damage occurs if nematodes and fungi are inoculated simultaneously. Tobacco varieties resistant to root knot do not show this response to the fungi. Tomato plants also respond similarly to such nematode-fungus interactions. In tomato, *R. solani* appears to be more aggressive than *P. ultimum*. The host tissues (galled as well as non-galled) in nematode infested plants become good substrate for colonization by the above two fungi. *Curvularia trifolii, Botrytis cinerea, Aspergillus ochraceus, Penicillium martensii* and *Trichoderma harzianum* are common soil inhabitants not parasitic on tobacco roots. When any one of these fungi is added to roots that have been previously exposed for three to four weeks to *M. incognita*, invasion is so severe that root destruction may be more than that caused by *R. solani* and *Pythium* under similar conditions.

Although *Meloidogyne* and *Heterodera* (*Globodera*) are similar in their host-parasite relationship except for the nature of nurse cell system, reports of *Heterodera* being involved in root disease complexes with fungi are not many. Some associations between *Heterodera* and species of *Fusarium, Thielaviopsis, Aphanomyces,* and *Colletotrichum* have been reported without convincing proof of interactions. Miller has reported interaction of *H. tabacum* and Verticillium and Fusarium wilts of tomato. On the other hand, in sugar-beet, when *H. schachtii* and *F. oxysporum* both were present there was antagonistic relationship. The damage was less than what only nematode was present. The complex of *H. schachtii* and *R. solani* (root rot) is sugar-beet is convincingly proved. Giant cells and adjacent areas provide a highly suitable substrate for extensive growth of the fungus.

(ii) *Migratory Endoparasites*: The role of nematode enzymes in fungal pathogenesis through change in host physiology was further demonstrated in the *Ditylenchus dipsaci- botrytis allii* complex of onion collar rot. When water containing enzymes excreted by the nematode was injected into the plant in the presence of the fungus 100 per cent plants showed collar rot against 50 per cent when only the fungus was present or when only plain water was injected. There is evidence that root decay phase of spreading decline disease of citrus (*Radopholus similis*) is accelerated when fungal pathogens such as *F. solani, R. solani, S. rolfsii* and *Thielaviopsis basicola* are present. In banana wilt (*F. oxysporum* f.sp. *cubense*) disease expression is aggravated and incubation period shortened when the plants are invaded earlier by *Radopholus similis. Pratylenchus penetrans*, the root lesion nematode, aggravates eggplant wilt (*V. dahliae*), which through a vascular wilt fungus, is not affected by the presence of root knot nematode. In this case the interaction appears be the dependent on root lesson and necrosis rather than the highly specialized host response in the case of root knot. Similar *Pratylenchus Verticillum* associations are reported in tomato, peppermint and *Impatiens balsamina*. Charcoal rot of sorghum (*Macrophomina phaseolina*), *Cylindrocarpon radicicola* on potato, carrot, and red clover, and black shank of tobacco are reported in increase when *Pratylenchus* is present. *Trichoderma* spp. are not important plant pathogens but in the presence of *P. penetrans* they can cause significant damage to alfalfa and celery.

(iii) *Sedentary Semiendoparasites*: In some nematode fungus interactions invasion by nematode selectively encourages certain fungal pathogens but not others. The citrus root nematode, *Tylenchulus, semipenetrans* aggravates the root decaying action of *F. solani* but does not encourage *F. oxysporum*. Increased severity of cotton wilt (*F. oxysporum* f.sp. *vasinfectum*) and okra root rot (*R. solani*) in the presence of *Rotylenchulus reniformis* has been reported. These studies have also shown that rhizosphere microflora influence the nematode-fungus interaction. The damage caused by *R. reniformis* and *V. dahliae* is less in unsterilized than in sterilized soil.

(iv) *Ectoparasitic Nematodes*: Ectoparasitic nematodes feed on cells of the root epidermis or in the outer cortex, or sometimes deeper. *Hoploalimus uniformis* is associated with *F. oxysporum* f.sp. *pisi* in early yellowing of peas. Singly the two pathogens cause only slight discolouration, lesion formation or cracks in the cortex. Jointly, they cause extensive dark brown discolouration along the roots and complete decay and the prior of inoculations. Rice plants infected with *H. indicus* are predisposed to infection of *Sclerotium*. The stunt nematode, *Tylenchorhynchus claytonii* causes only slight increase in black shank disease of tobacco cultivars resistant to the fungus but significantly increases Fusarium wilt in susceptible varieties. The severity of root rot of peas (*Aohanomyces euteiches*) is increased by *T. martini*. The increase is directly related to nematode population.

(v) *Saprobiotic Nematodes*: These nematodes are omnivorous, not feeding on living plants, and are rarely considered accomplices in plant disease complexes. However, some reports point to their indirect role in bacterial and fungal diseases. As their diet these nematodes ingest bacterial cells and fungal spores which remain viable and are protected from deleterious effects of fungicides and nematicides. These nematodes usually seek wounds, lesions or necrotic tissues as sites for feeding or microbes. In this process they deposit the living fungal spores and bacterial cells for infection of plants.

(vi) *Suppression of Nematodes by Secondary Pathogens*: In some interactions, the secondary invaders suppress the primary invaders. It depends on how early the secondary invader gains entry into the host. A gray sterile fungus is reported to reduce penetration of roots by *G. restochiensis*. In such combinations probably the destruction of the feeding sites by the fungus is the cause of suppression of the nematode. Some reports suggests that mycorrhizal fungi hinder penetration of roots by nematode. The bacterium *Psudomonas solanacearum* so heavily colonizes the giant cells induced by *M. incognita* that these feeding sites quickly degenerate and the nematode is starved.

(B) Nematode-Bacterium Associations

Investigations on nematode-bacterium interactions in plant diseases are comparatively fewer then the nematode-fungus interactions. Among root diseases major studies have been carried out with *Pseudomonas solanacearum* which causes bacterial wilt in potato, tomato, egg-plant, banana etc. The root knot nematodes, predispose potato, tomato, and egg-plant in infection by this bacterium. Similar increase in wilt is obtained if roots are cut and exposed to a thick suspension of bacterial cells. This suggests that the interaction is mainly the effect of wounding permitting easier entry of bacteria. In resistant varieties the only path of entry of bacteria is through injuries on roots. However, some reports suggests a role of structural and physiological changes also in these interactions. Sitaramaiah and Sinha have demonstrated that the egg-plant wilt caused by race 3 of this bacterium is more severe when *M. javanica* is present than when it is absent. The mass severe wilt membrane was observed when the bactenum was introduced two to three weeks after nematode inoculation. The bacterium showed a preference for the giant cells which degenerated following bacterial invasion.

M. hapla is also reported to increase the incidence of crown gall caused by *Agrobacterium tumefaciens* on raspberry and wilt of alfalfa caused by *Corynebacterium michiganense* subsp. *insidiosum*. Similar interactions between *D. dipsaci* and bacterial wilt of cotton and between *Helicotylenchus nannus* and *Pseudomonas solancearum* is tomato have been period.

Cauliflower disease or fasciation disease of strawberry is a good example of how the nematode-bacterium interaction is essential for full expression of symptoms. In this disease the interactions involves the foliar nematode *Aphelenchoides ritzemabosi* and the bacterium *Rodococcus* (*Corynebacterium*) *fascians*. The nematode alone produces localized feeding sites on the crown buds. When the bacterium is also present, proliferation of axillary buds on affected crown occurs and leaf galls are formed. It is believed that the nematode acts as a vector of the bacterium to carry it to the crown buds which it could not enter by itself.

One of the best studied interactions is the *'tundu'* or yellow slime or eat rot disease of wheat in which the nematode *Anguina tritici* and the bacterium *Calvibacter* (*Corynebacterium*) *tritici* are involved. The nematode causes ear cockle or seen gall disease. The bacterial yellow ear rot cannot develop unless the nematode is also present. Survival of the bacterium is carrying in soil, on surface or inside the galls. The nematode acts as a vector carrying the bacterium as the surface contaminant from the galls. There is involvement of an attractive forces of considerable magnitude between the cuticle surface of the nematode and the bacterial cell well.

(C) Mechanisms Involved in Interactions of Nematodes with Fungal and Bacterial Pathogens

A simple mechanism does not supply to all cases of interpathogen interactions. Each interactions has a mechanism of its own and distinct from another involving a different pathogen or a different host. Broadly, the mechanisms could fall under the following categories:

(i) *Wounding by the nematode (primary pathogen)*: The mechanism is simple. Where the secondary pathogen (bacterium or weak fungi) is not strong enough to break down mechanical barriers of the host, nematode punctures of the host epidermis not only provide avenues of entry, they also expose a greater root surface area for infection and colonization by the secondary pathogen. Indirectly, the root wounds increase root exudation which may alter the rhizosphere effect and the secondary pathogen may be stimulated or inhibit. Root wounding involves for more tissues than the epidermis alone. When cortical cells are damaged by the nematode, the dead cells or necroses may provide a good substrate for the secondary pathogens. The nematodes may externally carry the propagules of the secondary pathogen to these deeper tissues.

(ii) *Physiological changes*: During pathogenesis the nematode injects into the host cell several enzymes including proteases and β-glycosidases. Breakdown of proteins releases free amino acids the concentration of which increases in the host cell and also in the root exudates. The amino-acid tryptophan is metabolized to the auxin IAA and the resulting hormonal imbalance is not only confined to the feeding sites but also affects farther tissues. The physiological changes may affect pathogenesis by the secondary pathogen in one or more of the following ways:

(a) Characteristics of the rhizosphere microflora may change and the microflora antagonistic to the secondary pathogen, such as actinomycetes, may be suppressed. *Streptomyces* are effective antogonists of *Fusarium* species.

(b) Resistance to vascular wilts often lies in the endodermis which is normally avoided by fungi. The endodermis contains large quantities of antifugal substances like phenols, naphthols, and anthrols which, in later stages, give rise to benzoquinones, naphthoquinones and anthroquinones, compounds highly toxic to fungi and bacteria. The giant cell system of the root knot nematodes originating in the pericycle does not allow proper differentiation of the endodermis. The absence of the endodermis removes the barrier against invasion of xylem by the vascular wilt pathogens.

(c) The toxin-oxidizing or inhibiting mechanisms in the plant may be destroyed thus enabling the wilt fungus to break the host resistance.

(d) Tissues affected by nematodes or the nurse cell system and galls formed by nematodes are metabolically highly active containing denser cytoplasm. These tissues provide good substrate for nutrition of the secondary pathogens.

(D) Interactions between Nematodes and Plant Viruses

Although nematode-virus interactions are better known through the role of nematodes as vectors of soil-borne plant virus, there are reports of associations in which the viruses increase or suppresses nematode populations or results in enhanced damage. The exact mechanism are not known.

(i) *Stimulation or Suppression of Nematodes and Enhanced Damage*: Multiplication of *A. ritzemabosi* and *D. dipsaci* is increased in tobacco infected with arabis mosaic. Tomato mosaic virus increase multiplication of *M. javanica*. The tobacco mosaic virus suppresses population of *A. ritzemabosi* and *D. dipsaci* in tobacco. Increased damage due to *M. incognita* in the presence of leaf curl virus in tomato, maize mosaic virus in maize, potato virus Y in tobacco, and cowpea mosaic virus in cowpea has also been reported.

(ii) *Nematodes as Vectors of Plant Viruses*: Many virus diseases of plants were found to arise even when the plants were raised from clean propagation material and were protected from the visit of insect vectors. It was suspect that such viruses were soil-borne and were transmitted to plants from the soil. Cadman divided such viruses into two groups: those losing infectivity when the soil was allowed to dry at 20°C for a week or more, and those which remained infective on drying to soil for several weeks or years. These differences were found to be related to the transmitting agent, and not to properties of the viruses. The first group, in which the soil becomes non-infective on drying, contains the viruses transmitted by nematodes. Drying destroys the nematodes and hence no transmission of the virus. The second group contains the viruses transmitted by soil fungi. Drying cold not destroy resting spores of the fungi.

Hewitt and associates were the first to positively demonstrate that a nematode could transmit a plant virus when they found that the fanleaf virus of grapevines was transmitted by the nematode *Xiphinema index*. Now, four genera of nematodes, all from the same group, Dorylaimida, in the order Adenophorea, are known to be vectors of many plant viruses. These genera are *Xiphinema* and *Longidorus* of family Longidoridae and *Trichodorus* and *Paratrichodorus* of family Trichodoridae. The Longidoridae and *Trichodorus* and *Paratrichodorus* of family Trichodoridae. The Longidorids are relatively large nematodes and transmit the isometric (polyhedral) NEPO viruses (nematode transmitted polyhedral viruses). The Trichodorids transmit the tubular viruses of the Tobraviruses group, also known as NETU viruses (nematode transmitted tubular viruses).

Transmission of viruses by nematodes resembles the transmission by insect vectors. Viruses are picked up from the infected host and injected in the healthy host by the stylet during feeding process. There is acquisition feed time and inoculation feed time and the viruses are retained in the body of the nematode for different durations. Arabis mosaic virus in X. *diversicaudatum* and tomato ringspot virus in S. *americanum* can be acquired after one hour of feeding and transmitted in a similar period of feeding. S. *diversicaudatum* retains arabis mosaic virus for at least 31 days when kept in fallow soil and for at least eight months when kept on virus-immune cultivars of raspberry. The site of virus retention varies in different nematodes. In *Longidorus* the virus particles are found in the cuticular lining of the guide sheath of odontostylet. In *Xiphinema* the particles accumulate in lumen of odontosty and in the esophagus while Trichodorus has the virus particle in the entire oesophagus.

Specificity of the transmission also occurs. Different strains of the same virus are often transmitted by different species of the same nematode genus. For example, *Paratrichodorus christiei* transmits. American isolate of Tobacco rattle virus while the European isolate is transmitted by *P. anemone*. Site of retention or accumulation of the virus in the vector is responsible for some degree of specificity but more important is specific absorption of virus particles by linings of the nematode body.

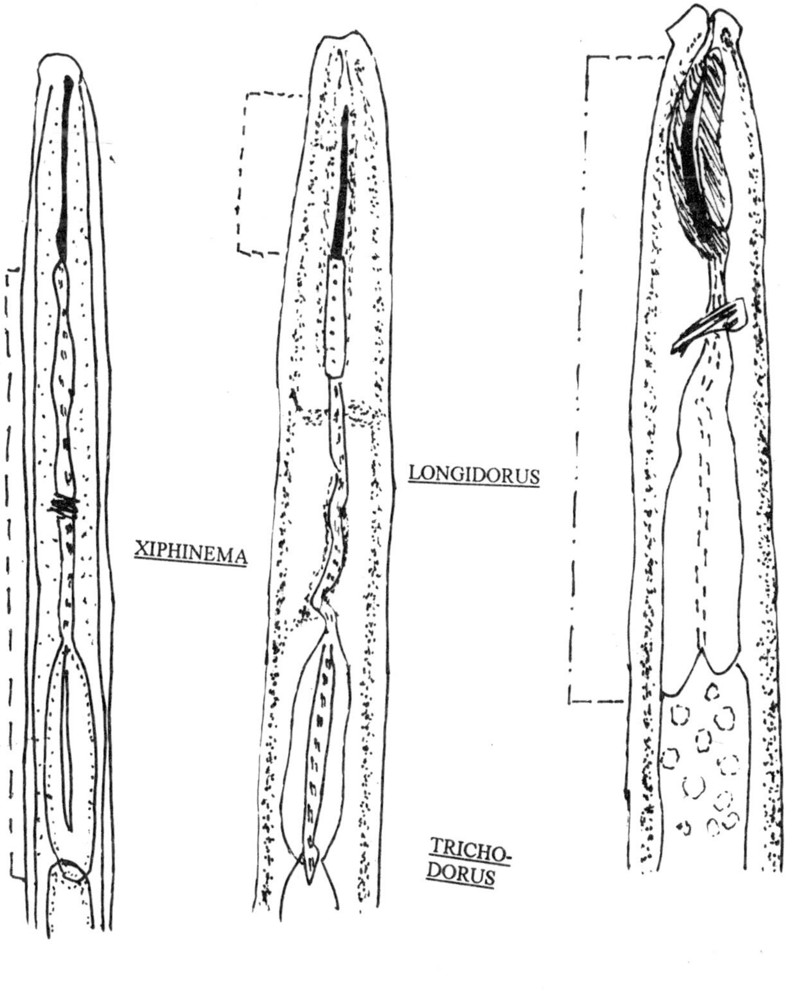

Fig. 15.4 Anterior portion of nematode vectors of plant viruses. Broken lines indicate part of alimentary tract virus particles are retained.

Table 1: Nematode species transmitting plant viruses

Nematode species	*Plant viruses transmitted*
1	2
Xiphinema americanum	Tobacco ringspot, Tomato ringspot.
X, coxi	Tobacco ringspot, Arabis mosaic (Type strain), Cherry leaf roll.
X diversicaudatum	Arabic mosaic (Type strain and Hop strain), Cherry leaf roll, Strawberry latent ringspot, Brome mosaic and Carnation mosaic.
X index	Grapevine fanleaf.
X italiae	Grapevine fanleaf.
X vuiltenezi	Cherry leaf roll.
Longidorus attenuatus	Tomato black ring (English and German strains).
L. elongatus	Tomato black ring (Scottish strain), Raspberry ringspot (English and Scottish strains), Carnation mosaic.
L macrosoma	Raspberry ringspot (English strain), Brome mosaic, Prunus necrotic ringspot.
Trichodorus cylindricus	Tobacco rattle (European strain).
T. primitivus	Pea early browning (English isolate). Tobacco rattle (European isolate).
T. minor	Tobacco rattle (European isolate).
T. similis	Tobacco rattle (European isolate).
T. viriliferus	Pea early browning (European isolate). Tobacco rattle (European isolate).
Paratrichodorus anemone	Pea early browning (English isolate). Tobacco rattle (European isolate).
P. allius	Tobacco rattle (American isolate).
P. christiei	Tobacco rattle (American isolate).
P. nanus	Tobacco rattle (European isolate).
P. pachydermus	Pea early browning (Dutch isolate), Tobacco rattle (European isolate).
P. porosus	Tobacco rattle (American isolate).
P. Teres	Pea early browning (Dutch isolate). Tobacco rattle (European isolate).

Chapter—16

Host-Parasite Relationship in Animals

All organism have evolved defence mechanisms for protection against invasion and establishment by micro-organisms and by metazoan parasites. Although ideally we should include the viruses and bacteria in our current discussion, space precludes such a full consideration. Before discussing immunity to parasites, it will be necessary to outline, in the briefest manner, the components of the defence systems—primarily of mammals.

The discipline of immunology has developed enormously over the last few decades, and it is often difficult to keep abreast of new developments. Although our knowledge of immunity to parasites is perhaps less complete than that in the areas of tumour and micro-organism immunology, a vast new literature is accumulating on the highly varied immune defence mechanisms adopted by mammals against parasitic infection. Nevertheless, many current textbooks of immunology pay little or no attention to parasite immunology, which perhaps reflects more the difficulties embodies in the subject, rather than the paucity of information. The immunology of parasitic infections is both an important academic subject, as well as a fundamental applied discipline. It is to the field of immunological achievement that many people from the Third World will be looking for relief from the major parasitic diseases that reduce the quality or length of their lives.

General principles of cellular and immunological defence systems

The essence of the immunological defence system of vertebrates is the recognition of self the ability to distinguish it from non-self, i.e. the host animal must be able to distinguish between its own macromolecules and those of an invading organism. On occasions this system of recognition can fail, giving rise to the so called auto-immune diseases, in which the host's defences act disastrously against its own tissue. Nevertheless, in the formal healthy animal, the ability to recognize non-self material and the development of a specific memory system associated with the recognition is the embodiment of the immune system.

Defence mechanisms of vertebrates against invading organisms are divisible into either specific or non-specific responses. Specific responses are associated with an immunological memory component and are normally directed against secondary and subsequent infections of an organism, whose primary infections initiates the response. Therefore, once a host has been infected with a particular parasite it retains an immunological memory of that infection and responds to eliminate or control any later infections by the parasite. Non-specific responses are more generalized in their nature, and are directed against any invading organism, with no involvement of a memory element. In some cases, specific and non-specific responses cooperative to form a combined defence against invasion.

Non-specific responses tend to have a cellular basis and involve the action of phagocytic cells (macrophages and polymorphonuclear leucocytes), which form the reticuloendothelial system and are present in the circulating blood and in the tissues of the host. Their function is to engulf invading organisms and digest them lysosomally. Engulfment and digestion are sequential processes: first, the invading organism becomes adhered to the surface of the phagoctyic cell; secondly, it is ingested by the formation of a vacuole, becoming trapped within a phagosome; thirdly, a laysosome fuses with the phagosome to form a phagolysosome; and finally, the engulfed organism is destroyed by an array of lyric factors, including enzymes. Phagocytosis is a common mode of defence against invading micro-organisms, and it can be enhanced by the action of antibodies known specifically as *opsonins*. The complex of serum factors, known as *component,* may also aid the process of phagocytosis by attracting phagocytic cells to the environs of an invading parasite. Other non-specific responses to parasites may take the form of inflammatory reactions and capsule formation. Inflammation is typified by increased capillary blood supply to a site of invasion, resulting in the accumulation of polymorphs around the invading parasite. Some polymorphs can transform to fibroblasts, which contribute to encapsulation and ultimately, calcification of the parasite; digenean metacercariae and pseudophyllidean plerocercoids are often encapsulated in this way in their intermediate, fish hosts.

Specific immune responses involve the formation of antibodies by the host as a result of stimulation by the presence of foreign material, termed antigens. Antigenic substances are usually polypeptides, proteins or polysaccharides, with a molecular weight in excess of 5,000 Daltons (lipids are only antigenic when combined with other substances, called haptens). Antigens derive from the surface coat of a parasite, i.e. surface antigens, or from secreted or metabolic byproducts, i.e. soluble or metabolic antigens. Substances are only antigenic if they exhibit a degree of foreigness to the host, which then responds by synthesizing antibodies, specific to each antigen.

Antibodies are soluble serum proteins that combine, specifically with antigens. They belong to the gamma-globulin fraction of serum proteins (figure 16.1), and are referred to as *immunoglobulins* (Ig). They are separable into five distinct classes, depending upon their structure, i.e. IgA, IgD, IgE and IgM (Table 16.1). The molecular structure of the immunoglobulins consists of two heavy peptide chains, linked by disulphide bridges, and two light chains, that are also cross-linked. Papain treatment cleaves the immunoglobulin molecule into two fractions, the F*ab* (antigen binding fragment) and the F*c* (crystallisable fragment) components. Antigen binding takes place at the F*ab* end of the immunoglobulin molecule, and thus, antibody function is retained after papain cleavage of the intact molecule.

Immune defence effected by antibodies is divisible into two distinct processes, known as *cell-mediated immunity* (delayed hypersensitivity) and *humoral immunity*. Both types of immunity arise from stem cells produced in bone marrow. In cell-mediated immunity, the stem cells are processed by the thymus gland to form a population of lymphocytes (T-lymphocytes) which become sensitized upon contact with antigen, in the paracortical region of lymph nodes. These T-lymphocytes are long-lived and carry firmly bound antibody on their surface membranes. In humoral immunity, the bone marrow stem cells are processed in lymphoid tissue (Bursa Fabricius in birds and Peyers Patches in mammals) to give rise to a population of B-lymphocytes, which are short-lived, after contact with antigen. These lymphocytes differentiate into plasma cells which synthesize and secrete antibody into the surrounding blood and tissues. Cellular co-operation may enhance an immune response, and such co-operation can involve contact between macrophages and lymphocytes or between B-cells and T-cells (helper T-cells). Cell mediated immunity is involved primarily in allergic responses and can take some time to develop (24 hours or longer), whereas humoral immunity is developed much more rapidly.

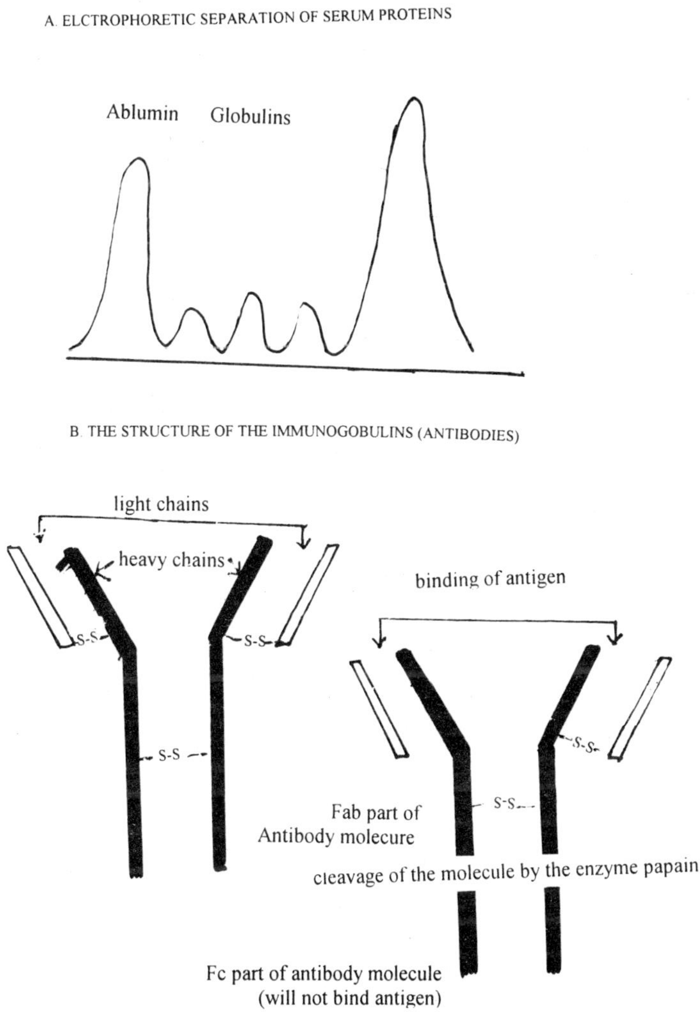

Fig. 16.1 : The identity and structure of mammalian immunoglobulins (antibodies).

S-S

Table 16.1 Major properties and functions of the immunoglobulin classes

Property	*IgG*	*IgA*	*IgM*	*IgD*	*IgE*
molecular weight	1.5×10^5	1.6×10^5	9×10^5	1.85×10^5	2×10^5
valency of antigen binding	2	2	5(10)	?	2
serum concentration (mg/ml)	8–16	1.4–4	0.5–2	0–0.4	0.017–0.45
carbohydrate content (%)	3	8	12	13	12
major known role	anti-micro-organisms	defence of body surfaces	agglutination: primary immune defence	found on lymphocyte surfaces	anti-parasite action: reaginic antibody (allergic reactions).

Antibodies react with antigen and effect lysis, agglutination, precipitation or enhanced phagocytosis of the cells producing the antigens. These reasons can form the basis of *in vitro* tests used to detect the presence of specific antibodies to parasite, and they serve to determine whether an individual animal has been exposed to a particular parasite (immuno-or sero-diagnosis).

Immunological defence is best developed and most often studied in mammals, and rather less is known for lower vertebrates and invertebrates. In non-mammalian vertebrates, humoral and cell-mediated immunity can be identified, but invertebrates appear to lack antibodies, in the mammalian sense, and rely, to a great extent on, phagocytosis as a defence against invasion.

Passive immunity is passed from a mother to her offspring and reflects in innate resistance to a particular

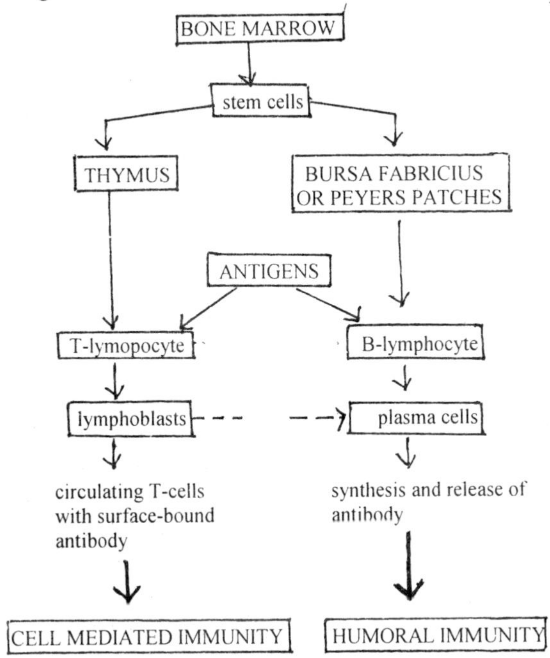

Fig. 16.2 : Lymphocytes and antibody production: A simplified scheme.

infection. Passive immunity can also be transmitted to the laboratory by the donation of serum or immunologically competent cells to recipient, naive animals. *Acquired immunity* depends upon the experience of a primary infection to initiate the specific immunological processes that render the animal insusceptible, or less susceptible, to a second or subsequent infection.

Immunity to Protozoa

Blood-dwelling Protozoa

In this section we shall examine briefly the immunity to malaria, babesias and trypanosomes, which together constitute the major blood-dwelling parasitic protozoans of vertebrates.

Malaria: The general picture of immunity to malarial parasites is remarkably complex due to a range of factors that include strain variation, antigenic variation within strains, and a life-cycle that involves intraerythrocytic stages and exoerythrocytic stages, both in the blood and in liver tissue. Acquired immunity to *Plasmodium* is variable, e.g. premunition is typical of avian malaria, whereas sterile immunity occurs in some rodent malarias, such as *Plasmodium berghei* and *P. vinckei*. In the former case, parasites persist in spite of the development of immunity, while in the latter case immunity is complete and no parasite survives. Premunition is thought to typify human malarial infections, in which relapses an recrudescences are not uncommon.

Protective immunity to malaria is recognized as a collaborative mechanism involving both cell-mediated reactions (T-lymphocytes) and humoral antibodies (B-lymphocytes). Immunoglobulin production increases considerably during the course of a malarial infection, but much of the antibody produced is not protective and does not serve to eliminate the parasite. Recovery from malaria requires protective antibody to the manufactured at a rate sufficient to control the multiplying parasite; drug induced immunosuppression in experimental malarias, can often result in a fatal infection developing from what would otherwise have been a non-lethal, self-limiting condition. There is considerable variation in the levels of the various immunoglobin classes, e.g. IgM levels rise slowly and are maintained for much longer periods. The roles of IgA, IgD and IgE are not clear in protection against malaria, and this function appears primarily with IgG. This is supported by the observation that, during pregnancy, IgG levels characteristically decrease, and there is often an enhancement of malaria in pregnancy.

Cell-mediated reactions, involving lymphokine production, the potentiation of antibody production from B-lymphocytes by helper T-cells and cytotoxicity, may all be involved in the development of immunity to malarias, but current knowledge remains incomplete. Surgical removal of the thymus (thymectomy) in neonatal rats and hamsters renders them more susceptible to *P. berghei* infections than intact controls. Depression of T-cell function by the administration of anti-T-cell serum, can also exacerbate malaria in experimental animals.

There is a number of diagnostic tools, based on immunoglobulin production in response to malarial infection, which serve to identify individuals infected with the parasite, where microscopic examination might prove negative. Such immunodiagnoses include the indirect florescent antibody test, the indirect hemagglutination test and various gel precipitation tests. The application of these methods to human populations can provide considerable and invaluable information on the epidemiology of malaria in endemic regions of the world. Serological monitoring is also proving to an important tool in the prevention of transfusion-induced malaria, which is an increasing problem related to the degree and ease of world travel that typifies the late twentieth century.

Malarial parasites can avoid the immune responses of their hosts to form residual populations, either in red blood cells or within liver tissue, and these residual populations of parasites are able to initiate recurrent relapses or recrudescences. The latter may result either from infected red cells that are situated at deep vascular sites, the plasmodia re-emerging as the effective immunity weakens, or from either the continued existence of merozoites in the circulating blood or sporozoites in the liver. Antigenic variation, whereby a sub-population of parasites of different antigenic structure from the original invasion, is developed, may also be important in this context. This new antigenic population can now withstand the effects of the antibodies directed against the antigenic variations represents an important problem, hindering the development of a vaccine.

The pathology of mammalian malaria is related, in part, to the immunological responses of the host to the parasite—hence the term immunopathology. Red cell destruction (haemolysis) and enlargement of the spleen (splenomegaly) are typical of advanced malaria. Haemolysis may be brought about by the phagocytic activities of spleen macrophages, and both infected and uninfected red cells are destroyed. Opsonising antibodies enhance this pathological process. Haemolysis in Blackwater fever (*P. falciparum* infections associated with quinine therapy) occurs in the vascular system and is not related to the splenic mechanism. Malarial kidney disease results from antigen-antibody complexes binding to the glomerular basement membrane, producing yet another aspect of malarial immunopathology.

An interesting feature of malarial parasites lies in their role as immunosuppressive agents for infections by other parasites. Laboratory animals infected with various species of *Plasmodium*, show a reduced immune response to non-living antigen, such as tetanus toxoid or sheep red blood cells, and in living antigens in the form of viruses, bacteria and several protozoans. It is not known whether this immunosuppressive effects is functionally related to cell-mediated or humoral immunity. There is god evidence for suppression of antibody production by B-lymphocytes, but the general mechanisms by which immunosuppression to unrelated antigens is effected remain to the elucidated.

Toxin production in malaria has been a much debated issue for many years. The key question of whether the parasite releases a toxic substance, which itself is pathogenic, has yet to be answered. There is some evidence that a transferable serum component will induce haemolysis in recipient animals, and that serum elements will induce haemolysis in recipient animals and that serum elements are capable of inhibiting mitochondrial respiration and oxidative phosphorylation in liver cells *in vitro*. Antitoxic immunity, however, has no effect on the levels of parasites in the blood, but may serve to alleviate aspects of the immunopathogenesis.

Babesia Members of the genus *Babesia* are intraerythrocytic parasites of vertebrates, occurring most commonly in mammals. Several species are of veterinary importance as they cause a variety of tick-borne fevers of domestic animals, e.g. *B. bovis*, red-water fever; *B. equi*, biliary fever. Immunity to many species of *Babesia* is of the premune type, existing only while the host remains infected by the parasite. For some species, however, complete sterile immunity is developed, e.g. *B. divergens*, *B. bigemina* and *B. rhodaini*. Immunity may be highly specific for strains of a single species, such that cattle, immune strain from one locality, can be susceptible to a different strain of the some species from other localities. In other, cases cross-immunity, between different species, is found e.g. *B. rhodaini* and *B. microti* in rodent. The extent of antigenic variation in babesias is not known.

Trypanosomes Extensive investigations have been made into immunity to both rodent and human trypanosomes. One of the best known model systems is *Trypanosoma lewisi*, a specific rat parasite, which provides a classical demonstration of acquired immunity. Additional interest in *T. lewisi* is generated by the occurrence of an apparently unique antibody, termed ablastin, whose functional significance has proved to be controversial. Typically, the course of a *T. lewisi* infection in a laboratory rat is characterized by an initial phase of rapid multiplication of the parasite, giving a rising parasitaemia. Parasite division is inhibited five days post-infection and a plateau in the number of trypanosomes occurs, but this is of short duration. Subsequently, a primary crisis takes place on or about the tenth day of the infection, during which most of the trypanosomes are killed. The few parasites that remain survive until approximately the thirty-fifth day, when a second crisis completely eliminates the infection. One explanation of these events involves the production of three distinct antibodies one that inhibits reproduction of the parasite (ablastin) and two lethal (trypanocidal) antibodies, which contribute to the two crises.

The name ablastin was coined by Taliaferro, more than fifty years ago, for the antibody that inhibited the reproduction of *T. lewisi* but lacked trypanocidal activity. Since the original experiment which demonstrated ablastic immunity were completed an interesting and controversial literature has arisen. Ablastin is found only in infection of two species of trypanosomes. *T. lewisi* and *T. musculi*, although ablastic-like responses are characteristic of many stercorarian trypanosome infections. Despite intensive

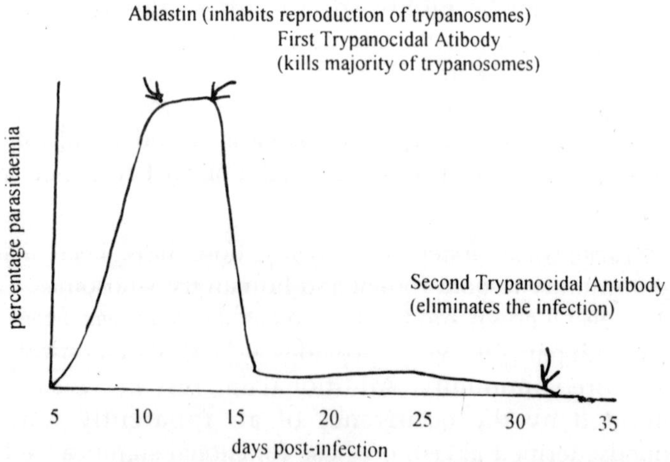

Fig. 16.3 : The course of an infection of *Trypanosoma lewisi* in a rat: one possible scheme of antibody action in eliminating the infection.

investigation, ablastin has not been unequivocally identified as an immunoglobulin, but it does possess physico chemical properties typical of IgG. However, ablastin is not adsorbed from infected rat serum by trypanosomes in *vitro,* a feature which throws some doubt on its role as an antibody. Furthermore, the trypanosome antigen (ablastinogen) that induces ablastin production, has yet to be identified, though there is some evidence that this antigen is released into the blood of the host during the course of a normal infection. Pregnant female mice are unable to limit infections of *T. musculi,* suggesting that ablastin belongs to the IgG class of immunoglobulins, which are known to decrease during pregnancy. Some authors, however, regard the apparent uniqueness of ablastin as sufficient reason, it itself, for discounting the presence of an antibody that is not lethal but inhibits only parasite reproduction. One alternative explanation that has been advanced, invokes the occurrence of two parasite antigens, one juvenile and the other adult, that stimulate the host to manufacture trypanocidal antibodies. Antibody against juvenile antigen eliminates juvenile trypanosomes by the tenth day post-infection, while adult parasites survive for longer periods, being—according to this hypothesis—less antigenic. Additionally, it has been proposed that ablastin may act as a metamorphosis-conversion-factor in the development of trypanosomes during an infection, enhancing the change from juvenile to adult form. The first crisis in an infection would then be attributable to the activities of a trypanocidal, anti-juvenile antibody, which may even be ablastin. These views contrast with the classical view of immunity to *T. lewisi*. Despite considerable speculation, the nature of a ablastin and ablastic immunity remains unresolved. Its apparent uniqueness as an antibody argues for generating research interest rather than for a denial of its unusual role.

Studies on immunity to the human trypanosomes, causing sleeping sickness, e.g. *Trypanosoma gambiense* and *T. rhodesiense,* are complicate by the appearance of *antigenic variation,* which typifies these haemoflagellates. Each species has a highly complicated antigenic structure, which gives rise to both strain and variant populations. These render attempts to immunize humans or domestic cattle against trypanosome infections almost futile. The earliest observations on antigenic, variations were made at the beginning of this century, when it was noted that *T. equinum*, isolated from the blood of a monkey, was unaffected by serum antibodies from the same animal in *vitro*, and it was concluded that the trypanosomes had undergone antigenic change during the course of the infection. Similar observations to this have been made on many subsequent

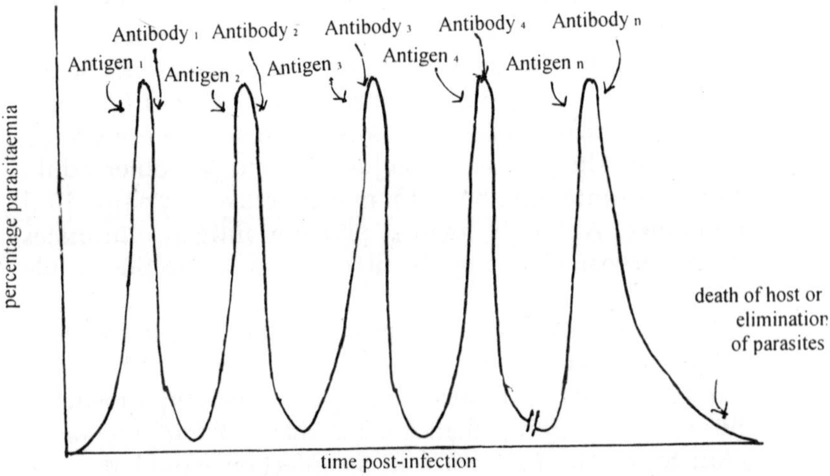

Fig. 16.4 : Antigenic variations and relapsing infections in trypanosomes. Antigens 1, 2, 3, 4.... n are variants produced by a population of trypanosomes in a single host. Each variant induces the production of a variant-specific antibody which eliminates the majority of the trypanosomes. A small number or parasites survive because they possess the new variant against which the host has not made antibody and they multiply to give a rising parasitaemia until there is sufficient variant-specific antibody circulating to eliminate this sub-population. The cyclic process can continue until the death of the host or the complete elimination of the parasites. This explains the characteristic relapsing nature of human sleeping sickness.

occasions. Antigenic variation can give rise to a *cyclic parasitaemia*, in which the numbers of parasites increase until a crisis occurs which eliminates the majority of individuals, except for a small number of antigenically variant forms which survive and divide to produce a second, rising parasitaemia, followed, in time, by a second crisis. This pattern of increasing and decreasing parasitaemia may continue until the host dies (figure 16.4). Experimental observations show the frequency with which new antigenic variants can arise, e.g. *T. congolense* produces a new variant every ninth day in sheep, and *T. vivax* develops new variants every seven days in sheep and calves. The maximum number of different variants arising from a single strain is 23, but this is thought unlikely to represent the upper limit. Only rarely will the same variant appear twice during a single infection. There is also evidence that antigenic variation is an organised process, with a particular strain of trypanosomes producing the same variants, often in a similar sequence, regardless of host species.

Two major subclasses of antigens have been identified from trypanosomes. The first are the bound antigens which remain stable during a relapsing infection. These antigens include enzymes and nucleoproteins and are only weakly antigenic. The second subclass contains the surface-coat antigens, which are unstable entities. The surface-coat antigens are the glycoprotein exoantigens that are present in the serum of infected animals and are responsible for inducing the production of protective antibodies.

Antigenic variation in trypanosomes may be regarded as an adaptive mechanism to overcome the immunological defence of the host. Its occurrence makes it most unlikely that successful immunisation will be accomplished in the forseeable future.

Tissue-dwelling Protozoa

Among the major tissue-dwelling protozoans of importance to man from a medical or veterinary point of view are the leishmanias, the coccidians and the amoebae.

Leishmania Various species of the genus *Leishmania* cause by unpleasant and often lethal diseases of man in many regions of the world. These include oriental sore (*L. tropica*), kala, azar (*L. donovani*) and New World cutaneous leishmaniasis (*L. mexicana* and *L. braziliensis*). The parasite is transmitted to man via the bite of infected phlebotomine flies (sand flies) and, after entry, the amistigote form of the parasite invades and inhabits the cells (such as monocytes and other phagocytic cells) of the reticuloendothelial system. Immunity to *L. tropica* develops after an initial infection, i.e. acquired immunity. The parasite invades the skin macrophages and a superficial sore develops. This sore will heal spontaneously, after which immunity to reinfection is complete. Should the sore be removed prematurely by surgery, however, the individual remains susceptible to reinfection. Circulating antibodies cannot be demonstrated in *l. tropica* infection, whereas humoral components are readily observed in *L. donovani* infections. In general, it is considered that the human immunological response to leishmaniasis is primarily via cell-mediated immunity, with circulating antibodies playing only a minor protective role. A population of specifically sensitized, thymus-dependent lymphocytes develops on contact with parasite antigens, and these cells destroy the infected macrophages. The actual mechanisms by which the parasite is eliminated is not fully understood, but it may occur by a number of processes, including phagocytosis by activated macrophages, cytoxic effects of lymphokines secreted by sensitized lymphocytes at the infected macrophage surface, or, perhaps by lytic antibody activity.

Some degree of cross-immunity occurs in leishmaniasis, e.g. *L. tropica* protects an individual against infection by *L. mexicana* and *L. braziliensis*, but not against *L. donovani*.

Coccidia Most domestic animals appear to develop immunity to reinfection by species of the genus *Eimeria*, which are obligatory, intracellular, but parasites. Sear of infected animals effectively prevents infection by the invasive stages of *E. tenella*, *E. bovis* and *E. meleagrimitis*, and the presence of circulating antibodies has been demonstrated by a variety of techniques. Rather surprisingly, passive transfer of immunity, by the injection of immune serum into uninfected hosts, has not proved to be particularly successful, and is only effective if the challenge infection is administered shortly after the inoculation of immune serum from the donor animal. Immunity to *Eimeria* is highly species-specific, so that there is no cross-immunity between the different species. The precise mechanisms of acquired immunity to *Eimeria*, either in chickens or laboratory rodents, is not clear. Humoral antibodies appear to be implicated in the development of protective immunity, while there is some evidence that cellular elements, polymorphonuclear cells, lymphocytes and macrophages, are also involved.

Protective antinodies are produced against another important intracellular gut parasite, *Toxoplasma gondii*, but it is thought that the cystic stage of the parasite is resistant to these antibodies.

Amoebae Entamoeba histolytica is normally a non-pathogenic inhabitant of the human, intestine and, under such conditions, is apparently non-immunogenic. However, the amoebae can occasionally penetrate the gut epithelium and spread throughout the body tissues. When this occurs antibodies are synthesized by a host, and serological tests can be used to identify their presence and thus serve to diagnose amoebiasis. At least two antibody systems have been identified, but it is not certain by which mechanism protective immunity is developed.

Immunity to helminths

Metazoan parasite differ fundamentally from protozoans in that the great majority do not multiply within the final host, unlike the Protozoa. Thus immunity to helminths will be directed primarily against the individual parasite, rather than its reproductive ability (Cf. *Trypanosoma lewisi*). There is increasing interest in the immunology of mammalian helminth infections, particularly with regard to those species that effect man's health of his economy.

While protozoan infections are frequently by an increase in the serum levels of IgG and IgM, helminth infections typically induce the synthesis of reaginic antibody, IgE, which is characteristically abundant in many helminthiases.

Digenea

Information on the immunology of digenean infections depends largely on studies of only two groups, the schistosomes and the liver flukes, e.g. *Fascoila hepatica*.

There is some doubt as to the occurrence of acquired immunity to *Fasciola* in cattle and sheep. Cell-mediated reactions may play an important role in the development of protective immunity in domestic animals. In the laboratory, experimental mice treated with peritoneal exudate cells from infected donors support fewer worms than control animals. T-cell responses may also be responsible for effecting repair of the tissue damage wrought by the parasite, since rabbits given antilymphocyte serum, to suppress T-cell responses, did of liver disease due to the parasite. Transfer of lymph node and spleen cells from infected donor rats to naive recipient rats offers a significant degree of protection against a primary challenge infection, but this is not repeatable using sheep.

The immunology of schistosome infections, particularly *S. mansoni*, has been examined in considerable detail using model systems in rats, mice and rhesus monkeys. Experimental studies in man are, naturally, not acceptable, so that evidence for acquired immunity in human infections derives mainly from epidemiological sources. In general, the young are more susceptible to infection than adult humans, and a premune condition develops, in which a primary infection renders the host immune to subsequent reinfection. In schistosomiasis, this condition is termed *concomitant immunity*, since the primary infection is not killed by the immune responses. This situation has led to the suggestion that it may be possible to immunize against infection using attenuated (non-developing) cercariae. There are, however, problems associated with this approach, so the search continues for a dead vaccine that will have fewer attendant risks and be more easily administered.

In experimental infections, acquired immunity of the concomitant type has been clearly demonstrated. Despite the development of a lethal immune response, the primary infection continues to thrive and produce eggs for considerable length of time. In rhesus monkeys, immunity to a challenge infection can be developed within sixteen weeks of administration of a primary burden of 100 cercariae per animal. Mice take much longer to develop resistance to a challenge infection. A *self-cure* type responses has been recorded in heavily infected baboons, in which the primary infection is completely eliminated; this does not accompany light infection doses.

The process of concomitant immunity has puzzled researchers for more than a decade, and several explanations for the observed effects have been put forward. It is generally accepted that concomitant immunity arises through the sharing of common antigens by host and parasite, but it is not clear whether the parasite synthesizes host-like antigens or merely binds antigenic material of host origin to its surface-coat. Support for the latter hypothesis comes from ingenious experiments in which schistosomes are grown to maturity in mice and then transferred to the hepatic portal system of rhesus monkeys that have been immunized against mouse red blood cells. Worms transferred in this way are killed within 24 hours, whereas mouse worms will live in non-immunized monkeys. Similar results have been obtained using worms grown, *in vitro*, in human serum and then transferred to monkeys immunized against human erythrocytes. These results support the notion that schistosomes bind host antigens to their surfaces, and by doing so become essentially disguised, rendering them indistinguishable from the host as far as the immune recognition system is concerned. Challenge infections are thought not to be able to bind host antigen at the rate sufficient to elude the antibodies produced against the primary infection. Lethal antibodies have been demonstrated in immune monkey serum, killing schistosomula in culture within three to four days.

Several antibodies have been isolated from schistosome-infected human and animal serum (many of which are used as sero-diagnostic tools), but it seems unlikely that any are directly involved in acquired immunity. Thus, while schistosomiasis is characterized by an increase in the serum levels of IgG, much of this is non-specific. The general inability to demonstrate the passive transfer of resistance to infection with immune serum has led some to believe that cell-mediated reactions are responsible for acquired immunity in schistosomiasis. Recent experiments on passive transfer have proved more successful when rats and mice are used, but not so with monkeys. Additionally the occurrence of a lethal serum component, capable of killing schistosomula *in vitro*, strongly suggests that humoral antibodies are, indeed, required for acquired immunity. It may well be that both B-cells and T-cells, working in a cooperative manner, will prove to be the elements involved, but further evidence is required to establish this hypothesis.

There is good evidence that cell-mediated reactions are concerned with the immunopathology of schistosomiasis. The disease, as seen in the model system of *S. mansoni* in the laboratory mouse, is characterized by liver and spleen enlargement (hepatomegaly and splenomegaly), by portal hypertension (increased blood pressure in the hepatic portal vein) and by oesophageal varices. Egg production by the worms is a fundamental feature in disease production. Typical hepatosplenic schistosomiasis does not develop in either unisexual infections of mice or in normal, bisexual infections of mice given drugs to inhibit the production of eggs by the parasites. The formation of granulomata (nodules of inflamed tissue) around eggs has been studied, experimentally, by examining pulmonary inflammation around schistosome eggs that are administered to mice via the tail vein and which become lodged in the lungs. These studies implicate cell-mediated reactions in ganuloma formation. Immunosuppressant drugs will inhibit the production of granulomata around eggs, but there is no evidence to suggest that humoral antibodies are involved in the inflammatory responses to the eggs of either *S. mansoni* or *S. haematobium*.

Partially successful attempts have been made to isolate the egg antigens responsible for initiating granuloma formation, which is thought to be due to the secretion of soluble egg antigens through pores in the egg shell. It is probable that soluble egg antigen stimulates the production and secretion of lymphokine by T-lymphocytes, and this attracts mononuclear cells to the egg. Soluble egg antigen includes the enzymes, secreted by the eggs to aid their passage through the host tissues, on their way to the external environment. In contrast with *S. mansoni* and *S. haematobium*, granuloma formation in *S. japonicum* is mediated by antigen-antibody reactions.

Cestoda

Tapeworms live as adults in the lumen of vertebrate intestine and as larvae in the tissues of either invertebrates or vertebrates. The immunological status of lumen-dwelling cestodes tends to be somewhat different from that of the tissue stages, since the latter are in much more intimate with the host's defence mechanisms than the former.

Acquired immunity to tissue-dwelling larval cestodes operates either at the transaction phase, during which the oncosphere hatches in the intestine of the host and invades the mucosa, or during the actual tissue phase itself. Liberated oncospheres are susceptible to immunological attack in the host gut, as shown by studies on *Taenia pisiformis, Hymenolepis nana* and *Echinococcus granulosus*. In immune hosts, the oncospheres of these worms either fail to hatch, or hatch but fail to penetrate the gut wall. Immune serum is capable of damaging the oncospheres of *T. pisiformis, in vitro.* Rats harbouring the cysts of *T. taeniaeformis* in their livers are resistant to challenge infections for several months, and a similar pattern of immunity occurs in other larval taeniids, e.g. *T. hydatigena, T. ovis, T. saginata* and *T. pisiformis*. Passive transfer of immunity is accomplished by injecting whole immune serum or lymph node cells from immune hosts. Thymectomy reduces protective immunity of mice to *H. nana*, a tapeworm whose cysticercoids can develop in the intestinal mucosa of the definitive host. Protection against reinfection is restored by implantation of thymus tissue. In *H. nana* infections, passive protection is also associated with serum antibodies, particularly IgG. Studies on the hydatid stage of *Echinococcus granulosus* suggest that a complete protective immunity does not develop, since challenge infections are numerically less successful than the primary, immunizing infection, but they still survive.

Although contact with host tissues tends to be less intimate in many of the lumen-dwelling adult cestodes, they are, nevertheless immunogenic. Current interest has focussed on *Hymenolepis diminuta, H. microstoma, Raillietina cesticillus* and *Echinococcus granulosus. Hymenolepis diminuta* develops very differently when grown in laboratory rats and mice. In the mouse, this parasite grows normally for ten days, then growth ceases and destrobilation occurs. Transfer of destrobilated worms to rates restores normal growth. A challenge infection given to immune mice (infected and cleared of worms by anthelminthic treatment) dies more rapidly than a primary infection. Additional evidence for the immunological basis of these events derives from the use of immunosuppressants, e.g. cortisone, methotrexate and antilymphocyte serum prevent both the rejection response and destrobilation. The rat, on the other hand, is the natural host for *H. diminuta* and will support low density infections, i.e. containing between one and ten worms, indefinitely. Destrobilation and worm loss is a feature of higher density infections, but it is not clear whether this is a function of the immune response of the host or if it is due to the physiological alterations associated with crowding of worms in the intestine. Recent evidence supports the former proposition; challenge infections of 50 or 100 worms survive less well than primary infections, an effect alleviated by cortisone treatment. Antibodies of various classes are bound to the tapeworm surface in the rat gut and it may be that this binding interferes with the absorption of essential nutrients. *Hymenolepis microstomoa*, which normally develops in the bile duct or mice, fails to grow or survive in the rat, but it has not been established that this is due to the immune response of the host. In the mouse, *H. microstoma* infections are accompanied by increased levels of circulating antibodies (IgA, IgE and IgG), but these do not appear to play a protective role.

In general, the early developmental stages of cestodes are more immunogenic than mature worms, so that the intermediate vertebrate host develops a strong protective immunity. Acquired immunity to lumen-dwelling adult cestodes may play a controlling rather than an eliminating role. Larval cestodes that develop in invertebrate animals are exposed to the haemocytic defences, which are either phagocytic or encapsulating processes. There is some evidence that tapeworm larvae, living in the haemocoel of insects, can either evade this defence or disguise themselves so that the host haemocytes fail to identify them as foreign, e.g., *H. diminuta* in four bettles.

Nematoda

Interest in the immunity of mammals to nematode parasites is well established, and has focussed particularly on the pathogenic species of economic or medical importance to man, or on convenient laboratory systems. Nematodes differ from other parasites in the presentation to the host of a cuticular surface, which is only weakly immunogenic. Furthermore, it is unlikely that antibodies or cellular defences will be able to exert any lethal effect by attaching to the nematode cuticle, and only the anal, oral and glandular openings of the parasite will be vulnerable.

Most nematode infections, and indeed helminth infections in general, stimulate the host to produce reaginic antibodies (IgE), but these antibodies do not appear to be exclusively responsible for limiting an infection. The parasite antigens that stimulate IgE production are termed *allergens,* and these have been identified in some nematodes. There is a relationship between IgE production, eosinophilia and immuno-pathology in many nematode infections, and this may be mediated via anaphylaxis, involving the synthesis and secretion of histamine from mast cells at the site of invasion or habitation of the parasite. The phenomenon called self-cure is probably related to the release of histamine in the gut of hosts infected with lumen-dwelling nematodes, and this causes leakage of plasma proteins into the gut, bringing about expulsion of the worms. Self-cure occurs in *Haemonchus contortus* infections of lambs and *Nippostrongylus brasiliensis* infections of rats, and may prove to be a common expulsive mechanism for gut-dwelling nematodes.

Haemonchus contortus is an economically important parasite, causing pronounced anemia and often killing young lambs. Self-cure of haemonchosis was first described almost 50 years ago. It is characterized by a sudden decrease in the numbers of parasite eggs present in the faeces of lambs between 12 and 13 weeks post-infection. This immunity is precipitated by the ingestion of additional larval nematodes from the pasture and occurs in lambs over four months old. Vaccination against *Haemonchus* is partially effective using irradiated third-stage larvae, that can migrate but do not develop to maturity. Variation in response by different breeds of sheep diminishes the usefulness of this approach to artificial protection. It has been proposed that the antigens responsible for stimulating self-cure arise from the exsheathing fluid of the third moult.

Self-cure is typical of *Nippostrongylus brasiliensis* infections in the laboratory rat. In heavily, infected rats, parasite egg production rises between six and ten days post-infection and then decreases rapidly, accompanied by the expulsion of the majority of the worms. The rats are not highly resistant to a challenge infection. By contrast, worms given to rats as repeated low doses, called *trickle infections*, fail to stimulate any protective immunity and these worms will survive and produce eggs for extended periods, being described as "adapted" and capable of surviving transfer to an immune host. Self-cure in *Nippostrongylus* was originally thought to be due to two distinct events; antibodies inhibiting worm acetylcholinesterase activity and immunopathological effects upon the rat gut epithelium rendering it an unsuitable environment for the worms. The latter is related to IgE production, stimulating histamine release from mast cells. Other immunoglobulins, including IgA and IgG, are thought to play some role in self-cure. It is apparent, on examining the immunological response of the rat to heavy infections of *Nippostrongylus,* that several aspects of the immune system must operate coincidentally to effect worm expulsion. Both antibodies are sensitized lymphocytes are required for the rejection process, whereas IgE, mast cells and eosinophils are not regarded as non-essential components. Protective antibodies belong to the IgG class of immunoglobulins but not to IgA. The former have been shown to interfere with worm metabolism and cause damage to the worm's alimentary tract.

Trichinella spiralis, causing trichinosis in man, is a good model parasite for the study of acquired immunity using mice, rats and guinea pigs as laboratory hosts. Trichinosis in man is not accompanied by a noticeable increase is serum antibody levels, but repeated infections of experimental animals results in a greatly increased immunoglobulin (precipitin) titre. In rabbits, IgM levels rise early in an infection, followed later by elevated levels of IgG. IgA remains consistently high throughout an infection, and forms a localized mucosal response to the parasite. Trichinosis is characterized by the production of homocytotropic antibodics, which stimulate the release of pharmacologically active substances from the cells to which they bind. This type of immediate hypersensitivity can be detected by active and passive anaphylaxis. In the mouse, homocytotropic antibodies are of two types; one is a reaginic antibody (IgE) and other is IgG_1. Inflammatory, cell-mediated immunity occurs in experimental trichinosis and it is passively transferable with the peritoneal exudate cells of infected mice. Acquired immunity to *Trichinella* is developed following a primary infection, so that the larvae of the challenge infection rapidly become eliminated in the immune host gut due to local inflammatory responses. Anti-inflammatory agents, such as cortisone, markedly depress the elimination of a challenge infection, as does whole-body-irradiation of hosts. Both anti-lymphocyte and anti-thymocyte sera reduce the resistance of immune mice to a challenge infection, implicating T-cells in acquired immunity.

Although the exact nature of the mechanisms of acquired resistance in laboratory animals to *Trichinella* remains to be elucidated, current opinion is that the parasite is expelled from the gut of the host by specific inflammatory mechanisms. It is possible that non-specific processes that alter the environment of the gut, rendering it unsuitable for the worms, are also involved. In trichinosis, protective, immunity is directed essentially against invading larvae rather than established worms.

Ascaris suum in pigs and laboratory animals serves as a convenient model for human ascariasis (*A. lumbricoides*), the immunology of which is poorly understood. Immunity to *A. suum* possibly involves both humoral and cell-mediated elements, since passive transfer of immunity can sometimes be accomplished with anti-serum, but cannot be effected with lymphoid cells. Again, protective immunity acts against invading larvae, though self-cure has been recorded for adult ascarids in pigs. During an infection, IgM levels are the first to increase, followed by IgE, IgG and IgA. It is not known to which class of immunoglobulins the protective antibodies belong. In human infections with *Ascaris*, by contrast, the immunoglobulin titre does not always increase. Experimental ascariasis is typified by both inflammatory and cell-mediated responses directed, particularly, against larval worms in the tissues of the host. Granulomata are formed around parasite eggs, probably as a result of cell-mediated reactions. The third-stage larvae are the most active immunogenically and some characterization of parasite antigens has been accomplished by gel precipitation techniques.

Rather little is known about immunity to other nematodes that parasitize humans. *Ancylostoma caninum*, a dog hookworm, has been studied experimentally and is similar to *Nippostrongylus brasiliensis* in the rat, in the immune responses it invokes. Immunity to the human hookworms has not been examined in any detail. The immunology of human and experimental filarial infections has received some attention using *Litomosoides carinii* in cotton rats, *Dirofilaria immitis* in dogs are *Brugia pahangi* in cats as model systems. Infection in these systems produces increased IgE and IgG levels in the respective hosts. There is a significant degree of common antigenicity between filarial worms of different genera and other helminths.

Many nematodes like some protozoans and platyhelminths, are capable of living in a host which has responded to the parasite by humoral and cellular reactions. The avoidance of the immune responses is a characteristic feature of many parasitic infections, but little is known about the mechanisms involved. Additionally, there are unexpected interactions between nematodes and micro-organisms at the immunological level, e.g. mice infected with *Plasmodium berghei, Trypanosoma brucei*, on various babesias, lose their protective immune responses to *Trichuris muris* and *Trichinella spiralis*. No explanation of these events is, at present, forthcoming.

Immunization against parasitic diseases

The development of vaccines to combat any of the major parasitic diseases of man and domestic animals is of fundamental importance to the medical well-being and economy in much of the world, particularly the tropical and subtropical regions. Unfortunately, vaccine production has not yet met with any real success since so many parasites are capable of evading the immune responses of the host (antigenic variation in malaria and trypanosomiasis, and concomitant immunity in schistosomiasis are example of the problem). Nevertheless the search continues and it is a widely held view that a complete knowledge of the immune responses to each parasite is a prerequisite for the development of artificial immunization. As we have seen already in this chapter our depth of knowledge is far from complete for the immunology of any parasitic disease, but despite this, there have been encouraging results with immunization against some parasites.

A variety of different approaches has been adopted to produce immune hosts by artificial treatment and these include the use of attenuated parasites, homogenized parasite tissue, dead parasites, soluble parasite antigens, heterologous protection, controlled live infections and non-specific measures. Perhaps the most successful immunization yet achieved has come through the use of attenuated larvae to protect cattle against the nematode lungworm, *Dictyocaulus vivaparus*, and there is now a commercially available vaccine of irradiated larvae which confers a long-lasting immunity. Similar success has been achieve with *D. filaria* in sheep and *Ancylostoma caninum* in dogs. The use of larvae attenuated by irradiation has been found to offer protection in experimental infections of a number of nematode parasites, including *Brugia malayi, Dirofilaria, immitis, Capillaria obsignata, Haemonchus contortus, Litomosides carinii* and *Trichostrongylus colubriformis*, and in some schistosome species (*S. mattheei*). An effective live vaccine has been developed against *Babesia argentina* in Australia. The parasite is obtained from splenectomized calves and therefore has a lowered virulence and will not infect the tick vector. The use of killed parasites has not found much success and it has proved more efficacious to use soluble parasite antigens to induce immunity. Highly immunogenic substances have been isolated from a number of parasites, e.g. *Ascaris suum, Trichinella spiralis, Trichostrongylus colubriformis* and larvae of

Taenia saginata and *Taenia ovis*. One of the more promising recent development in the quest for suitable anti-parasite vaccines derives from tumour immunology. Following treatment with BCG (*Bacillus Calmette Guérin*) vaccine, non-specific immunity is developed for a wide range of parasites, e.g. experimental success has been achieved with rodent babesias, malarias, *Trypanosoma, cruzi, Leishamania tropica* and *L. donovani, Èchinococcus granulous* and *E. multilocularis*. It is remains to be seen whether the potential of non-specific immunization with BCG vaccine and other such substances can be realized as a practical approach in the field.

Pathogenesis of parasitic infections

Immunopathology

It is well recognized that the pathology due to infection with many species of parasite is related to the immunological of the host to the parasite. Immunopathology is classified into four subclasses, that together constitute the allergic mechanisms often responsible for the production of disease symptoms.

Type I mechanisms: This class of allergic reactions includes immunopathology due to immediate hypersensitivity, in which parasite antigens react with host tissues that have been passively sensitized by antibodies, thus stimulating the release of pharmacologically active substances, which can, among other effects, increase blood supply and stimulate smooth muscle contraction in the host. The general term *anaphylaxis* is applied to the effects of these substances. Antibodies capable of sensitizing tissues for anaphylaxis are called *homocytotropic* antibodies and belong to the IgE class; IgG can also be involved. Included among the pharmacologically active substances that are released are histamine, serotonin, "slow reacting substances of anaphylaxis" and peptides called kinins (e.g. bradykinin).

Immediate hypersensitivity reactions have been identified in a number of parasitic infections including *Leishmania braziliensis, Trichomonas foetus,* and *Trypanosoma cruzi*, as well as in many helminth infections, the latter involving both short-lived (IgG) and long-lived (IgE) homocytotropic antibodies. The prevalence of immediate hypersensitivity reactions in helminth infections provides a useful immuno-diagnostic tool for the identification of infected individuals, e.g. skin reactions, of the wheal and flare type, are often regarded as diagnostic for *Ascaris, Trichinella, Echinococcus, Schistosoma*, hookworms, filarial worms and many others. In ascariasis, pathological conditions, including bronchitis, bronchial asthma and pulmonary eosinophilia, are associated with type I reactions. In trichinosis, the reactions include fever, skin, rash, oedema and eosinophilia. Penetration of schistosome cercariae into sensitized mammalian skin is accompanied by urticaria ("nettle rash"), subcutaneous oedema, leucocytosis, eosinophilia and, occasionally bronchial asthma. Infection with filarial worms is characterized by eosinophilia while local swellings arise a result of worm migration.

Type I reactions, occurring as they do in individuals with a previous experience of an infection (necessary to cause sensitization), are probably not involved directly in protective immunity. They may, however, affect the growth and development of parasites and the ability of the parasite to migrate within the host.

Type II mechanisms This class is of reactions involves the, usually, lytic effects of antibodies against antigens that are fixed to cell or basement membranes. The antigen may be an integral part of the membrane or it may simply be adsorbed to it. Complement and lymphocytes are both involved in type II reactions. It is thought that haemolytic anemia in malaria results from this class of allergic reactions, as indeed may the anaemias associated with schistosomiasis, trypanosomiasis, leishmaniasis and babesiasis.

Type III mechanisms Immunopathology associated with this class of reactions is either inflammatory (arthus) or systemic (serum sickness), both of which result from the formation of antigen-antibody complexes. Complement is also involved in type III reactions. Arthus lesions are characterized by oedema, infiltration of neutrophils, necrosis of arteriole and venule walls, and thrombin containing platelets and leucocytes. Serum sickness, due, for example, to the treatment of humans with animal serum, is rare nowadays. Pathological processes associated with animal serum, is rare nowadays. Pathological processes associated with this class of events include glomerulonephritis in malaria and schistsomiasis, and the nephrotic lesions in some malarias.

Type IV mechanisms. These reactions include cell-mediated responses or delayed hypersensitivity. Delayed skin reactions have been demonstrated in human trypanosomiasis, leishmaniasis, toxoplasmosis, trichomoniasis, ascariasis, toxocariasis and schistosomiasis mansoni. In laboratory animals, many parasites induce delayed hypersensitivity reactions. The sores developed in cutaneous leishmaniasis (*Leishmania donovani*) and hepatosplenic disease (due to the formation of granulomata around tissue-bound eggs) in schistosomiasis are classical examples of type IV reactions.

General pathology

Apart from the immunopathology that we have discussed briefly above, many parasites cause damage to their hosts, resulting in the production of disease symptoms. Protozoan parasites multiply within the final host, whereas helminths do not. Therefore, any density-dependent effects of the former in the process of pathogenesis will depend upon the rare of division or reproduction and in the latter on the rate of invasion, longevity and size of the parasites. These factors are of particular importance in pathogenesis. It must be made clear, however, that a great many parasites do not, under natural conditions, appear to cause their hosts any ill effects, i.e. an undetermined number of wild animals may be parasitized without ever showing outward or overt physiological signs of infection. It is rare, in the author's experience, to find unparasitized fishes, amphibians, birds or small mammals, yet these animals are living, until capture, apparently in harmony with their parasite fauna and it seems probable that it is only under certain circumstances that disease conditions arise from an infection. Such conditions would include the intensive monoculture of fishes, domestic fowl or cattle. Hence the view that a parasite is, by definition, a disease-causing organism is patently not true, and we must exercise care when defining the pathogenic nature of parasites, since only a relatively small number of species are disease-causing in natural environments. It has been an understandable tendency among parasitologists to focus attention on the pathogens and relegate the innocuous species to a lower order of significance.

Nevertheless man, domestic and wild animals can be seriously affected by their parasite fauna. The ways in which pathology can arise are varied and are briefly outlined below.

Mechanical injury. There are a number of different ways in which parasites inflict mechanical injury to their hosts, e.g. during penetration and migration through the tissues, during feeding, through the action of the organs of attachment and cellular injury wrought by the presence of intracellular parasites within host cells. Many of the skin-invading digeneans produce localized damage at the site of entry, along with allergic reactions. Mechanical damage due to invasion and subsequent migration is often related to the number of invading larvae. The feeding habits of helminths, particularly nematodes, are responsible for considerable local damage to host epithelia, e.g. hookworms bite into the intestinal mucosa to obtain a blood and tissue meal and in doing so may leave open wounds with resultant bleeding and loss of tissue fluids. Mechanical damage attributable to parasite attachment organs is usually associated with the digeneans and cestodes. The unarmed tapeworm scolex, lacking hooks (e.g. Pseudophyllidea) may or may not inflict damage to the host mucosal tissue, whereas the presence of hooks on the scolex tends to be more injurious. In general, attachment of tapeworms to the intestinal mucosa is associated with inflammatory lesions and characteristic fibrosis. Digeneans, such as the liver flukes, produce pathological conditions by the mechanical disruption of host tissue, attributable to the body spines, attachment organs and the physical presence of the parasite in the narrow ducts of the liver. Anemia, resulting from the ingestion of host red blood cells (e.g. *Schistosoma mansoni, Fasciola hepatica* and *Haemoncus contortus*) or from haemolysis (malaria) is not a common form of mechanical injury. Heavy infections of duct-dwelling parasites may cause occlusion of the ducts occupied, with its associated pathology, e.g. *Fasciola* in the bile duct, malaria in hepatic venules, and cestodes in the intestine. The severity of mechanical injury due to parasitic infection will usually be directly related to the number of parasites present, the size and behaviour of the parasite, or its multiplication rate if it is a protozoan.

Toxic effects. The most potent toxins known to man are produced by bacteria and include tetanus, diphtheria and botulism. By contrast, the majority of parasites do not release toxic substances into the blood and tissues of their hosts. Although toxins are thought to the produced by *Ascaris*, some digeneans (*Schistosoma*), some cestodes (*Moniezia*) and several protozoans (*Plasmodium* and *Sarcocystis*), there is no unequivocal evidence to support this notion. In malaria, for example, a toxin is postulated to inhibit mitochondrial respiration and oxidative phosphorylation in host liver cells, but this substance has not been isolated or identified.

Effects on host cell growth. Several parasites are responsible for inducing altercations in host cell and tissue growth patterns. Such effects are most commonly hyperplastic, i.e. cell growth without cell proliferation, or neoplastic, i.e. new cell growth, as in tumour production. Hyperplasia accompanies a number of infections, including *Eimeria, Schistosoma, haematobium, Paragonimus westermani, Fasciola, hepatica* and many adult tapeworm infections. Neoplasia, or tumour development, has been described for infections with *Taenia taeniaformis* in rodents, and *Schistosoma haematobium* in man.

Effects on the reproductive system of the host. There are a small number of well documented cases in which the host's reproductive system is altered, or its tissue destroyed, by the presence of a parasitic infection. The general term for this phenomenon is *parasitic castration.* The best known examples are *Sacculina parasitica* (a cirripede crustacean) that causes sex reversal in male crabs and sterility in females, and *Ligula intestinalis* (a pseudophyllidean cestode plerocercoid) which causes gonad regression and sterility in cyprinid fishes. In neither of these examples is the mechanism of castration understood. It seems probable that the parasites release compounds that mimic the activity of the host sex hormones, but extensive investigation has proved negative to date. The larvae of some digeneans are also known to castrate their molluscan intermediate hosts, e.g. *Himasthla secunda, Cercaria emaculens,* and *Cercaria lophocerca* in *Littorina littorea; Cercaria milfordensis* in *Mytilus edulis;* and *Leucochloridium paradoxum* in *Succinea putris*. The mechanisms involved here are also known.

Metabolic effects. Many parasites may bring about altercations in the various metabolic pathways of their hosts, but few cases have been documented. Among the best known are the pernicious anemia and vitamin B_{12} deficiency characteristic of some human infections with the cestode *Diphyllobothrium latum,* and the growth-stimulating effects of larval *Spirometra mansonoides* (Pseudophyllidea) on laboratory rats and mice. In the latter case, the growth effects is due to the parasite releasing a substance (known as sparganum growth factor) that mimics the activities of mammalian growth hormone. This has questionable significance, however, since the larval parasite, the sparganum or plerocercoid, normally infects water snakes and is grown in rodents as a laboratory model only. Similar growth-stimulating effects are found in natural infections of molluscs with larval digeneans, and the phenomenon of molluscan parasitic gigantism is well-known. Molluscs parasitized by some larval digeneans may show the reverse effect, i.e. stunting. Explanations for these interactions are not yet forthcoming.

Chapter—17

Control of Nematodes

The main objective of the control of phytonematodes is to improve the growth, quality and yield of crop plants. Phytonematode control is achieved mainly through management of their populations because even with the best anti-nematode agricultural chemicals hundred per cent elimination of their numbers is not possible and the remaining population multiplies rapidly to injurious levels in the presence of a suitable host. Therefore, their population management involves attempts to keep the populations low at a safe level where they do not cause economic loss. This can be achieved by a single method or by a combination (integration) of several methods. The latter approach is always preferable. Whatever the nematode control programme is used, it must be profitable to the grower. It is important to evaluate the cost:benefit ratio of each or combination of methods to make the approach effective.

Regulatory Measures

Regulatory control involves certain measure, self-imposed or enforced by state law, which prevent entry of any serious pest or disease agent into a geographic area, to which it is exotic, from outside that area, within or outside the country. These regulatory measures fall under the category of "quarantine regulations." A country undergoes certain expenses and inconveniences to enforce these regulations so that it can save on continued expenses and losses that could occur if the unwanted crop enemy got itself established. Some quarantine measures have been highly successful while others have not.

The term 'quarantine' is derived from the latin word *quarantum* meaning forty. It was originally applied to the period of detention of ships arriving from countries where epidemic fevers occurred. The passengers and crew of the ship were compelled to remain isolated on board giving enough time to permit latent cases of disease to develop. When compared to human quarantine plant quarantine is of recent origin. The first plant quarantine legislation in the world was enacted in the year 1877 in Indonesia. India adopted plant quarantine measure in 1914 by introducing the Destructive Insects and Pests Act.

(i) *Necessity of Regulatory Measures:* Easy transportation of food materials (plants, seeds, etc.) and necessity of such materials in many areas of the world has created the problem of easy transport and entry of plant parasites also. In many countries, including India, intensive developmental research work to increase crop productivity is going on. This involves import of germ-plasm of important crops, especially from areas of their origin. In many instances the parasites of such crops are also established in these areas.

Associated with these exotic introductions has been the danger of entry of some of the most serious plant parasitic nematodes so far not known in India. Interception of several destructive nematodes such as *Heterodera schachtii, H. goettingiana* and *Ditylenchus dipsaci* from imported planting material were recently reported from this country. These nematodes are not known to occur in India [70]. The presence of sugar-beet cyst nematodes (*H. schachtii*) in inert seen material such as dried root bits, plant debris, and in accompanying soil clods suggests that there is every possibility of these cyst forming nematodes gaining entry into the country along with imported seed, provided the crop is grown widely. This strongly emphasizes the importance of examining introduced seen or plant material for nematode infestation in order to prevent or at least restrict the introduction of serious nematode pests into the country. Thus, plant quarantine should function as the first line of defence in plant protection programmes. The modern concept of plant quarantine is that it should function as a filter but not as a barrier.

There are some essential requirements for implementations of quarantine measures. The regulations should clearly specify methods and procedures for inspection, treatments and certification, etc. The quarantine stations should be equipped adequately to carry out the duties. Treatment schedules should be clear so that the exporting country and the inspector at the point of entry are able to follow them with reasonable resources.

(ii) *Quarantine Methods and Their Application*

(a) Pre-entry precautions are to be ensured by the exporting country as well as by the importing country. The latter should avoid, as far as possible, importing plant materials from countries where dangerous nematodes exist but are not existing in the importing country. On their part, the exporting country should ensure that the material being exported has been produced in properly inspected fields and proper laboratory testing and treatments have been provided before export was made.

(b) *Post-entry control*: These involve inspection of the material at the first point of entry to detect and refuse delivery of anything which may show infestation. The inspection may involve laboratory testing, treatment if necessary and also quarantine of the material, especially plants, for specific duration in isolated area.

The necessity for mandatory inspection at the port of entry, in spite of the fact that consignments are accompanied by phytosanitary certificates from the exporting country, can be seen by the following example:

About 112 metric tons of potato tubers were imported from Scotland, Ireland and West Germany through Madras Seaport in 1965 and 1968. The consignments were accompanied by the mandatory phytosanitary certificates and were additionally certified to be free from soil as well as potato golden nematode cysts. The examination of about 80 to 90 kg of soil accompanying the consignment, although import of soil is prohibited, yielded 16 cysts of the golden nematode and one cyst of sugar-beet cyst nematode. This necessitated destruction of the entire consignment by dumping into the sea.

(iii) *International Cooperation in Plant quarantine*: Full benefit of plant quarantine can be achieved only when countries situated in the same geographic area, having similar agricultural interests, take common action in preventing the introduction of pests and diseases. Realizing the need for cooperation in enforcing quarantine laws, governments coordinate their activities under the guidance of F.A.O. Eight international plant quarantine commissions were set up both and without F.A.O. participation. The main objectives of these commissions were to prevent the introduction and spread of destructive pests and diseases in their respective areas through appropriate plant protection measures with mutual cooperation among countries in their zone. Asia and Pacific Protection Commission is one of these commissions. There are 22 countries, including India, in this commission. F.A.O. plays a major role in organizing and supervizing as well as disseminating information regarding pest and disease incidences, regulations regarding prevention of introduction etc.

(iv) *Administration and Organization of Plant Protection in India*: In India the first law, an act called Destructive Insects and Pests Act, 1914 (Act 11 of 1914), was passed on third February 1914. Under this act various notifications have been issued from time to time prohibiting or restricting the import of certain plants, plant materials, insects and fungi, by air or sea. These regulations are called Foreign Quarantine Regulations. Regulations have been imposed prohibiting transport by rail, road, sample post etc. of various types of plants and propagating materials from infested areas in a state or Union Territory. These are called Domestic Quarantine Regulations and draw their authority from Agricultural Pests and Disease Acts of the respective States.

The effectiveness of quarantines depends on complete information regarding the biology of the nematode (life history, dispersal methods, environmental relations, and host range), present distribution in the concerned area, probability of introduction in the absence of quarantine and probability of establishment and relative importance. There should be periodic review of the measures being followed and if and when necessary, modifications should be made in the existing regulations. Objective evaluation of effectiveness of quarantine measures is not possible since there is no check for comparison. However, the long-term benefits compensate for inconveniences and cost incurred by the government and the growers Under special circumstances the government may pay compensation to the growers for the financial loss suffered by them in the enforcement of the regulations.

Cultural and Land Management Practices

The aim of these practices, based on the principle of attrition, is to reduce nematode populations to a low level before a suitable crop is grown satisfactorily. Usually the cultural operations are carried out in conjunction with normal crop production practices, at little or no extra cost, and are aimed at a particular nematode pest. The art of weakening the parasite to the point of exhaustion by constant harassment, breaking down or wearing down from repeated attacks or constant diminution of the parasite is known as attrition and is an important principle of plant disease control. The nematodes are gradually weakened to the point that they are incapable of causing invasion and multiplying or spreading.

Plant parasitic nematodes are obligate parasites and do not feed, reproduce or multiply in the absence of their host plants. When the infested soils are kept free from host plants, reproduction and multiplication of nematodes is limited. There are several other practices which causes premature death of nematodes. If these practices are followed for several seasons the populations are starved due to lack of food and killed. These methods are practices can be briefly listed as below:

(i) Land management practices such as flooding, fallowing, deep summer ploughing, and drying of soil.

(ii) Crop rotations.

(iii) Growing antagonistic crops and cover crops.

(iv) Prevention of spread and multiplication by destruction of host weeds, selection of nematode free, healthy planting material and field sanitation.

(1) *Crop Rotation*: Crop rotation, sequence or arrangement of crops grown in a particular, is an ancient agronomic practice used by farmers throughout the world to avoid crop failure due to soil problems. It continues to be one of the most

problem component of any cropping system to combat the problem of soil-borne plant pathogens including nematodes. Mention has been made earlier that more than 100 years ago Julius Kuehn had recommended crop rotation as the best method of control of the sugar-beet cyst nematode, *Heterodera schachtii*.

Extensive work on the effects of crop rotations on phytonematodes has been done in other countries but in India such studies are limited. The value of crop rotations in the control of root knot nematodes, *Meloidogyne* spp. cyst nematodes, *Heterodera* spp. and migratory ecto and endo-parasitic nematodes is well established. There are specific beneficial crops which, when they follow a susceptible crop, eliminate or reduce the population of phytonematodes; for example, barley following cotton or groundnut following tomato in *M. incognita* infested soil, wheat following rice in *Pratylenchus indicus* infested soil, or beet following cereals in *p. penetrans* infested soil.

Control of phytonematodes by crop rotation is based on the fact that being obligate parasites these nematodes can feed and multiply on their host plants. They must have a suitable living host to complete their life cycle. Better the host for feeding larger will be the number of generations of the nematodes in one season. By introducing non-host crop plants or resistant/immune cultivars between the susceptible crops or their cultivars, there is significant decrease in the number of generations and, finally especially in long rotations, the population of the nematode declines to insignificant or innocuous level.

Field trails carried out by Chawla and Prasad and Mukhopadhyay and Prasad have shown that when a fallow land, with insignificant numbers of phytonematodes, was put under one crop (wheat) the total numbers of Tylenchids (including *Tylenchorhynchus, Hoplolaimus, Pratylenchus,* and *Helicotylenchus*) almost doubled. Taking fallow population as a unit, the Tylenchid population build up in one crop (wheat), double cropping (maize-wheat) triple, cropping (mung bean-maize-wheat), and four cropping (mung bean-maize-potato-wheat or mung bean-maize-*toria*-wheat) was 16.5, 27.1, 29.8 and 10.2, respectively. In Punjab the field trials conducted by Sharma and associated to determine the effect of *kharif* cropping sequences (groundnut, sesamum, soybean, tomato) and four rabi corps (wheat, mustard, chick-pea, tomato) in different combinations on the population of *M. incognita* have shown substantial increase in populations in tomato monoculture and in sequences having tomato in combination with chick-pea and soybean. The population was reduced considerably under groundnut-mustard rotations. Monoculture of tomato, brinjal, okra, chilli, sponge gourd or rotations in all combinations with these crops brought about manifold increase in the population of *M. incognita*. A decrease in the nematode population was obtained when marigold, spinach and bottle gourd were grown or when the field was left fallow. In the U.S.A. the barley root knot nematode, *M. nassi* was effectively controlled by growing a non-host, potato, for two years.

For controlling the potato cyst nematode there is a law in the Netherlands prohibiting growing growing potatoes or other crops susceptible to the nematode more than once in three years. A four to five year rotation is usually sufficient to ensure that the population left after harvest of susceptible potato has fallen to a safe level before the next susceptible crop is grown. Populations decline about 30 per cent annually in the absence of the host crop. If long rotations of five to seven years are followed loss in yield can be minimized.

The cereal cyst nematode, *Heterodera avenae*, is highly host specific, being restricted to wheat and barley among the common cereals in India. There is, therefore, good scope for crop rotation as a method of control for this nematode. In the presence of the host crop the nematode multiples five to six times. When a non-host is sown in the infested soil there is about 60 per cent decline in nematode populations. The average decrease in population after lucerne has been found to range from 45 to 65 per cent. In the U.K., the damage by this nematode to oats, the preferred host there, is considerably reduced when oats are avoided for two years in the infested soil. Thus, if non-host crops are grown for two to three years there is good possibility of controlling this pest of wheat and barley. Crop rotations for four-year duration with such non-hosts as chick-pea, mustard and carrot have shown 50 to 75 per cent decline in *H. avenae*. Mustards have special advantage being a good source of cash returns and also because the mustard roots have adverse effect on cyst nematodes.

The mechanism of control phytonematodes with crop rotation has not been studied in detail. The general belief is the populations of these obligate parasites in the absence of the host and reduced because they are unable to multiply and produce new generations. However, there could be other possibilities such as changes in soil physico-chemical and biotic environment caused by different types of crops, chemically different crop residues of the rotation crops releasing decomposition products harmful for the nematodes, and better nutrition and root growth of the plants. There are many reports of the decline being attributed to parasites and predators of the phytonematodes in specific rotations. This biological control is discussed later. Roots of some plants such as marigold and some cucurbits produce toxic substances harmful to nematodes as discussed later.

In spite of the fact that crop rotation is without doubt an effective method of disease control, it has certain limitations, especially in countries where land is in short supply and choice of crops is limited. A nematode having a wide host range will not favourably respond to crop rotation within the narrow choice of crops unless the whole cropping system is changed. Root knot nematodes of vegetable crops

(*M. javanica, M. incognita*) is an example. These nematodes have very wide host range and attack almost every vegetable crop in India. Vegetable-cereal rotation effectively reduces the incidence of root knot. However, vegetable cultivation in India is a specialized cropping system and normally the vegetable growers will not put the land under cereals. In situations where, for economic reasons there are difficulties in changing the cropping system, cultivar rotations, instead of crop rotation can be followed. In cultivar rotation, the crop remains the same but rotation includes resistant or immune cultivars of the crop(s) in place of non-host crops. In such rotations susceptible cultivars may be included occasionally to prevent emergence of new races or biotypes of the nematodes. Even in the case of host specific phytonematodes such as *H. avenae,* farmers may be reluctant to practice long rotations of three to four years in areas where this nematode is widespread because the choice of alternative crops is limited. It is not only choosing a crop but also the consideration of cash return from such crops that matters. Farmers always prefer to follow prefer to follow a rotation in which each crop yields a reasonably good cash return.

The success of crop rotation, thus depends on a variety of factors which such as selection of crops in the sequence, relative susceptibility of the crops, identity of the nematode(s) predominantly present, races of the nematode present, host range, initial inoculum density, length of life cycle and other aspects of population dynamics. The length of rotation is related to the magnitude of initial population level and to the rate of population decline during the presence of the rotation crops.

(2) *Fallows*: Fallowing is the practice of keeping the land free of all vegetation for a specified period by occasional ploughing of soil, especially during not summer months. Fallowing may have to be adopted out of sheer necessity, such as because of

limited water supply or failure of a crop soon after planting and no time left for replanting or planting of a substitute crop. It can also be adopted by choice in order to get specific benefits. In fallowing out of the necessity the field should be kept free from weeds and repeated turning of the soil should be done to expose the nematodes to adverse elements of weather. Fallowing by choice is largely an economic consideration, taking into account revenue from crops that could be grown during the fallow period, alternative methods of reduced nematode populations in the soil and benefits of the fallow on the succeeding crops. Even where fallowing is known to give definite advantages in disease control the farmers have not accepted it because of economic consideration.

Fallows are of different types such as dry fallows, wet fallow and flood fallow. Dry fallowing for nematode control is generally restricted to extensive farming where other methods of disinfestation of soil are not economical. It is combined with specific crop rotations, soil amendments, and fertilizer and tillage practices. Clean fallow deprives nematodes of their food source and exposes those in surface soil layers to extremes of weather, especially heat and desiccation. While most non-cyst forming nematodes are considerably reduced by this treatment, even cyst forming nematodes such as *H. avenae* which are highly susceptible to desiccation are also adversely affected. Root knot nematodes may be killed by drying soil for two days in 2.5 cm depth, for four days in 5 cm depth and for 32 days in 25 cm depth. The nematodes present in deeper layers can also be killed by giving frequent turning of soil. Considerable reduction in the population of cereal cyst nematode can be achieved by ploughing the fields during summer. It is, therefore, a good practice to keep the fields free from vegetation during hot summer months (May-June), give one early irrigation and then two to three ploughings to turn and expose the soil to hot sun and desiccation.

There are some objections to the practice of including a fallow in crop rotation or in a cropping system. From an economic viewpoint, in countries where due to pressure on land maximum crop production per unit area is the aim, keeping the land free from crops may not be acceptable. However, during May-June when water scarcity makes crop cultivation expensive fallowing can be practised to advantage. From an agronomic viewpoint, frequent turnings of the soil without vegetation cover causes depletion of nutrients. This may necessitate use of cover crops.

(3) *Cover Crops, Trap Crops, Antagonistic or Enemy Crops*: Cover crops usually have not much economic value and are grown to provide vegetation cover to the land and for fodder etc. Such crops are highly resistant to phytonematodes. Growing cover crops in nematode infested soil has been advocated in some countries to reduce nematode populations. Cover crops can be harvested as fodder crop or used as green manure. *Crotalaria* spp. have been used for green manure and reduce root knot nematode damage.

Trap crops are plant highly susceptible to a nematode parasite of the major crop and are planted so as to get infested by the nematode at a particular stage of growth of the plant and both the crop and nematode can complete its life cycle. This reduces the primary inoculum for the susceptible major crop that follows. When trap crops are grown in root knot or cyst nematode infested soil, the second stage juveniles enter the plants but before they reach adulthood to lay eggs the plants are destroyed. Oats have been used in England as a trap crop for cereal cyst nematode. The crop is harvested for fodder before the nematode produces the next generation of cysts. *Solanum nigrum* stimulates hatching of eggs of *G. rostochiensis*. The second stage larvae enter the roots as on normal host but cyst development is poor.

Enemy or antagonistic plants produce some toxic compound that destroys the nematodes in soil. Some grasses, varieties of mustard, marigold, species of *Crotalaria,* and asparagus have been listed as enemy plants. The relationship between marigold (*Tagetes* species) and a number of root infecting nematodes was studied by Oostenbrink and his associates. They found that by growing marigold, populations of *Pratylencus* could be reduced by 90 per cent. Later, it was found that these plants reduced populations of *Paratylenchus, Tylenchorhynchus,* and *Rotylenchus* also. The effect is due to root exudates of marigold that contain terthienyl compounds toxic to nematode. Prolonged stand of the plant in the soil is essential before sufficient concentration of the toxic compound is reached in the soil. Since marigold is not a commercial crop in most places its cultivation for at least for months in the field may not be feasible.

The effects of potato root exudates, essential for hatching of cyst nematode, are neutralized by growing mustards along with potato. This prevents the larval release from cysts. Later, it was found that mustard oils contain allyl isothiocyanate toxic to nematodes. Ellenby reported that mustard oils reduced the severity of cyst namatode infestation and increased the yields of potato. Populations of *Trichodorus christiei* are not supported by *Asparagus officinalis*. The fleshy roots of the plant contain a nematicidal agent, asparaguisic acid being one component. The compound also suppresses *Pratylenchus curvatus*. Root exudates of *Brassica nigra* and *B. hirla* neutralize the effects of potato root exudates on cysts. The root exudates of certain grasses contain pyrocatechol that is toxic to nematodes. While the effectiveness of these antagonistic plants against several phytonematodes has been demonstrated in laboratory and glasshouse tests their practical utility on a field scale is not certain.

(4) *Flooding*: Submergence of land under water can be a part of fallow (food fallow) in which the crop-free land is submerged in water for a specific period or it can be a part of crop culture such as wet rice culture in which the soil is flooded with water for most of the duration of the crop. While the former practice can be followed where the availability of water is not a problem and the land is well levelled, the latter is a normal practice with the crop grown during rainy season when only occasional watering of the field may be required.

Flooding of the land can be employed as a means of reducing certain phytonematode populations. It was first reported for control of root knot nematodes of Florida, the U.S.A. by Watson. Many reports have appeared on the effectiveness of this method in nematode control. Periods from seven to 40 days are recommended to get decreased infestation of root knot nematodes. Compared to continues submergence, alternate flooding and drying increases the destruction of root knot nematodes. Thomas and Stoner had reported reduction of root knot infestation in vegetable crops taken after wet land rice culture.

The mechanisms(s) involved in the control of phytonematodes by flooding are not clearly understood. It is presumed that flooding eliminates the desired host plants and the nematodes die from starvation. Exhaustion of second stage juveniles at the expense of their body reserves of energy is hastened by increased mobility in water. High soil moisture hastens the decay of nematode galls exposing the eggs and larvae to adverse conditions. Although there is lack of aeration and free oxygen in flooded soils free oxygen is not a limiting factor. Swarup and Pillai have reported that there is no marked difference in the total number of juveniles hatched when eggs were kept submerged or under free oxygen conditions. Experimental evidence presented by Hollis and Johnson had shown that non-sterile flooded soils exhibited more reduction in nematode populations than sterilized flooded soils. This points to the role of microbial activity in the anaerobic conditions created by flooding or submergence. Anaerobic bacteria such as *Clostridium* sp. are known to release highly nematotoxic substances during decomposition of organic matter in flooded soils. Nematicidal concentrations of fatty acids are also formed within four days in corn meal amended saturated soils. The role of hydrogen sulphide in wet rice soils has also been pointed out by Rodriqucz–Kabana and associates and Rhoades. In laboratory tests *Tylenchorhychus martini* was killed by hydrogen sulphide in five to ten days at concentrations similar to those found in flooded soils.

(5) *Organic Amendments and Fertilizers*: The value of decomposition of organic matter in soil in the reduction of nematode damage was first demonstrated by Linford and his associates in 1938. Since then a large number of research reports have confirmed that addition of a variety of decomposable organic matter to nematode infested soil results in the reduction of population of several phytonematode species.

(i) *Organic Manures*: A brief review of the effects of manuring on nematode populations in soil was given by Oostenbrink. Addition of farm-yard manure and compost reduces the infestation level of *G. rostochiensis* in potato roots. These manures hampered the development of the nematode in plant roots where they developed more slowly than in plants grown in unmanured soil. There seemed to be some type of biochemical resistance imparted by the manure to the plants. Deep ploughing during hot summer months followed by application of farm-yerd manure at planting spots increases the yield of tomato in root knot infested soil. Reduction in the citrus nematode populations by application of steer manure and chicken manure has also been reported. Organic nitrogen in sewage sludge has been found more effective then ammonium nitrate for reducing populations of *Belonolaimus longicaudatus* and *Hypsoperine graminis* in turf grass.

(ii) *Green Manuring*: Green manuring is a conventional practice in which green plants are ploughed into the soil and allowed to decompose to provide nutrients to the crop grown subsequently. The decomposition of organic matter brings about important physical, chemical and biotic changes in the soil. Linford and associates had demonstrated that the incidence of root knot could be significantly reduced by incorporating chopped pineapple leaves into the soil. In India, chopped leaves of *Pongamia glabra, Crotalaria juncea, Azadirachta indica, Melia azadirachta* and *Sesbania aculeata* to infested soil have reduced root knot incidence.

(iii) *Cellulosic Soil Amendments*: Excellent success in control of phytonematodes has been achieved with the use of cellulose and sawdust. Miller and Edgington reported of chopped paper and white pine sawdust at the rate of 2 g per 500 g soil. Similar control of *Heterodera tabacum* with a 1 per cent amendment of paper and white pine sawdust was observed by Miller and Wihrheim. Addition of fertilizers accentuated the effect of cellulose. Singh and Sitaramaiah, in extensive studies, have shown that sawdust amendment of soil supplemented with nitrogen in the form of urea gave a highly significant control of root knot caused by *M. javanica* in okra and tomato. Similar reduction in the activity of citrus nematodes by soil amendment with cellulose is also reported. Chitin has been found a very effective cellulose material against many nematodes.

(iv) *Mature or Dry Crop Residues*: Application of mature crop residues such as straw is not a conventional manuring practice. However, substantial quantities of dry crop debris are annually ploughed into the soil as a regular agronomic practice. The amount of such residues depends on the type of crops included in the rotation being followed. The effect of decomposition of such residues in suppression of nematodes in soil has been studied by many. Johnson found that every type of crop residue introduced into the plots infested with root knot nematodes reduced the number of galls per plant. Maximum reduction was obtained when the residue was allowed to decompose for 30 days before the host was planted. Later, Johnson and associates confirmed that when mature dry residues of lespedeza, alfalfa, oats and flax were chopped to about 3 mm particles and incorporated into field plots infested with *M. incognita* the incidence of root knot of tomato was significantly reduced. A 10-ton/acre dose was more effective than 5-ton/acre dose. Longer durations allowed for decomposition gave better results than shorter durations. Subsequently, Johnson found that application of 1 per cent (w/w) oat straw with a fertilizer mixture was effective in controlling *M. incognita*. Similar observations were reported by Miller and Wihrheim. Morgan and Collins got a very high degree of control of *Pratylenchus penetrans*, as good as D-D fumigation, by application of rye straw into soil in summer fallow. Other such residues are cotton waste, cocoa bean hull, timothy hay, lucerne hay, rice straw etc. which have suppressed such phytonematodes as *Tylenchulus semipenetrans*, and *Belonolaimus longicaudatus*.

(v) *Meals and Oils-cakes*: Fertilization of soil with meals and oils-cakes has been a practice in parts of the world. In Florida farmers believed that the application of tungnut meal to soil reduced losses from root knot. Lear reported significant in populations of phytonematodes when large quantities of castor bean pomace were mixed with soil. Populations of the lesion nematode are reduced by soil amendment with soybean and corn meal. However, soybean meal has no effect in sterilized soil. Use of oil cakes of margosa, castor bean, and many other oil bearing plants were extensively studied by Singh and Sitaramaiah who found significant reduction of root knot of tomato and okra in field trials. Earlier, Singh had, in 1964, demonstrated in pot experiments almost complete suppression of root galls of tomato (*M. javanica*) with oil cake of *Pongamia glabra* (*karanj*). The results of use of sawdust and oil cakes in nematode control was subsequently confirmed in studies conducted in India and many other countries.

(vi) *Oils and Sugars*: Vegetable oils and carbohydrates such as sugars and sugar related products like molasses are not normal soil amendment materials but have been shown to reduce phytonematode populations in soil. Adverse effect of black mustard oil that contains allyl isothiocyanate on the potato cyst nematode was reported by Ellenby as early as 1945. In his field trials 0.1 cwt/acre oil increased potato yields by 100 per cent and decreased the emergence of larvae from cysts. Spraying of 1.5 per cent water dispersible fish oil on soil surface is reported to reduce populations of *Radopholus similis* in citrus roots. Corn oil, cotton seed oil, groundnut oil and soybean oil when applied to soil at the rate of 1 per cent (w/w) reduce populations of lesion nematodes, corn oil being the most effective.

Osmotic destruction of nematodes in the presence of high concentrations of sugar in soil and in solutions has been reported by many workers. As little as 1000 ppm of dextrose is nematicidal. Sugarcane bagasse and molasses with 0.2 per cent sucrose content give 50 per cent larval mortality in *M. javanica*. Most workers have used such substances in very large dosages which are not practicable on the field scale.

Mechanisms Involved in Nematode Suppression by Organic Amendments: Organic amendments using sawdust, dry or green crop residues, oil-cakes and other easily decomposable materials have proved highly effective against nematodes in subtropical and tropical regions. Intense microbial activity and quick decomposition of the added materials is hastened in tropical climate. The effects of such soil amendments vary from soil to soil and region to region. The mechanisms involved are not clear and require detailed studies. The available literature suggests the following mechanisms.

(1) *Parasitism and Predation of Phytonematodes by Nematophagous Fungi, Bacteria and soil Microfauna*: The intense microbial activity during decomposition of organic matter improves the chances of increase in numbers of microflora and fauna that can destroy phytonematodes. This is a form of have been reviewed by Mankau, Sayre, Singh and Singh and Sitaramaiah. Till the sixties the attempts on biological control of nematodes were mostly observational records of parasites and predators attacking nematodes. After 1960 detailed studies on the nature and ecology of these enemies of nematodes were intensified.

1. The Nematode Destroying Fungi

Most investigators believe that nematode destroying fungi or nematophagous fungi are potentially most useful biocontrol agents. These fungi are found in all types of environments and soils. Their ecological study may provide means to encourage development of their populations to the level where they can become successful destroyers of the nematodes. Most of the nematophagous fungi belong to the order Moniliales of Fungi-Imperfecti but some are from Zoopagales in Phycomycetes. These fungi can be divided into two groups: (a) the trap forming fungi and (b) the non-trap forming endozoic parasites.

The Nematode Trapping Fungi: These fungi develop special structures from their hyphal branches in the presence of nematodes to trap or catch the latter. Most of them are ordinary saprophytes than can be cultured on synthetic media where they do not form traps. But when nematodes are added to the medium trap formation is initiated. Death of the nematode occurs due to mechanical damage by the trap and also due to release of certain nematotoxic secretions. Nutritional conditions are the primary factors affecting their growth and trap formation. The active morphogenic substance believed to induce trap formation of *Arthrobotrys* is known as "nemin". The predaceous activity of nematode trapping fungi is often independent of nematode populations but dependent on the release of certain soluble carbohydrates.

The trapping devices may be grouped as below:

(i) *Sticky branches*: Short lateral branches, often only a few cells long, are formed. These may anastomose to form loops which never form complex network.

(ii) *Sticky network*: The branches curl around and anastomose with similar branches or with parent hyphae. These loops produce complex network. The adhesive surface of the network helps to hold the prey.

(iii) *Sticky knobs*: Small spherical to subspherical lobes on one or two celled lateral hyphae are formed. Only the terminal cell is sticky.

In the above sticky traps (*Arthrobotrys oligospora, Dactylella lobata, D. cinophaga, D. ellipsospora*) the wandering nematodes are caught in the loops and held thereby an adhesive fluid secreted by the fungal cells. The trapped nematode is immobilized within one or two hours by which time fungal hyphae have already penetrated the nematode cuticle. Inside the nematode body the fungus produces swollen bulb-like structures from which hyphae radiate throughout the lumen of the nematode body and absorb the body contents.

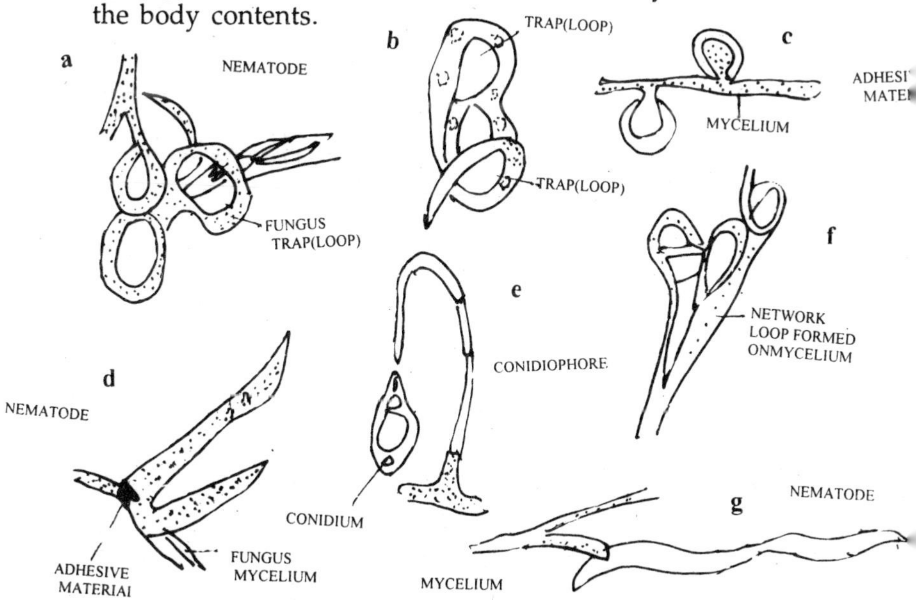

Fig. 17.1 : Nematophagous fungi, Trapping structures.

(iv) *Constricting rings*: The trap is formed when a short hyphal branch curls back on behalf and anastomoses, forming a three- or four-celled ring. The ring is borne on a slender one- or two- celled stalk. When a nematode enters the ring and contacts the inner walls of the ring cells, the cells bulge inward filling the lumen of the ring and holding the prey fast. Such constricting rings are seen in *Dactylella bemicoides*.

(v) *Non-constricting rings*: The non-constricting rings develop as the constricting rings but are neither constricting nor sticky. The nematodes get entrapped by the rings which have a diameter less than the diameter of the nematode body. In attempts to get away the namatode gets further entrapped by the snares. In constricting as well as non-constricting traps the method of feeding on the nematode body is the same as for the fungi producing sticky branches, rings and knobs.

The Endoparasitic Fungi: The endoparasitic (endozoic) fungi produce conidia as infection agents. They are obligate parasites and form vegetative mycelium only within the nematode body. There is little or no mycelial development outside the host. Conidia germinate and form an infection tube that penetrates the nematode cuticle and causes infection. The condia are generally sticky and adhere to the body of nematodes coming in contact with them. Hyphae ramify within the nematode body and absorb the body contents. Latter, spore bearing hyphae emerge from the body and help in dispersal of the fungus.

The Phycomycetous endoparasites of nematodes are more common in soil and have studied in some detail. The examples are some species of *Catenaria*. These have been especially found attacking free-living and predatory nematodes (Dorylaimids and *Mononchus*). The zoospores of these fungi stick to nematode body, germinate, and the germ tube penetrates the nematode cuticle causing infection. Within the body the ramifying hyphae produce encysted zoospores. Sporophores come out through the cuticle.

The thallus of *Catenaria vermicola* develops within the nematode body. Fine strands of separate hyaline hyphae running along the length of the body develop. The rate of infection of plant parasitic nematode genera is slow. It has been observed that nematodes of genera in the Dorylaimoidae are more easily attacked by parasitic fungi with mobile zoospores than are the Tylenchida. This is especially evident among *Xiphinema* spp., which are members of the Dorylaimoidea. Certain phycomycetes attack eggs of nematodes in the soil and destroy them. Some endozoic parasites such as *Nematoctonus heptocladus* and *N. concurrence* kill the nematodes on contact before penetrating the cuticle. *Curvularia pallescence and Fusarium* spp. have also been isolated from females of *M. javanica* extracted from root galls of tomato grown in soil amended with oil-cakes. The fungus destroys the body contents leaving only an empty sac or the infected females fail to lay eggs.

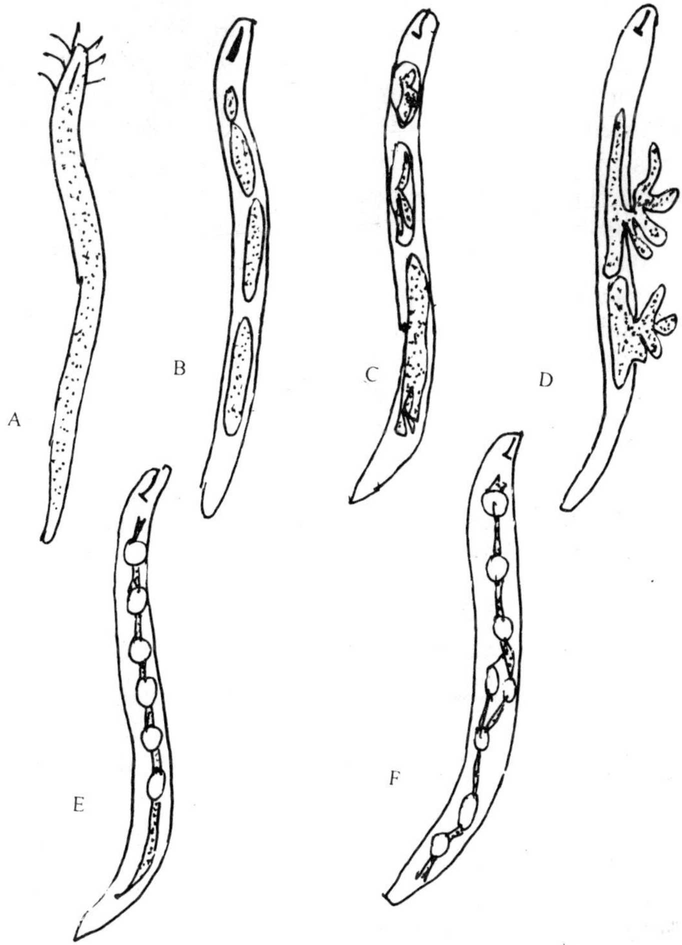

Fig. 17.2 : An endozoic parasitic fungus (*Protoaseus* sp.)

A. Fungus spores attached to the ant, end of the nematode.
B. Naked protoplasts of the fungus inside the nematode.
C. Segmentation of fungus thallus.
D. Nature species coming out directly through cuticle of nematode body.
E, F. Formation of chlamydospores of the fungus in the nematode body cavity.

There are many reports of parasitization of eggs and cysts of phytonematodes by fungi. The viability of eggs of the potato cyst nematode is known to decline over a time cysts are left in the soil for long. The fungus *Verticillium chalmydosporum* and many other non-sporulating fungi are reported to the responsible for the decline of viability of eggs in cysts of *H. schachtii* and *H. avenae*. *Cylindrocapon destructans, Acremonium strictus* and *Fusarium oxysporum* have been isolated from cysts and are capable of destroying eggs of sugar-beet cyst nematode. The fungus *Phialophora heteroderae* enters the cysts of the potato cyst nematode, *G. rostochiensis*, through oral and vulval openings and penetrates the egg shell reducing the number of viable eggs in the cysts. Eggs of *M. incognita* are infected by *Paeciolmyces lilacinus*. The fungus invades both egg masses and females of *Meloidogyne* eggs. This fungus is capable of limited saprophytic growth and is a specialized hyphomycete parasite of nematode eggs. *Nematophthora gynophila* parasitises eggs of *H. avenae* and is the major factor in limiting the natural multiplication of this nematode.

2. Predatory Nematodes

Predatory nematodes can be considered as effective enemies of other nematode. Because of their size larger than the size of some of the destructive phytonematodes such as *Meloidogyne* which are not only small and slender but sluggish in movement the predatory nematodes can prove effective biocontrol agents if suitable soil conditions can be created for their multiplication. The predatory nematode *Seinura* feeds voraciously on nematodes. Some members of the superfamily Dorylaimoidea such as *Discolaimus, Dorylaimus,* and *Actinolaimus* pierce the nematode body and eggs and suck out the contents. These predatory nematodes can be conveniently grouped in categories based on their feeding habit:

(a) The nematode prey is ingested *in toto* such as by *Tripyla* spp. and *Monohystera* spp. These nematodes have a very flexible buccal cavity and plain esophagus. The stoma is usually unarmed and often small.

(b) Some nematodes feed by puncturing a hole in the nematode body with fixed or moveable stomal teeth. Example is *Mononchus papillatus*.

(c) Predatory nematodes like *Seinura christiei* and *S. tenuicaudata* are equipped with stylet with which they feed on other nematodes in the same manner as phytonematodes feed on their host plants.

Prey preferences of the predaceous nematodes in water agar cultures have also been observed. The lesion nematodes,

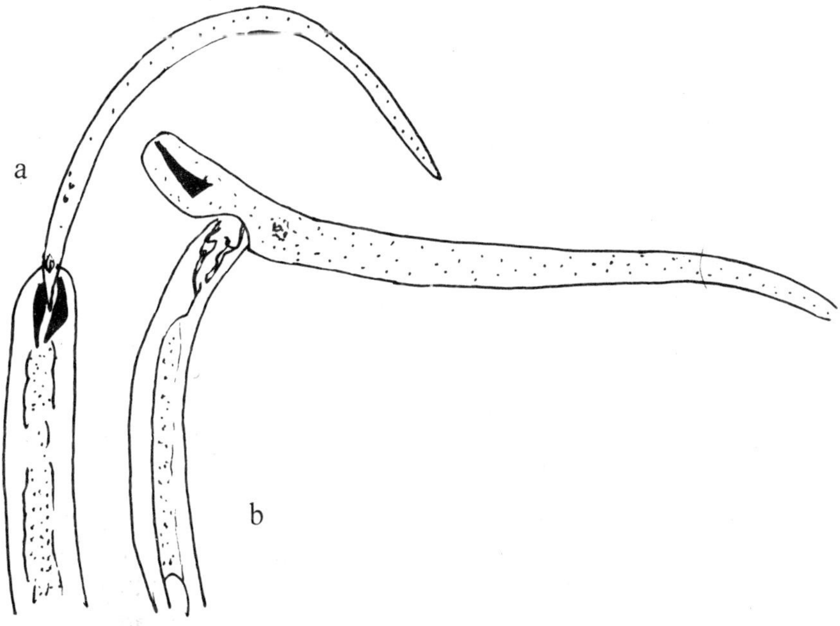

Fig. 17.3 : Feeding in predatory nematodes.

(a) A predatory nematode (*Mononehus* sp.) devouring another nematode.

(b) A predatory nematode sucking body contents of a plant parasitic nematode.

Pratylenchus penetrans, P. vulnus, Paratylenchus curvitatus and *Meloidogyne* spp. are susceptible to predation by several Dorylaimida whereas *Hoplolaimus tylenchiformis, Belonolaimus longicaudatus* and *Criconemoides* spp. are fairly resistant to predation. Despite the apparent success of predaceous nematodes they probably do not occur in soil in sufficiently large numbers in the crucial early phase of population build-up of the phytonematodes to be efficient natural regulators of nematodes.

3. Viruses and Bacteria

Unconfirmed evidence suggests that plant parasitic nematodes do suffer from virus infection. However, conclusive proof of pathogenicity and the presence of virus particles in the infected nematode are lacking. In *M. incognita incognita* it was reported that juveniles showing virus infection had a very sluggish motion and that the sickness was transferable to healthy juveniles when they came into contact. After eliminating the possibility of involvement of bacteria and fungi it was shown that the filtrates of a suspension of sick nematodes cold include the disease in healthy juveniles.

In another case the pnenomenon of swarming (aggregation of nematodes in masses) was also reported. Electron microscopic studies of the cuticle of *Tylenchorhynchus martini* revealed abnormal and disruptive morphological changes in the cuticle layers that suggested a diseased condition. Virus-like particles were observed in the hypodermis and muscle layers, digestive and reproductive systems and on the surface of the cuticle. Normal healthy nematodes exhibited intact tissues without any suspected virus particle. However, in these studies neither the virus could be isolated nor the disease could be transmitted to healthy nematodes.

Bacteria have been detected in the body of free-living nematodes but it is not clear whether they are saprophytes or parasites. *Bacillus penetrans* is an obligate parasite of plant parasitic nematodes. It was first detected in *Pratylenchus pratensis* and described as a protozoan, *Duboscquia penetrans*. In addition to *P. pratensis* it also attacks *P. crenatus, P. penetrans, Rotylenchus robustus, Tylenchorhynchus dubius* and *Meloidogyne arenaria*. Spores of the bacterium adhere to second stage juveniles in soil and germinate after the nematodes enter the host roots and start feeding. Germ tubes penetrate the cuticle, colonies develop in the pseudocoelum and ultimately some of the vegetative cells differentiate into spores which fill the body cavity. Infected nematodes do not reproduce. All stages in the process of infection of *M. javanica* by *B. penetrans* were favoured by temperature optimum for the nematode. At 30°C the parasite proliferates extensively in females before they reach maturity. Tomato roots containing large numbers of *Meloidogyne* females infected with *B. penetrans* were obtained by inoculating tomato plants with second stage larvae infected by the bacterium. Roots were then air-dried and powered. The preparation had the potential of microbial nematicide. However, this method cannot be a practical method of large scale production of the nematicide. The major problem in large-scale utilization of *B. penetrans* is, thus the inability to culture the bacterium *in vitro* on any of the standard bacterial culture media. Neither vegetative cells nor spores of the bacterium can be harvested in sufficient quality for extensive tests.

Intracellular bacterium-like organisms have been reported also in *Heterodera goettingiana, H. glycines,* and *Globodera rostochiensis*.

Protozoa and Soil Microfauna

(a) *Protozoa*: A large amoeboid organism, *Teratromyxa weberi* (Proteomyxa, Vampyrellidae), frequently observed ingesting nematodes, is considered of practical importance in the control of phytonematodes. It has seen engulfing larval stages of *G. rostochiensis*, *Meloidogyne* spp. and *Pratylenchus* spp. *Urostyla* spp. have also been found to feed on nematodes. Other soil protozoa probably have incidental predatory relationship with nematodes.

(b) In addition to predatory nematodes many other soil fauna are known as enemies of nematodes. These include tardigrades, collembola, turbellarians, mites and enchytraeids. Although occasionally seen destroying nematodes females, larvae and eggs in soil especially where organic matter is decomposing, there is not much information on their practical use.

Tardigrades are known to feed on several nematodes and other soil organisms, *Hypsibius myrops*, a taridgrade, has been cultured on *Panagrellus redivivus* in a laboratory for several weeks and can reduce populations of *D. dipsaci*. Turbellarians such as *Adenoplea* spp. have also shown good promise as nematode controlling agents. They consume significantly high numbers of nematodes in certain soil types although under natural conditions they are likely to be present in sufficient numbers to regulate nematode populations.

Collembola: These organism are also active predators of nematodes feeding rapidly especially on eggs. *Onychiurus armatus, Isotoma viridis* and *Hypogastrura* spp. destroy cysts of *Heterodera cruciferae*. Mites such as *Pergalumma* spp. are reported as predators of *Tylenchorhynchus martini*. The enchytraeid worms *Fridericia* and *Enchytraeus* feed on sugar-beet cyst nematodes and in pot experiments were able to reduce the population of this nematode.

The above account of the parasites and predators of nematodes leaves no doubt that these enemies of nematodes naturally exist in soil but are not effective in controlling the phytonematodes because of their low population. Their exploitation as biocontrol agents are reasonable cost can be achieved through manipulation of soil environments in their favour. This may increase chances of increase in their numbers to desirable level. The modification of soil environment as an approach to achieving desirable populations of biocontrol agents in soil is designed in full accordance with the ecological principle that soil population at any time will be determined by habitat conditions and that the populations can, therefore, be changed in any direction by making appropriate changes in soil conditions. Major emphasis to bring about these changes in soil conditions at the microsite level has been given to the exploration of the effects of decomposition of crop residues and organic waste especially those commonly available at low cost in the area. This aspect of soil amendment has been discussed in the preceding section of this chapter.

Chemical Control of Plant Parasitic Nematodes

Soil disinfection for control of phytonematodes is used mainly when cultural and biological methods fail to reduce the nematode populations or when these methods are not economical for the growers. Chemicals are also used when there is very heavy infection of the soil and proper resistant varieties are not available to be grown satisfactorily. Under these situations chemical soil disinfestation is the most effective method to reduce nematode populations to a threshold level. These measures allow the farmer flexibility in choice of crops to be grown on nematode infested soil and the farmer can grow what he wants, where he wants and when he wants. The main objectives of chemical soil treatment are to protect plants from nematode damage, prevent or restrict nematode multiplication, improve the growth of plants, and finally improve the quality of the produce. One should not aim at complete eradication of nematode populations since it is impossible to achieve this in a complex biological environments of the soil. The population should be brought below the injury or harmful level so that profitable crops can be grown in the infested land.

Nematicides have been on a large-scale use since 1945 in Europe and North America and their use has been increasing every year. However, the nematicides are now generally used even in these countries only for cash crops which give high cash returns. The chemicals are also used in nursery beds to obtain nematode-free plants. The earliest soil fumigants used against nematodes were formaldehyde, carbon disulphide and calcium cyanide. These were mostly used in glasshouses. At present these are not used on a large scale. Perhaps, the earliest report on the application of carbon disulphide as soil fumigant is that by Kuehn, who tried to control the sugar-beet cyst nematode with carbon disulphide in Germany. In 1919 Byars used hydrocyanic gas to control root knot. The nematicidal property of chloropicrin was discovered in the same year and this chemical was widely used in the controlling root knot nematodes in pineapple fields in Hawaii. Methyl bromide is widely used as an insecticide and nematicide for nursery beds and in glasshouse soil. Hurst and Franklin and Smedly reported the nematicidal properties of cynamide and isothiocynate compounds, respectively. D-D was discovered in 1943 by Carter and was the first soil fumigant used on a large scale to control many species of plant parasitic nematodes. The discovery of ethylene dibromide (EDB) followed soon. Later, Mc Beth reported, 1, 2 dibromo-3 chloropropane (DBCP) as a potential nematicide. In India, the first serious attempt to demonstrate nematode control by soil fumigation with D-D were undertaken by Burmah Shell in 1953. Their efforts were primarily directed at nematode control in tea gardens and nurseries.

Types of Nematicides

Nematicides in commercial use basically belong to two groups, fumigants and non-fumigants. The fumigants consist of compounds belonging to halogenated hydrocarbon group and isothiocynate group. The non-fumigants consists of organo-phosphates and carbamates.

A. Fumigant Nematicides or Volatile Soil Fumigants

The fumigant nematicides possess a high vapour pressure are diffuse through soil pore spaces where nematodes live. They are dissolved and dispersed in soil moisture in concentrations high enough to kill the nematode larvae. These nematicides are not very specific. They can be used for the control of many kinds of nematodes and soil insects and, at high dosages, even fungal and bacterial pathogens of plants. There are two major types of fumigants in the market.

Halogenated hydrocarbons such as methyl bromide, ethylene, dibromide, Telone, D-D mixture, Chloropicrin and methyl isothiocynate (MIT) and MIT releasing soil fumigants such as metham sodium or Vapam and Dazomet, Mylone or Basamid.

B. Contact and Systematic Nematicides

These nematicides have little fumigant action do not kill the nematodes directly. They include organophosphates such as Phenamiphos, Fensulphothion, Thionazin, Mocap, Diazinon, VC-13, Phorate etc., and Oxime N-methyl carbamates such as Aldicarb, Oxamyl and Carbofuran.

The Soil Fumigants

(i) *D-D mixture* ($CH_2Cl.CH$ $CHCl$ + CH_2Cl $CHCl$ CH_3)

D-D is a mixture of 1, 3-dichloropropene and 1, 2-dichloropropane, in 2:1 ratio, with some other hydrocarbon impurities. Ever since its discovery in 1943 this fumigant has been widely used to control plant parasitic nematodes in different parts of the world. It kills nematodes, soil, insects and wireworms and also some fungi at high dosages. The nematicide is sold as undiluted or as an emulsifiable concentrate in drums. D-D is the trade name of the product of the Shell Chemical Co. The dosage rate is 400 to 500 lit per ha as a pre-plant soil fumigant. The best results of the treatment are obtained at soil temperatures between 10° and 27°.

D-D has no permanent detrimental effect on beneficial soil organisms but there is a temporary reduction in the number of nitrifying bacteria and inhibition of nitrification results in high levels of ammonia. It is, therefore, desirable that nitrogen in the form of nitrate should be added to the soil, especially if it contains high levels of organic matter. Crops like sugarcane and pineapple can tolerate ammonia; hence, in their case there is no need to apply nitrates after fumigation with D-D. The nematicide is also sold under the names of Vidden-D, Nemafume and Nematox.

(ii) *Telone* (1, 3-dichloropropene. $Ch_2Cl..CH=CHCl$): This nematicide is a mixture of *cis-* and *trans*-isomers of 1, 3-dichloropropene with small quantities of related hydrocarbons. The chemical was introduced in 1956 by Dow Chemical Company. The *cis*-isomer is more toxic than the *trans*-isomer. Dorolone is a mixture of EDB and Telone. The dosage rate of the nematicide is 200 lit/ha depending upon the soil type.

(iii) *E D B* (1, 2-dibromeothane or ethyl dibromide): Christie had reported that this chemical had excellent nematicidal properties. It kills nematodes as well insects and has been used extensively in tobacco. Since EDB contains bromine it may cause phytotoxicity in bromine sensitive plants. The biological activity is similar to that D-D mixture. Rate of application is 18 to 120 lit/ha depending upon soil type and the crop. The nematicide is not effective when the soil temperatures are below 10ºC. This nematicide also reduces nitrification.

(iv) *D B C P* (1, 2-dibromo 3-chloropropane, $CH_3Br.C$ $Br.CH_2$ Cl): Under the name of Nemagen this chemical had been found very useful as pre-plant and post-plant treatment of established plantation crops. Due to its low phytotoxicity it was also used through irrigation water in citrus nursery soils. Rate of application was 20 to 50 lit/ha depending upon the crop and soil type. The use of Nemagen has now been restricted because of the residue problem.

(v) *Methyl bromide* (CH_3 Br)

This fumigant is very effective against nematodes, soil insects, fungi, rodents and weeds. The material is compressed into liquid and marketed in metal containers. Since methyl bromide is toxic to humans and is odourless a warning agent such as chloropicrin is usually added to it. For seen and nursery beds the treated surface is immediately covered with polythene sheets. In tobacco nursery beds methyl bromide is applied at the rate of 500 g per 5 to 20 sq m surface. The bed is covered with a polythene sheet to aid diffusion of the chemical. For large-scale field treatment methyl bromide is applied at 15 to 20 cm depth. This nematicide also destroys nitrifying bacteria and problems may occur in ammonia sensitive crops. It is marketed under the name of Dofume MC-2, Meth-O-gas and Terabol, etc.

(vi) *Chloropicrin* (Trichloronitromethane. CCl_3NO_2): Chloropicrin was developed primarily as a soil fungicide to control fungal wilts of plants but was found extremely useful for controlling root knot nematodes. Sclerotia forming fungi are not affected. In addition to nematodes and fungi chloropicrin kills bacteria and weeds also. Rate of application is 500 lit/ha. Chloropicrin is also known as pic-fume and tear gas.

(vii) *Dazomet* (3, 5-dimethyl-tetrahydro 1, 3, 5.2 H Thiadiazine Thion): On application Dazomet breaks down into methyl aminomethyl dithiocarbamate which in turn, yields isothiocynate. It is a good fungicide and nematicide. Rate of application is 20-35 g/sq m as a pre-plant soil fumigant. Other names of the nematicide are Mylone are Basamid.

(viii) *Metham sodium or Vapam* (sodium N-methyl dithio-carbamate. CH NH.CS.SNa.2HO): Vapam releases methyl isothiocynate which is the active component. The chemical is used against fungi, nematodes, weeds and sometimes soil insects. Rate of application is 700 to 1000 lit/ha. The soil temperature should be reasonably warm (15° to 30°C) at the time of application.

(ix) *Trapex* (methyl isothiocynate. CH.NCS): This nematicide is a 20 per cent (w/w) product of an organic solvent, generally xylol, with or without an emulsifier. In addition to nematodes, soil insects and weeds it has been found effective against cyst forming nematodes also.

Methods of application of soil fumigants: There are several methods of application of fumigant nematicides to the soil for good results. Application of fumigants to the entire filed is known as area treatment. The field is prepared well in advance of the treatment into fine seed bed condition. Marker lines are drawn at 30 to 35 cm distance in either direction. The nematicide is injected into the soil to a depth of 22.5 to 30 cm at the intersecting points of lines. Hand-operated injector guns are used for the purpose. In case of large size areas (several acres) the operation is carried out with the help of tractor-mounted applicators and the nematicide is injected at a depth of 30 cm. Area treatment is undertaken where maximum control is desired especially in commercial crops like tobacco and vegetables.

When the crops are planted in rows of 60 cm or more with, row treatment can be used. The rows are marked and the nematicide is applied under the planting row. Row treatment is widely used in vegetable crops and tobacco to get reasonably good control of nematodes. Strip treatment is commonly used when perennial crops are planted in strips. The planting row of 2 to 3 m are treated by hand injector guns or a tractor applicator. Site treatment is done when individual trees are to be planted. The area around the trees can be treated with nematicide. This method has been used for large-scale treatment of citrus trees as pre-plant and post-plant treatment with Nemagon.

Criteria for success of soil fumigation: The fumigant nematicides are usually injurious to seeds or plants if they are planted immediately after soil fumigation. In order to avoid the injurious effects, generally two to three weeks waiting period is required depending upon the type of nematicide used. A good fumigant must disperse and diffuse through the soil particles to kill nematodes. Newhall had discussed the theoretical considerations of soil fumigants. The initial stage is relatively short and characterized by radial diffusion from injection points. This is followed by transition stage and then by a uniform distribution of the fumigant through the soil. The efficacy of fumigant nematicides in soil, therefore, varies with vapour pressure of the chemical, its molecular weight, soil conditions such as porosity, moisture content, temperature and organic matter content. For maximum efficiency the soil must be properly prepared for the treatment. It is also important to remove all weeds are undecomposed plant material including roots of the previous crop. Their presence interferes with normal application of the nematicide and may also protect the nematode stages present inside the plant tissues where the chemical cannot reach.

Mode of action of fumigant nematicides: The distinguishing feature of the action of fumigant nematicides form the non-fumigant nematicides is that when there is sufficient exposure of the nematodes to the fumigant nematicide they are killed whereas non-fumigant nematicides usually incapacitate them in one or more functions at field concentrations and the effects are reversible.

Penetration of the nematode body by the chemical is a necessary prerequisite to physiological reactions associated with death. Chemical may enter the nematode body directly through the cuticle or through such orifices as mouth, anus, and vulva. EDB is known to enter the nematode body two and half times faster than water and produce a narcotic effect on the body. Eggs are less sensitive than juveniles and adults to the effects of nematicides. The first reaction of the nematode to the entry of chemicals in the body is hyperactivity followed by gradual decrease in activity, leading to eventual paralysis but this narcotic effect may be temporary and the nematode may recover to complete its life cycle. Within the nematode body the chemical may act in many ways such as by precipitating proteins, blocking nerve may act in many ways such as by precipitating proteins, blocking nerve endings, destroying nerve sheaths, cellular membranes and amphids. The nematicidal activity has been suggested to involve alkylation of nucleophilic sites on proteins or oxidation of iron (Fe II) porphyrine and haemoproteins. There is no information on the mode of action of methyl isothiocynate or their generators or nematodes although it would be expected that isothiocynate molecules have an affinity to a variety of relativity nucleophilic sites within cells.

Beneficial effects of soil fumigation: There is plenty of information in developed countries on the impact of soil fumigation on food production through control of nematodes. Improved plant growth in fumigated soil is the general effect observed in most crops. The increase in yield is reported to be as high as 87 per cent. Although it can be expected that nematode control is the major factor involved in this increase, there are many other associated factors that account for the yield increase. Fumigation reduces the losses from disease complex. Biologically bound nutrients are released after the effect of fumigation on soil microflora and fauna and are utilized by the plants. Plants in treated plots show distinctly improved and uniform growth with extensive root system and feeder roots. The increased root system enhances uptake of nutrients and water resulting in better yields. The healthy plans growing in treated plots are also less susceptible to other plant diseases and can tolerate adverse growing conditions better.

Side effects of fumigant nematicides: All soil fumigant nematicides are general biocides. Halogenated compounds can temporarily upset normal soil nitrification process which results in increased ammonia accumulation with the result that plants may suffer from ammonia toxicity or nitrate starvation. Brominated compounds like methyl bromide or EDB have residue problem. Bromine residues have been found in a number of crops like citrus, carrots, potatoes, groundnut and tomato following fumigation with these nematicides. Increased uptake of bromine or chlorine by some plants has been reported. Increased halogen content in tobacco leaves reduces leaf quality and adversely affects the burning quality of tobacco grown in fumigated fields. Nematicides with fungicidal properties can adversely affect nematophagous fungi and other parasites and predators of nematodes. Normal infection and development of mycorrhizal fungi are also suppressed by nematicides having fungicidal property. Citrus seedlings grown in fumigated soil are commonly stunted and chlorotic. Heavy fertilization partially overcomes stunting. For many years this problem was attributed to soil toxicity developed by fumigation. Kleinschmidt and Gerdemann later showed that stunting was caused by inadequate nutrition due to destruction of vesicular arbuscular mycorrhizal fungi. Inoculation of fumigated soil with VA-mycorrhizae partially solved stunting problem and reduced need for frequent fertilization.

Little information is available on the long-term ecological effects of frequent and repeated use of soil fumigants in cultivated fields. Investigations at the University of California in the U.S.A. had indicated that EDB, DBCP and D-D mixture are not persistent in soil and have no lasting adverse effects on the physical and biological properties of the soil, on the environment and on the nutritional value of the crop grown in fumigated fields. The fumigants are quickly degraded by normal biological activities in the soil.

Contact and Systemic Nematicides

The introduction of granular contact and systemic nematicides in the seventies has provided very effective chemicals for nematode control. Some of these chemicals have replaced the fumigant nematicides in several countries. These non-volatile, non-fumigant nematicides are relatively less phytotoxic and can be applied before or at the time of planting or even on standing plants. They are effective at much lower dosages than the fumigants. Most systemic nematicides are taken up by roots from soil application and translocated up in the plant system giving the plant protection against nematodes and many insects including white flies, but not against soil-borne fungal plant pathogens. The applications of these nematicides does not require any special equipment. Commercial contact and systemic nematicides belong to the organophosphate and carbamate groups.

(i) *Phenamiphos* [0-ethyl-0 (3-methyl thiophenyl) isopropyl amidophosphate]: Sold in the market under the name Nemacur, phenamiphos is a highly effective systemic nematicide against nearly all the economically important phytonematodes. The active ingredient simultaneously affords good protection against sucking insects for six to eight weeks. The chemical is readily absorbed by roots as well as by leaves and translocated both up and down in the plant system.

(ii) *Phorate* [0, 0-diethyl-S (ethyl thiomethyl) dithiophosphate]: Commonly sold in the market as Thimet, phorate is a systemic organic phosphate soil insecticide and nematicide. Absorbed by roots it is translocated in the plant system are persists for several weeks protecting it not only from nematodes but also sucking insects.

(iii) *Fensulphothion* [0.0-diethyl-0(4-methyl sulphinyl) monothiophosphate]: Sold as Terracur P and Dasanit this non-systemic organo-phosphate can be used as an insecticide as well as nematicide. The active ingredient disperses in soil via the water system so that the chemical easily comes in contact with nematodes. The chemical easily comes in contact with nematodes. The chemical remains effective in soil for about three months at a temperature of 20°C.

(iv) *Thionazin* [0,0-diethyl-0 (pyrazinyl) monothiophosphate]: This organophosphate contact nematicide and insecticide was introduced in 1961 as soil pesticide bv American Cynamid Co. Its other names are Nemafos, Cynem and Zinophos.

(v) *Mocap* [0-ethyl S,S-dipropyl dithiophosphate]: Also known as Ethoprophos and Ethoprop, mocap is a phosphatic compound with a broad spectrum of activity as nematicide and soil insecticide. The chemical is non-fumigant, non-systemic and has efficient contact action with good soil movement and residual properties. It can be applied immediately before or at the time of planting or used for treatment of established plants. The chemical should be either applied with irrigation water or if applied direct it should be mixed with soil.

(vi) *Diazinon* [0,0-diethyl-0 (2-isopropyl-4 methyl) pyriidinyl]: Diazinon (Basudin, Neocidol) is an organic phosphate insecticide, nematicide, and acaricide with contact action.

(vii) *VC-13* [0,0-diethyl-0 (2-4 dichlorophenyl) monothiophosphate]: This is also an organic phosphate contact nematicide-insecticide applied to soil.

(viii) *Oxamyl* [methyl-N-N-methyl-N (methyl carbomoyl) oxy-1 thioxamidate]. Oxamyl is an oxime carbamate and sold as Thioxamyl, Vydate and Du Pont 140.

(ix) *Aldicarb* [2 methyl-2(methyl thio) propionaldehyde-0-(methyl carbomoyl) oxime]: Aldicarb is usually sold under the name Temik. It is a strong systemic insecticide-nematicide and acaricide. The chemical was introduced in 1965 by Union Carbide Corporation. As a versatile insecticide-nematicide to control several nematode parasites of important crops it is used extensively.

(x) *Carbofuran* (2, 3-dihydro-2,2-dimethyl 7-benzofuranyl N-methyl carbamate): Carbofuran is sold as Furadan or Curatter and is a highly effective soil nematicide-insecticide which has systemic action is plants.

Some of the organophosphate and carbamate nematicides are systemic but are only absorbed by roots and translocated up to the foliage. Few, such as oxamyl, phenamiphos and carbofuran have the capacity to move from foliage down to roots and from roots up to foliage under certain conditions in some plants and provide protection against nematodes. The movement of these nematicides in plant is, however, faster from roots to foliage than from foliage to roots.

Methods of application of systemic nematicides and response of nematodes: The non-volatile nematicides can be applied to the soil at planting time and also around established plants without any phytotoxic effect. Application can be by spot treatment at the time of planting young seedlings in the main fields or by row treatment. Chemicals like oxamyl and phenamiphos can be used as foliar sprays also or through watering of the seedlings at the time of planting. This method saves the amount of chemicals and provides protection against initial infection. The rate of application may vary depending upon the method of treatment, nematode species to be controlled, depth of application necessary to reach the target nematode , soil conditions, environments and cultural practices.

(i) *Soil Application*: Broadcasting the granules or spraying the emulsifiable concentrates on the soil surface followed by through mixing with the soil is the normal method of application for achieving maximum protection of crops. Soil application places the nematicide in the top few inches of the soil and then the chemical moves downwards and reaches greater depths. In field trails conducted by Whitehead and his associates effective control of potato cyst nematodes was obtained when oxime carbamates (oxamyl, Aldicarb) were mixed with soil at the rate of 11 kg a.i./ha. Aldicarb affects the behaviour of the nematode by disrupting their orientation to potato roots. The chemical also causes a marked increase in the male/female ratio. Development of potatoes is reported to be much improved by soil application of fensulphothion at the rate of 10 kg a.i./ha. A high degree of control of various species of *Meloidogyne* (root knot) was reported under field conditions with less than 10 kg a.i./ha of phenamiphos. According to Brodie and Good soil application of systemic nematicides has often proved better than fumigants in the control of root knot and in increasing tobacco yields.

The method of soil application influences the performance of systemic nematicides. Reddy compared different methods are rates of application of granular nematicides for the control of root knot of tobacco. Spot application (at 1.0 kg a.i./ha) and row application (4.0 kg a.i./ha) of aldicarb 10 G, fensulphothion 5G or carbofuran 3 G to tobacco seedlings planted in soil infested with *M. incognita* and *M. javanica* increased yield of cured tobacco leaves and decreased root galling. Yields were 83 to 135 kg/ha higher in treated than in untreated plots. Nematode control and yields from spot treatment were as high as those from row application and were higher for fensulphothion and aldicarb than for carbofuran. Spot application of aldicarb, carbofuran and phorate is reported to be superior to row treatment or broadcasting in the control of root knot of okra and tomato. Sitaramaiah and Vishwakarma obtained increased tomato yields by applying the nematicides to the soil surface and then mixing into the soil better than the row or spot treatment. Time factor is also important in application of granular nematicides. Application of aldicarb, phorate, diazinon and fensulphothion seven days before planting tomato seedlings gave better control of root knot nematodes than the soil treatment immediately before transplanting the seedlings.

(ii) *Root Treatment*: Roots of young seedling or nursery planting material infested with endoparasitic nematodes can be disinfested by root treatment with systemic nematicides. Mostly bare root dips for specific durations have been tested. Depending upon the type of planting material the duration can be for a few minutes to hours. Nursery stock of perennials can be exposed to nematicide solution for as many as six hours while roots of tender seedlings require 10- 30-minute exposure. Although longer durations are more effective, there may be phytotoxicity.

(iii) *Seed Treatment*: Some systemic nematicides have been tested as seed dressing material to reduce the cost of treatment without any deleterious effects on the plant growth. Among such nematicides carbofuran and aldicarb have been found effective against several plant parasitic nematodes. Seeds are uniformity coated with gum or starch (2 per cent) as stickers and the required quantity a wettable formulation of the nematicide is sprinkled over the seeds and mixed thoroughly to give a uniform smooth coating. The treated seeds and dried in shade before sowing. This method was used by Prasad and Mathur to treat sugar-beet seeds with carbofuran and aldicarb sulfone. Sugar-beet seeds coated with carbofuran at the rate of 1 per cent a.i. and aldicarb sulphone 2 per cent a.i. reduced *M. incognita* populations in pot tests and improved the growth of plants. Sivakumar and associates tested carbofuran as okra seen treatment for the control of root knot and observed that such treatments do not give absolute protection but could be effectively employed to reduce the severity of infection. Potato tuber dip in phenamiphos at the rate of 4.75 a.i./lit water is reported to eliminate *M. hapla* from tubers. The duration of dip was 10 seconds only. There was no phytotoxicity.

(iv) *Foliar Treatment*: Bunt had reported control of the lesion nematode, *Pratylenchus penetrans*, by foliage spray with oxamyl. Foliar application of promising nematicides could also be used in conjunction with soil application of granular nematicides which could provide initial seedling root protection until the foliage was sufficiently expanded to be sprayed with systemic nematicides. Alam and associates found the highest reduction of root galling by *M. incognita* when the seedlings were immersed in oxamyl for 20 to 30 minutes followed by five foliar sprays at weekly intervals. Spraying alone provide poor control. Prasad and Rao also fond foliar sprays with oxamyl at 100 to 2000 ppm effective in reducing populations of the stunt nematode. *Tylenchorhynchus* spp. Foliar treatments of plants for nematode control are not yet practical on a field scale and hence the traditional soil application in the common method.

Mode of action of contact and systemic nematicides: It is generally accepted that at field dosages the organo-phosphate and carbamate nematicides do not kill the nematodes directly. They act by impairing the nervous system of the nematodes, reducing the rate of hatching, motility, orientation to host roots, feeding and development. These nematicides primarily act by inhibiting the enzyme acetyl cholinesterase at cholinergic synapses in the nematode nervous system as is known in the case of mammals and insects.

Soil application of thionazin reduces the fecundity of *M. incognita* females while aldicarb has no such effect. Larvae from eggs produced in thionazin and aldicarb treated plants do not show any adverse effect of the treatment. Similarly, hatchability of *Meloidogyne* eggs is not influenced by these nematicides when they are subsequently transferred to water. For thionazin only the highest concentration of 1000 ppm tried has been found lethal to *M. incognita* but at the recommended field dosages a five-day exposure has no effect on the hatched larvae from egg masses. The gelatinous matrix of the egg sac acts as a physical barrier against the penetration of nematicides. The sedentary endoparasites such as root knot and cyst nematodes can be most effectively controlled only by treating the soil before the juveniles enter the host roots.

Prevention of attack on roots and feeding on the host by the nematodes in treatments with systemic nematicides has been reported by many investigators. Oxamyl and carbofuran are known to disrupt orientation and attack on roots by *M. incognita* and *Pratylenchus penetrans*, respectively. On reaching the roots the nematodes did not feed. Mc Leod and Khair used a precise test to determine the effect of nematicides on development of nematodes with tomato seedlings as indicators. Nematodes were allowed to invade the roots over a 24-hour period, the roots were then washed and the plants placed in tubes containing sand and nematicide solutions. The number of juveniles which had developed to mature saccate female stage was recorded after 10 days. Aldicarb and thionzin at 2 ppm completely prevented development, methomyl, ethoprophos, phenamiphos and phorate were progressively less effective.

Experimental chemicals other than commercial nematicides: Plant growth regulators, phenolics, aromatic acids, amino acids and fatty acids have been tested either as foliage treatment, soil drench or as bare root dip treatment to reduce nematode infestation of plants. These chemicals, though effective, are still in experimental stage and no commercial formulations have been proposed.

Foliar application of 2, 4-D and Cycocel reduced the egg mass production in *M. javanica* and *M. incognita*. The ability of grape fruit seedlings to grow well in the presence of the burrowing nematode, *Radopholus similis*, was increased by application of exogenous phenols. Cinnamic acid and p-caumeric acid sprayed on susceptible potato plants prevented susceptible responses to *Globodera rostochiensis*. Exposure of roots of susceptible tomato seedlings to phenols, aromatic acids and fatty acids imparted some resistance in invasion of juveniles of *M. javanica*.

Resistant Varieties for Nematode Control

Soil treatment with chemicals and employing practices such as crop rotation are common methods of nematode control. However, in a nematode control programme, or in any plant protection programme, these methods should not be the first choice to consider. A most effective and economic method is the use of resistant cultivars. Growing resistant varieties of crops in crop rotations is economical for the farmer for reducing nematode in populations gradually in the infested fields. According to Oostenbrink the cultivation of a resistant variety may suppress a nematode's population to 10 to 50 per cent of its harmful density. With reduced multiplication and population density through successive cultivation of resistant varieties for sometime, it is possible to grow even a susceptible cultivar at frequent intervals without extra cost on use of nematicides. The use of nematicides can be directed to high value crops or where the farmer wants to grow a particular crop continuously in the same field year after year. The use of nematode resistant cultivars and chemical soil treatment should be considered complementary methods of control and not as comparative measures.

Considerable progress has been made in the U.K., Europe and the U.S.A. in breeding crop varieties resistant to specific sedentary and migratory endoparasitic nematodes. Resistant varieties have been developed in cotton cowpea, tobacco, beans, soybean, tomato, grapes and peaches against the root knot nematodes in alfalfa for resistance to *Tylenchulus semipenetrans*.

Nematode resistance has been discovered in both cultivated and wild plants. In most of the economically important crops like potato, tomato, barley and oats, resistant genes to several endoparasitic nematodes have been identified. Resistance to the potato cyst nematode has been found in *Solanum verni, S. multidissectum, S. oplocense, S. spegazzinni S. kurtzianum* and *S. tubersum* s.sp. *andigena*. Screening for resistance to potato cyst nematode was studied extensively in the U.K. and the Netherlands. Several genes for resistance (R genes) to many pathotypes of *Globodera restochiensis* have been found and used for developing commercial cultivars. Resistance to root knot nematode has been found in *Lycopersicon peruvianum, L. pimpinellifolium, Gossypium barbadense* and several species of *Nicotiana*. Gilbert and Mc Guire studied the inheritance of resistance in tomato to root knot and found the presence of a single dominant resistance in tomato to root knot and found the presence of a single dominant resistance gene Mi which was effective against all species of *Meloidogyne* except *M. hapla*. Subsequently, two against all species of *Meloidogyne* except *M. hapla*. Subsequently, two dominant and one recessive genes for resistance were designated as . L Mi Rl, L Mi R2 (cultivars Nematex and Small Fry-1) and L Mi R3.

High resistance to the cereal cyst nematode, *H. avena,* was found in species of *Avenue* (*A. sterilis, A. byzantina*). These species possess resistance to several pathotypes of *H. avenae* and were effectively utilized to develop resistence in cultivated oats (*A. sativa*). Resistance in barely against this nematode has been detected in barely varieties Primus, Svanhals, Chevalier, etc. Two pathotypes (Race 1 and Race 2) were originally known in *Heterodera avenue*. Resistance to Race 1 is found in cultivar Drost. At present there are at least six resistance genes identified in various barley cultivars which have been designated differently in different countries. High resistance to *Heterodera schachtii* has been found in *Beta petellaris, B. procumbens* and *B. webbiana*. These sources of resistance are useful for incorporation of resistance in the cultivated species.

Types and measure of resistance have been briefly described in the Chapter on "Host-Parasite Interaction in plants".

The resistance in plants to nematodes can be due to the presence or release of preformed antinematode compounds or it can be due host reaction to invasion wherein the host does not allow development of juveniles to adult eggs laying stages. Both these aspects have been discussed in some detail earlier under "antagonistic plants and under "Host-parasite Relationship-Resistance and Susceptibility." Interesting information regarding the antinematode compounds produced by antagonistic plants is available but such compounds have not been identified in commercial crop varieties.

The antinemic compound in *Asparagus officinalis* var. Mary Washington that kills the larvae or disrupts reproductive progress is a glycoside with a low molecular weight aglycone. It is toxic to *Trichodorus christiei, Paratylenchus projectus, Tylenchorhynchus claytoni, Helicotylenchus nannus,* and *Belonolaiamus gracilis*. The compound is reported to interfere with acetyl cholinesterase in the nervous system. The nematicidal properties of marigold (*Tagetes erecta* and *T. patula*) are attributed to the presence of terthienyl and bithienyl compounds which are highly toxic to *Pratylencus* and many other nematodes including root knot. Pyrocatechol in *Eragrostis curvula* is toxic to *Meloidogyne* spp. Similarly, the alkaloids nimbidin and thionemone in *neem* tree (*Azadirachta indica*) are reported to have antinemic action.

In most crop varieties possessing resistance to nematodes, there is hardly, well in resistant as well as susceptible plants. It is only after entry that the resistant plants exert their influence on nematode activity, mostly by not allowing the larvae to feed properly and develop into adult egg laying females. This could be, as discussed earlier, due to suppression of giant cell and nurse cell development which causes starvation of the nematodes. There is no strong evidence to suggest the resistance in plants to attack and invasion of certain nematodes is due to an absence of a particular nutrient for the nematode in the plant.

The foregoing account of possible approaches to solve the problem of nematode control discusses regulatory measure (quarantine, etc.), cultural practices including biological control, use of chemicals and growing of resistant varieties. How far these approaches can be easily adopted by the farmers in India is a difficult question to answer. The majority of farmers in India are interested in maximum return from their limited holdings with minimum investment. He is not voluntarily willing to obey certain strict regulatory laws that affect his annual income or modify his methods of crop culture, or invest in costly antinematode chemicals. It may be due to his need for short-term planning. There is hardly any work in India to impress upon the farmer the need for long-term planning to avoid serious losses in future or losses already occurring in his field.

There is no doubt that the farmer can be provided with resistant varieties of commonly grown crops, he will readily accept them because he has not to invest extra money. Therefore, availability of resistant varieties in the first and foremost suggestion that can be practical. Unfortunately, there is insufficient work in this direction. Because of intricacies involved in breeding for resistant cultivars of a crop some of such varieties introduced from abroad have failed in India. Further, resistance breeding is a time consuming process. Indigenous and exotic varieties/lines resistant to specific nematodes are listed in the respective chapters.

In the absence of resistant varieties, practical and chemical methods need emphasis. Flood fallowing, use of decoy or trap crops or growing of antagonistic plants are not practical methods under Indian conditions. Crop rotation is a time-tested successful method. However, its limitations have been pointed out. Choice of crops with the Indian farmer is very limited. For example, in north India, rice in *kharif* and wheat in *rabi* season are common crops except where a crop like sugar cane predominates. With vegetable growers the problem is more acute. Almost all the vegetables grown for the market are susceptible to root knot. The grower cannot switch over to cereal cultivation in place of vegetables. The old system of a legume following cereals is no more in practice. In the absence of a wide range of crops for choice the best alternative could be the use of resistant cultivars of the main crop in the rotation but, as stated above, resistant cultivars, with a desire yielding quality, are not available. However, there is a good scope for using the crop-free period during summer as summer fallow with frequent ploughings to expose the nematodes to extremes of weather. Using one *kharif* season occasionally for green manuring is also possible. The advantage offsets the loss of one crop. There is a need for developing techniques for quick decomposition of crop residue in the field. This could encourage indirectly the biological control of nematodes.

Finally, chemicals remain the only direct methods of nematode control. But the prohibitive of chemicals and lack of the knowledge about their proper use excludes a majority of the farmers who own limited areas of land. It, therefore, requires a good deal of work on the part of scientists of develop well-designed package of practices that could be integrated with the normal crop growing practices followed by the farmer.

Chapter-18
The Host: Immune Reactions Against Parasites

When a parasite or other 'foreign' body (e.g. a splinter) invades a host, the latter is able, in some way not understood, to recognize that 'non-self' material is present, and respond accordingly. How this recognition is carried out is not understood, but presumably recognition must occur at the molecular level on or near the surface of susceptible cells and requires contact between the invading material and the recognizing cell. In some cases, possibly associated with a phenomenon termed *molecular mimicry* the host is unable to 'recognize' the parasite as 'non-self' and consequently does not respond to its presences; this phenomenon is discussed later.

Vertebrate hosts respond to the presence of 'non-self' material by bringing into action mechanisms which can be broadly grouped into two kinds; these compliment each other to some degree:

(*a*) *non-specific responses*: these are selective to the extent that they can differentiate 'non-self' from 'self' but they are not dependent on *specific* recognition (i.e. of some very specialized type of 'foreign' molecule);

(*b*) *specific responses*: these depend on the prior exposure to 'non-self' material and the subsequent recognition and reaction to this material specifically, during later invasions.

Immunity (resistance) and insusceptibility

It is usual to speak of specific responses as *immune* or *allergic* responses and the field of study incorporating these phenomena as *immunology*. In many situations, these responses do not actually confer immunity on the host, and it has been recommended that responses or reactions should not be designated 'immune' until shown to be concerned with immunity the 'uncommitted word "allergie" is much better and presupposes nothing as to their role or significance'. Ideally, it would be desirable to clarify usage of the terminology in this way, but the terms 'allergic' and 'immune' have been widely used as being synonymous. The term immune is generally used throughout this text, since the word allergic is derived from the noun *allergy*, often considered to be synonymous to hypersensitivity.

In a similar manner, the terms 'immunity' and 'resistance' have been widely used as synonymous, although the former term is now, perhaps, more frequently used. Theoretically, the word 'resistance' implies that the host is not completely successful in repelling invading organisms, whereas 'immunity' suggests complete rejection of such organisms.

Insusceptibility. Although the words 'resistance' and 'immunity' are widely used, in many instances an alternative term, *insusceptibility* may be more appropriate. Insusceptibility implies that certain features related to the morphology, behaviour, ecology, physiology or biochemistry of the host are unsuitable for the establishment of a particular species. For example a parasite with enzymes adapted to a cold-blooded host is unlikely to establish itself in a warm-blooded host. Again, the physiological conditions in a gut—such as pH, PCO_2, PO_2 and Eh—may be such that it provides an unfavourable site in relation to the metabolism of particular species of parasite.

In the gut too, the size and shape of the villi and/or the composition of bile may be important. For example, it has been shown that the tegument of the cestode *Echinococcus granulosus*, a parasite of carnivores is rapidly lyzed by deoxycholic acid but not by cholic acid. Since the former occurs in high concentration in herbivore bile and the latter in carnivores, this would be one factor in restricting the host specificity of this parasite. Williams (1960) has shown that the structure of the gut in fish is an important factor in the host specificity of Tetraphyllidea. Numerous other examples involving behavioural or ecological insusceptibility could be quoted.

Advances in our understanding of the immune process evolve with bewildering rapidity and it is not possible here to give more than the most generalized account of the current position. Useful general texts on immunology are those of Bellanti (1971), Gold & Peacock (1970), Humphrey & White (1970), Roitt (1971), Ward (1970); a valuable glossary of immunological terms (and their origins) is also available (Halliday, 1971). Immunity to animal parasites in particular has also been the subject of a major (two volumes) review by Jackson, Herman & Singer (1969—1970) and several symposia.

Natural and acquired immunity

If a parasite makes contact with a potentially susceptible host but fails to become established in it, the host is said to possess *natural* or *innate* immunity or resistance to that parasite. If an initial infection induces a degree of resistance to subsequent infections of a parasite, the phenomenon is termed *acquired* immunity or resistance. Acquired immunity may develop in two ways:

(*a*) *naturally*, as the result of a natural infection with a parasite of the same (i.e. homologous) species or a different (i.e. heterologous) species. Some degree of immunity in newly born mammals can occur due to transfer of antibodies from the mother to the offspring either via placental passage or in milk.

(*b*) *artificially*, as a result of the deliberate inoculation (= *vaccination*) of the live or dead organism or some material associated with it (i.e. an extract, homogenate, secretion, excretionor metabolic product). Passive immunity may be induced artificially by injection of antiserum i.e. serum containing antibody.

These various types of immunity are discussed further below.

Cell types in the immune system

In vertebrates, the effector system concerned with immune mechanisms is known as the *lymphoid-macrophage* or *lymphoreticular* system. The cellular elements are found throughout the body but especially in the thymus, lymph nodes, spleen, bone marrow and in the respiratory, gastrointestinal and urinogenital systems. The cells chiefly involved are the *macrophages, granulocytes, plasma cells* and *lymphocytes;* they all share the property of being activated by contact with non-self material. Lymophoctes, in particular, play a central role in the specific immune response and an understanding of their origin and properties is central to an understanding of immunology. In the non-specific response, on the other hand, the phagocyte is the major cell involved.

Non-specific responses

Although the reactions below are considered under the heading of 'non-specific' responses, they are all capable, under certain conditions of being influenced by, or interacting with, certain components involved in the specific immune response, such as antibody or complement.

Phagocytosis

The first reaction of the host to the invasion of small particles of foreign material, such as bacteria, is to attempt to engulf them by phagocytosis. Two types of phagocytic cells are involved in this process, *macrophages* and *polymorphonuclear leucocytes* sometimes referred to as *microphages.* Macrophages in the circulating blood are known as *monocytes,* those in the connective tissues as *histiocytes,* and those in the spleen, thymus and lymph nodes as reticulum cells. *Sinuslining* macrophages occur in liver sinusoids, capillaries etc. Phagocytes belong to the *reticulo-endothelial system*; the latter is now often called the *lymphoid-macrophage system*.

The role of phagocytes is to engulf particulate matter and digest it, where possible, by lysosome action, or if undigestible (e. g. carbon particles) to store it so that it no longer acts as an irritant. Although phagocytoiss can occur in the absence of antibody, the process can be greatly facilitated by the action of *opsonizing antibodies* or *opsonins* (especially, IgG and IgM antibodies) which, in some way not understood, make the antigen more readily ingested. Phagocytosis is also enhanced by a serum component known as *complement* in the presence of antibody.

Macrophages can also pass on antigens or their products to lymphoid cells and thus can act as an important link between the non-specific and specific immune responses.

Inflammatory response

If the amount of foreign material invading the host is small it may be surrounded by phagocytic cells and gradually immobilized by the deposition of collagenous tissue around it. If the amount is large, the reaction is more severe and

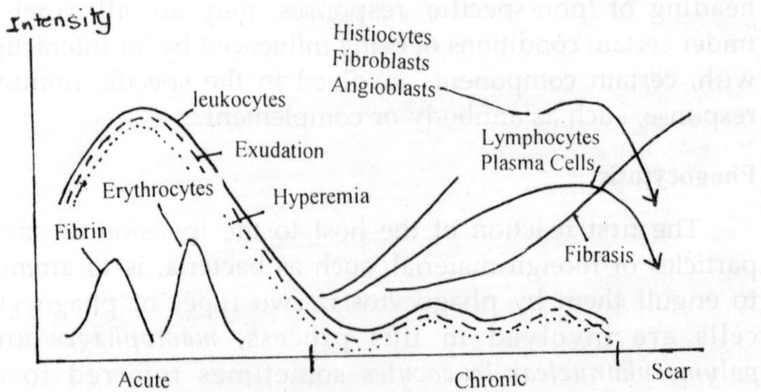

Fig. 18.1 : Schematic representation of the tissue reactions to inflammatory stimuli.

inflammation occurs. This shows itself as a swelling (*oedema*) of the tissue brought about as a result of a localized dilation of the capillaries (*vasodilation*) which results in an increase in the blood supply to the affected area. This is accompanied by the passage of protein from the blood into the tissue fluids together with polymorphonuclear leucocytes which collect (by chemotaxis), in large numbers, at the invasion site.

The onset of vasodilation is brought about by a number of factors which include (*a*) a local nervous reaction and (*b*) the release of an active substance termed *histamine* from mast cells. In addition to histamine, other pharmacologically active substances, 5-hydroxytryptamine (=serotonin in USA), bradykinin and SRS-A (slow reacting substance of anaphylaxis) are released as a result of antigen-antibody reactions. The role of these substances is not fully understood (see *hypersensitivity*).

The passage of leucocytes is followed by a migration of lymphocytes, some of which may transform into mononuclear cells or *fibroblasts*. The latter also take part in the formation of the fibrous *capsules* which surround many issue parasites — especially larval helminths (e.g. *Cryptocotyle lingua, Trichinella spiralis*). If capsules persist for long periods, calcification may occur.

Abnormal growth responses to parasites

One of the most interesting features of tissue reactions, often associated with the presence of parasites, is a change in the growth pattern of parasitized tissue. This change of pattern may be reflected in a number of ways as follows:

(*a*) *Hyperplasia*, that is an increase in the rate of cell division. In this condition, the cells increase in number but not in size. This is seen frequently during repair, when the reparative processes have been carried out in excess of the actual requirements. It is often produced as the result of irritation (e.g. the liver of a rabbit infected by *Eimeria stiedae*;. Note that *hypertrophy* is an increase in cell *size* but not number.

(*b*) *Metaplasia,* which is the transformation of one tissue type into another. It is not commonly associated with the presence of animal parasites, but is known to occur, for example, in the lungs of mammals infected with the fluke *Paragonimus westermani.*

(*c*) *Neoplasia,* which is the growth of new kind of tissue from existing cells, and such growths are commonly referred to as *tumours.* Any definition of a tumour can only be regarded as temporary and provisional. 'A neoplasm or tumour may be defined as a new growth arising from pre-existing tissue, independent of the needs of the organism, and subserving no useful purpose, but on the contrary, often acting deleteriously. It is autonomous, i.e. it follows its own line of growth, and is not regulated by those governing the tissues of the body in which it is growing. . . the life history. . . of a neoplasm appears to have no typical termination, the tumour cells seeming to be endowed with the power of continuous, aimless and unlimited proliferation.' A malignant tumour is popularly referred to as a 'cancer.'

'The term cancer is a convenient label or pigeonhole for a number of pathological tissue conditions exhibiting certain common characteristics, but all unique in certain other respects. Indeed, it is not a disease at all, but an assemblage of many different diseases which have certain symptoms in common. A century ago, medical men spoke of "fever" as a disease. Now they no longer do so. Something of this sort is happening with cancer during the present century.'

In medical terminology, however, the term 'cancer' is used synonymously for a *carcinoma,* a malignant tumour of epithelial tissue. The term *sarcoma* refers to a malignant tumour of connective tissue.

It is significant that helminth parasites are amongst the very diverse agents associated with the production of neoplasia; the literature has been reviewed by Schwabe

(1955). A number of species, has been incriminated. Of these, only the larvae of the cestode *Hydatigera taeniaeformis (Cysticercus fasciolaris)* in the liver of rodents, and the adults of the nematode *Spirocerca lupi* (see below) in the oesophagus of dogs have been unequivocally shown to be associated with sarcoma formation in these sites. The other cases listed are doubtful ones. The presence of *Schistosoma haematobium* has long been associated with cancer of the bladder in man, and this question has received much attention. The evidence has been reviewed by Jordan & Webbe (1969) who conclude that *S. haematobium* is apparently associated with bladder cancer in some areas of Africa but not in others. It has been suggested that this variation in effect in different areas may be nutritional in origin or possibly related to the absorption of a carcinogenic substance from the different cooking pots used in these areas!

Table 18.1. Parasites believed to be associated with tumour formation (neoplasia) in mammals. Only in the case of those marked* has the relationship been unequivocally established.

Group	*Parasite*	*Host*	*Site of tumour*
Protozoa	*Eimeria stiedae*	rabbit	liver
Trematoda	*Schistosoma mansoni*	man	intestine
	Schistosoma haematobium	man	bladder
	Schistosoma japonicum	man	intestine
	Paragonimus westermani	tiger	lung
	Clonorchis sinensis	man	lung
Cestoda	**Hydatigera taeniaeformis*	rats	liver
	Echinococcus granulosus	man	lung
Nematoda	*Gonylonema neoplasticum*	rat	tongue
	**Spirocerca lupi*	dog	oesophagus

Cysticercus fasciolaris

Hepatic sarcoma in the liver of the rat induced by the presence of the larva of *Hydatigera taeniaeformis (Cysticercus fasciolaris)* has received much attention since its initial discovery by Borrel in 1906. To date, more than 7500 cysticercus tumours have been developed experimentally.

The majority of these have been sarcomas, but a few have been of other histological types. In general, malignant tumours arise only following an infection of from 250 to 675 days' duration. The tumours appear to arise not from the liver tissue itself, but from the fibrous capsule formed by the host around the larva. Attempts to isolate a specific active agent from *Cysticercus fasciolaris* have been unsuccessful; there is some evidence that it may be associated with the calcareous corpuscles normally present in the body fluid of cysticerci.

The extraordinary difficulties inherent in attempts to obtain unequivocal results in experimental work on tumour formation are reflected in the case of *Gogylonema neoplasticum*. This nematode occurs in the tongue, oesphagus and forestomach of rodents. In these sites, it was thought to induce malignant tumours and an intensive research programme was carried out on this organism, especially by Fibiger and his co-workers. Critical experiments have shown that, although this nematode can act as chronic irritant, it does not induce malignancy. It was shown that the so-called maligancy described by early workers was related to the unintentional use, in the original experiments, of Vitamin A deficient diets for the experimental rats used. The parasite was first encountered in rats in a sugar refinery where the abundance of sugary food available resulted in the natural diet being a vitamin A deficient one, although this was unrecognized at the time. When the infections were established and maintained in laboratory rats, it happened that a vitamin A deficient diet consisting of white bread *and water* (not milk) was again used. Examination of the records of the millers suggests that the bread used in Norway at that time was deficient in vitamin A. Thus, by a series of coincidences, the abnormal diet conditions occurring in the sugar refinery were essentially reproduced.

Spirocerca lupi

The association of this nematode and the occurrence of sarcomas in dogs has been much studied and presents an interesting problem in host-parasite relationships. The adult worm lives in a nodular mass in the wall of the oesophagus of dogs and other carnivores; there is a small passage from the lumen of the nodule through which eggs pass. Numerous species of dung beetles act as intermediate hosts and animals such as birds, hedgehogs, lizards, mice and rabbits serve as transport hosts. Infective larvae ingested by a dog are freed in the stomach, penetrate the mucosa and eventually, after a migration, end up in the oesphageal wall. A reactive granuloma develops around the worms and, under conditions not understood, becomes transformed into a sarcoma; the genetic disposition of a dog to the neoplasm is probably an important factor. Research on this intriguing problem has been reviewed by Bailey (1963, 1972).

Specific immune responses: basic concepts of immunology

Antigens and antibodies

In addition to the reaction of phagocytic cells described earlier, the invasion of a vertebrate host by a parasite or other 'foreign' (i.e. non-self) material is associated with a number of serological (or humoral) reactions, so-called because they occur primarily (but not always) in the serum of the infected host. A substance which invokes these reactions is said to be *antigenic* and the molecules it releases or exposes to the host are known as *antigens*. The introduction of antigen provokes the appearance, in the blood and tissue fluids, of a soluble protein entity known as *antibody* which has the property of combining with that specific antigen. The nature of antigens and antibodies is considered further below.

Antigens. Antigens are macromoleules of protein or complex polysaccharides. In order to act as an antigen, a substance must (*a*) have a minimal molecular weight (rarely less than 5000), and (*b*) exhibit 'foreignness', i.e. must contain chemical grouping or configurations which are foreign to the animal. Lipids are not directly antigenic but may act as haptens, i.e. 'partial antigens,' which contain at least one determinant group of an antigen. When coupled to proteins or complex polysaccharides they can dictate specificity of the coupled 'antigen.' The term *allergen* has recently been introduced for antigens which provoke *allergic* reactions (see hypersensitivity).

Antibodies. As mentioned above, the introduction of an antigen results in the appearance in the serum of a soluble protein entity termed *antibody*. Antibody has the property of combining specifically with the antigen which provoked its production, but with that antigen only, so that the molecular structure of antibody is in some way related to that of antigen. *In vitro,* the presence of antibodies in serum is revealed by the ability to react with their corresponding antigens. A number of complex antigen-antibody reactions are known and these are discussed later.

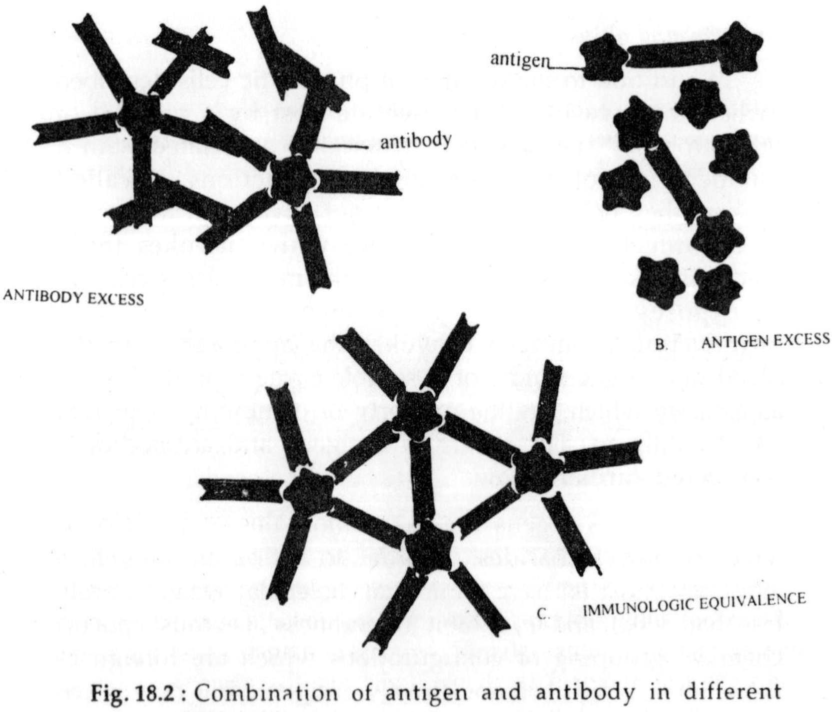

Fig. 18.2 : Combination of antigen and antibody in different proportions.

Theories of antibody synthesis

Although antibody appears to originate in cells of the lymphoid series, or their derivatives (e.g. plasma cells, the method whereby antibodies are actually synthesized is one

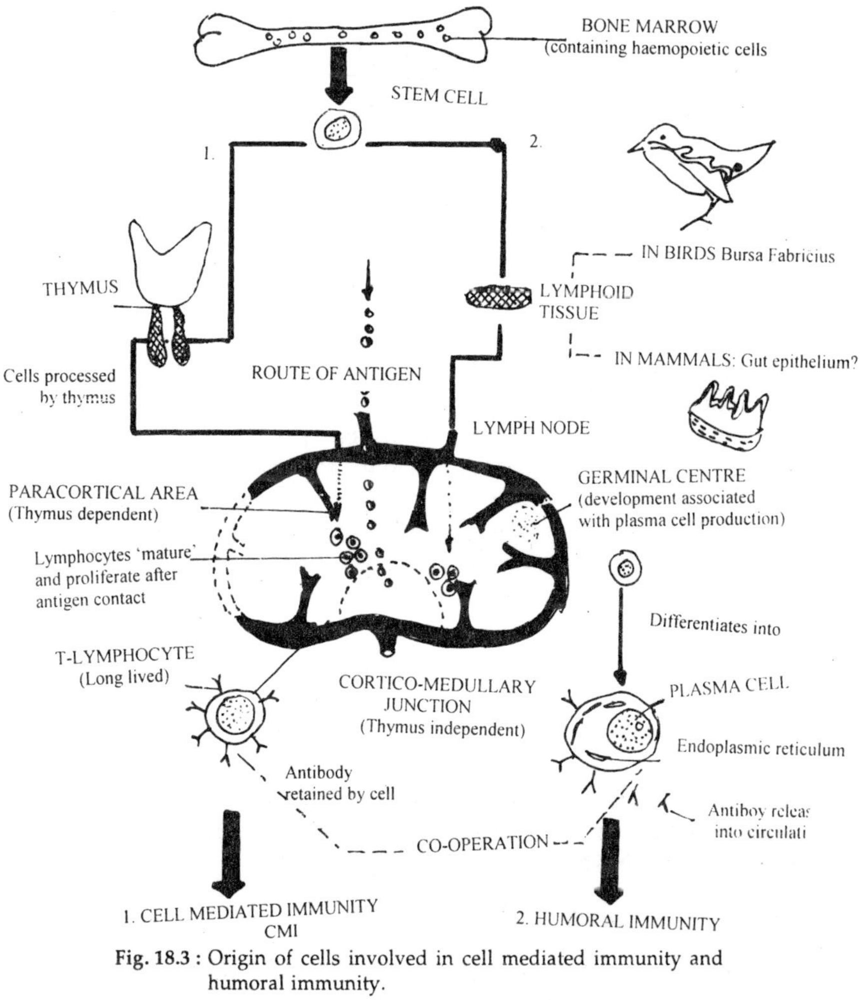

Fig. 18.3 : Origin of cells involved in cell mediated immunity and humoral immunity.

of the most controversial problems in immunology. Detailed consideration of this problem is beyond the scope of this book; the position may briefly be summarized as follows.

There are two main theories: (*a*) the instructive theory, and (*b*) the selective theory; several modifications of these are known.

The instructive theory postulates that antigen enters the site of serum protein synthesis and acts instructively as a template around which an immunoglobulin chain could be moulded. This then takes on a configuration which is complementary to one of the determinant groups of the antigen. On separation from the template, the antibody molecule would thus have a specific antigen-combining site.

The selective theory was first put forward by Jerne (1955) and later modified by Burnet (1959) who offered a 'clonal selection theory.' This postulastes that before lymphoid cells reach maturity large numbers of variants arise each capable of producing one specific type of antibody. When an antigen enters the body it makes contact with the cell (s) which have the corresponding reactive sites on their surfaces and combines with them. This stimulates the cell to multiply by repeated cell division, producing a clone of similar cells all producing the same antibody. The validity of this theory in the light of modern work is discussed by Gold & Peacock (1970).

According to either of these views, for an antigen to stimulate the formation of a reacting antibody, its specific molecular pattern must not be destroyed or altered to any significant extent. This probably explains why some antigens used as vaccines (e.g. bacteria) are only active in a living state, while others may remain active after having been killed by formalin or other treatment.

Cellular processes in antibody formation

Specific immune responses are brought about by two types of effector mechanisms: (*a*) *cell-mediated immunity* (CMI) or *delayed hypersensitivity*, mediated by specially sensitized lymphocytes, and (*b*) *humoral immunity*— mediated by antibody. Cells of the lymphoid series which are concerned with these immune responses are said to be *immunocompetent* and such cells have been shown to engage in a remarkable 'cellular traffic' throughout the body.

Cell mediated immunity. A tentative scheme showing the origin of the components involved in CMI is given in Fig. 18.3. The basic cells from which immunocompetent cells arise are the *stem cells* in the bone marrow. These involved in CMI are 'processed' by the thymus and (in some way not understood) pass to the paracortical region of various lymph nodes, where, after stimulation by antigen, they proliferate to produce a population of 'sensitized' lymphocytes (thymic dependent cells or *T-cells* or *T-lymphocytes*); these carry surface-bound antibodies directed against the antigen which stimulated their proliferation. T-lymphocytes are especially active against 'fixed' antigens or substances (such as drugs) which persist and which are themselves not immunogenic but become so through combination with tissue protein. They are thus largely concerned with so-called *allergic* reactions. The cellular events involved in CMI are slow to develop, usually requiring at least 24 hours - hence the expression 'delayed hypersensitivity.' In contrast, humoral reactions (see below) can occur within minutes (*anaphylaxis*) or hours. T-lymphocytes are long lived.

Humoral immunity. The production and secretion of serum antibody is dependent on a population of B-lymphocytes which, like T-lymphocytes, develop from stem cells in bone marrow (Fig. 18.3). These B-lymphocytes develop into cells known as *plasma* cells which in birds are 'processed' by the Bursa of Fabricius, a lymphoid organ connected to the cloaca. Plasma cells are characterized by possessing a well developed rough endoplasmic reticulum and they are actively engaged in the synthesis of protein which is secreted into the bloodstream. The equivalent of the avian Bursa of Fabricius has not been identified in man although it is thought possible that gut-associated lymphoid tissue, such as Peyer's patches, the tonsils and the appendix may act as the processing sites.

Structure and function of antibodies

The proteins in plasma may be separated by electrophoresis into five groups—albumin, and four globulins a_1, a_2, β and *y*. Antibodies were formally thought to be associated with the *y* globulin fraction, but this is now recognized to be an oversimplification, for some antibody has been detected in other components also. The more general term *immunoglobulins* is now used to describe the closely related proteins involved in antibody activity.

In each vertebrate species, immunoglobulin molecules can be subdivided on the basis of the structure of their 'backbone'. In man, for example, five classes of immunoglobulins have been recognized. These are referred to by the letters G, A, M, D and E after the symbol *y* (indicating their electrophoretic mobility as *y* globulins), e.g. *y*G *or* after the abbreviation Ig (indicating their immunoglobulin function), e.g. IgG; the latter terminology is used here. IgG, IgM and IgA are the major immunoglobulins and may be detected by immunoelectrophoresis. The properties of the various immunoglobulin classes are summarized in Table 18.2, and are briefly commented on.

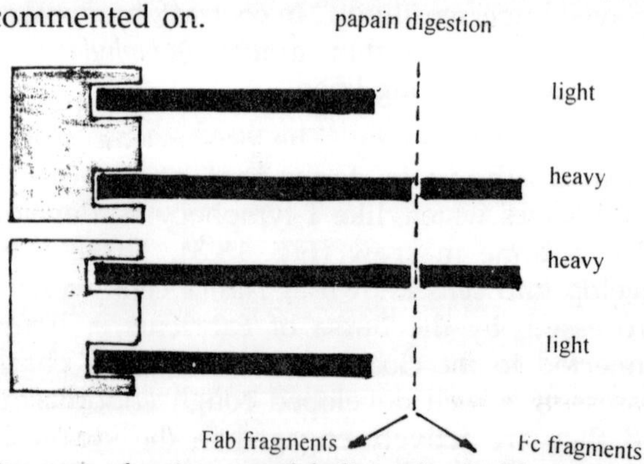

Fig. 18.4 : Structure of an immunoglobulin molecule. There are two heavy and two light chains with di-sulphide linkages. Papain digestion splits the molecule into two fragments, **Fab** which retains its antibody activity, and **Fc** which shows no antibody activity but is crystallizable. The binding of antigens to antibody is schematically represented.

IgG. This is the most abundant of the immunoglobulins and is believed to provide the bulk of immunity in the tissues to invading organisms including bacteria, viruses and fungi, as well as parasites. It accounts for more than 85 per cent of the total. It has a molecular weight of about 150 000 and a sedimentation constant (in Svedberg units) of 7S. It can cross the placenta and can fix antigen. It achieves significant concentration in the internal body fluids, particularly in extravascular sites. Most antibodies concerned in serological tests for helminths occur in this class. The complexes of bacteria with IgG antibody are able to attach to the phagocytic cells because some of these cells have specialized receptors for sites on the heavy chain of the Fc portion of IgG.

Table 18.2. Some properties of human immunoglobulins. (Data from Halliday, 1971 and Roitt, 1971)

	IgG	*IgA*	*IgM*	*IgD*	*IgE*
Mean serum concentration (mg/100 ml)	1240	280	120	3	0.03
Sedimentation constant ($S_{20,}$ W) (Svedberg units)	6.5-7	7-15	18-19	6.2	8
Molecular weight	150 000	170 000	890 000	150 000	1960 000
Electrophoretic behaviour	γ_2 (slow)	γ_1 (fast)	γ_1 (fast)	γ_1 (fast)	γ1 (fast)
Heavy chain designation	γ	α	μ	δ	ε
Light chain designation	$K_1\lambda$	$K_1\lambda$	$K_1\lambda$	$K_1\lambda$	$K_1\lambda$
Antibody activity	Crosses placenta Complement fixation Anaphylaxis Antitoxins 'Late' antibody' valence = 2	Secretary antibod Antitoxins Some reagins	Agglutinins Lysins Opsonins Bactericidins 'Early' antibody; valence = 5	None known	Reaginic antibody

IgA The second most abundant immunoglobulin. It appears especially in the external sero-mucous secretions such as tears, nasal fluids, saliva and secretions of the lung and alimentary canal and serves to protect these surfaces against attack by foreign material. IgA is synthesized locally by plasma cells and secreted as a dimer. One type of IgA is non-secretory.

IgM. Immunoglobulins in this have a higher molecular weight (890 000 or more), a higher sedimentation constant (19S) and a greater electrophoretic mobility than the others. They appear early as a response to antigen, and, on account of their large size, are confined to the bloodstream. They efficiently agglutinate particulate antigens such as red blood cells and bacteria. In elasmobranchs IgM appears to be the only immunoglobulin produced.

IgD. Occurs in very small quantities in normal serum. No antibody activity has yet been associated with IgD.

IgE. Present only in trace amounts in serum; known as *reagin, reaginic* or *cytotropic antibody*. It has the ability to remain firmly fixed when injected into human skin, probably being bound to mast cells. Contact with antigen leads to release of mast cell granules (degranulation with release of vasoactive amines, e.g. histamine hypersensitivity reaction, Type III. It has been suggested that the release of histamine initiated by parasites coming in contact with mast-cell bound IgE antibody facilitates the rejection of parasites (e.g. *Nippostrongylus brasiliensis*, by the gut. Reagins probably play an important role in the resistance to several helminths.

Subclasses. Immunochemical analysis of the above main classes of immunoglobulins has revealed that they can be further divided into subclasses such as IgG_1, IgG2 etc. Further consideration of these subclasses is beyond the scope of this book.

Serum of an immunized animal (containing antibodies) is termed *antiserum* or less aptly *immune* serum and if some of this serum is injected into a nonimmunized animal, this second animal also will usually exhibit immunity. Immunity produced by injection of antiserum is termed *passive immunity* since the recipient animal has not taken part in the development of the immune state. Duration of passive immunity is short, perhaps a few weeks, and depends partly on the amount of antibody injected. In contrast, the kind of immunity developed after the injection of antigen is termed *active immunity,* since the infected animal body is itself actively responsible for its production. The degree and duration of active immunity depends on a number of factors; in some it may last for years; in others it may wane sharply after the antigen has disappeared.

Antibody-antigen reactions

General principles. In vivo. At the cellular level, combination of antibody and antigen may cause (*a*) agglutination (clumping), or lysis of cells containing antigen or (*b*) more ready ingestion of such cells or antigens by phagocytes or (*c*) damage by the antigenatibody complexes forming precipitates.

In vitro. Many of the above reactions can better be studied *in vitro* (see below). For laboratory methods and detailed descriptions of these reactions see Campbell *et al.* (1963), Crowle (1961), WHO/UNESCO (1964), Ackroyd (1964), Weir (1973) and Kwapinski (1972). The interactions between antibodies and antigens are.

Antigens (multivalent) combine with the antibodies (generally bivalent) to form three-dimensional lattices which aggregate and precipitate. However, (at least *in vitro*) the proportions in which antibodies and antigens are mixed have a marked effect on the composition of the resulting precipitates:

(*a*) *Excess antibody*. The resultant precipitate contains more antibody than antigen; the complexes formed are insoluble.

(*b*) *Immunological equivalence*. The combining sites are used to the full and virtually all the antigen and antibody precipitate together and neither can be detected in the supernatant.

(*c*) *Excess antigen.* Resultant precipitate contains more antigen than antibody and the precipitate tends to dissolve due to formation of solube complexes.

Tests for antigen and antibody

(*a*) *Titre.* Antibody concentration is expressed as the *titre* of a serum sample. Titre is the reciprocal of the highest dilution of serum which gives a detectable antibody-antigen reaction and is expressed as the number of (arbitrary) antibody units per unit volume of undiluted serum.

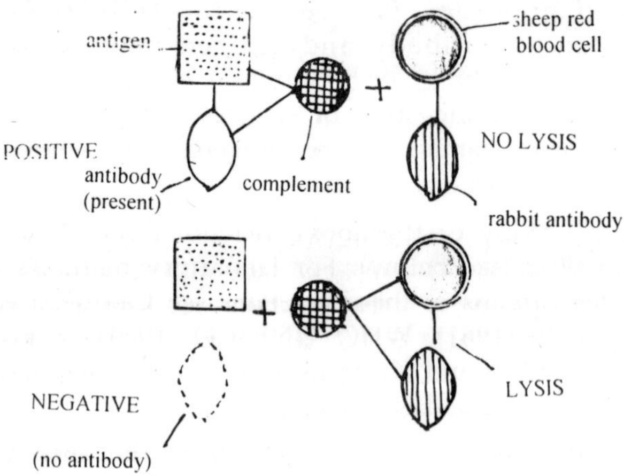

Fig. 18.5 : The complement-fixation reaction.

(*b*) *Precipitation.* In the simplest test, antigen is carefully overlayed on the antiserum in a tube. A white line of precipitate forms at the interface if an antibody-antigen reaction has taken place.

(*c*) *Agglutination.* The basic mechanism of agglutination is essentially the same as that of precipitation except that the antigens are larger and either particulate or cellular (e.g. bacteria, yeasts, erythrocytes). Agglutination of erythrocytes is known as *haemagglutination.* The use of the haemagglutination reaction has been extended to produce a delicate test for antibody in serum.

Erythrocytes are treated with tannic acid after which antigens can be absorbed on their surface and used to test for the presence of antibody. Other particles—such as polystyrene, latex, sephadex or bentonite can similarly be used as 'carriers' and coated with antigen. Particles or cells treated in this way are exceedingly sensitive and can detect as little as 0.003μg of antibody (measured as nitrogen).

Complement-dependent reactions. The term *complement* was originally applied to a component of normal serum which, when acting upon an antibody-coated cell, led to lysis of that cell (cytolysis). This activity is destroyed by heating serum to 56°C for 30 minutes. Guinea-pig serum is particularly rich in complement. Reactions which are largely dependent on complement include cytolysis (cytotoxicity), phagocytosis-promoting activity (opsonization), chemotaxis, and altered permeability.

Although 'complement' was originally thought to be one substance, it is now known to consist of at least nine functional entities (or eleven discrete proteins) which act in sequence. When complement is first 'activated' by an immune complex (e.g. antibody bound to a red cell) it develops the ability to activate the next component, which in turn activates the next and so on. The process involves a series of complex reactions. Hence the 'triggering' of a single C_1 molecule can lead to activation of many thousands of the lat components, the whole process thus producing an *amplification* effect. This effect manifests itself in lysis of the cell probably through the activation of a phospholipase. For a detailed account of the 'complement sequence' see Roitt (1971).

Lysis. This occurs as the result of the effect of specific antibody directed at a surface antigen and mediated via the complete complement sequence. The reaction is limited to the destruction of erythrocytes (haemolysis), certain gram-negative bacteria (bacteriolysis) and the vegetative forms of protozoans.

Phagocytosis, chemotaxis, altered permeability. The combined action of antigen, antibody and complement greatly enhances

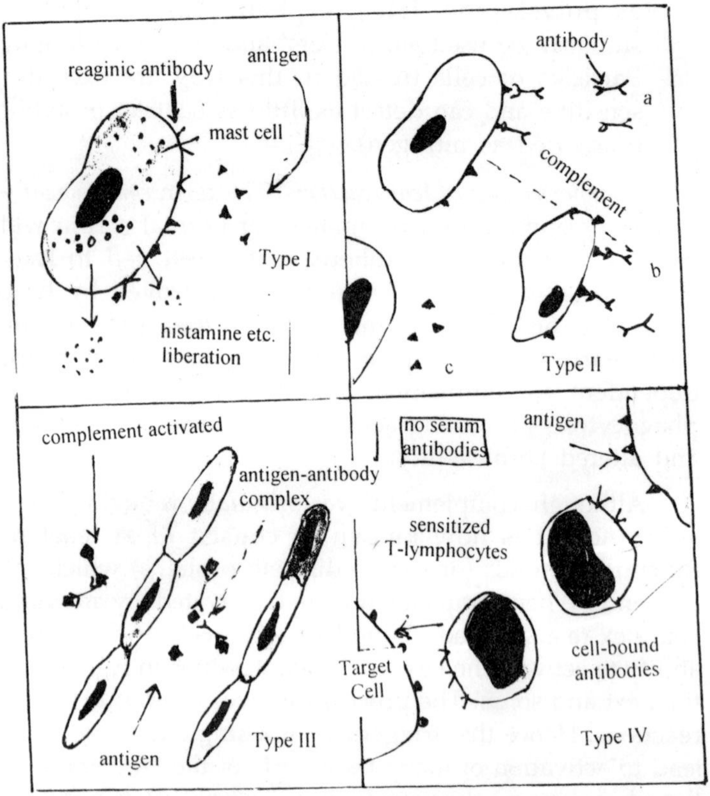

Fig. 18.6 : The four chief types of hypersensitivity reactions; Type V omitted. Sites of involvement of complement are indicated by dotted lines.

Type I. *Anaphylactic-type hypersensitivity (reagindependent)*. Initiated by antigen reacting with reaginic antibody (produced elsewhere) which has bound to mast cells and basophils. Histamine and other pharmacologically active substances are released.

Type II. *Cytotoxic-type hypersensitivity.* Initiated by antibody directed against antigenic components of cell surfaces. Cell death may occur due to (*a*) complement-dependent lysis; (*b*) adherence reactions, followed by phagocytosis; (*c*) direct attack by certain lymphoid cells.

Type III. *Complex-mediated hypersensitivity.* Formation of antigen-antibody complexes initiate (via activation of the complement system and the Hageman factor) a series of complex reactions including release of vasoactive amines and kinins and attraction of polymorphs. In mammalian pathology these reactions result in Arthus-type damage.

Type IV. *Cell-mediated (delayed-type) hypersensitivity.* Sensitized T-lymphocytes bearing specific antibody on their surface transform into blast-like cells capable of killing target cells bearing the appropriate antigen. Serum antibody does not appear.

the activity of phagocytic cells. This is due to the appearance of chemotactic factors, through the activation of components of the complement sequence, especially that known as C_3, an event which results in the accumulation of leucocytes at the site. A small cleavage product of C_3 (C_{3a}) brings about the attraction of neutrophilic granulocytes in a unidirectional manner. C_{3a} also has *anaphylatoxin* activity which results in histamine release from mast cells. As well as causing smooth muscle contraction, histamine release results in increased vascular permeability. The overall effect of the latter, together with neutrophil reaction, is to produce *acute inflammation*.

Phagocytosis is also enhanced by the action of *opsonins*—antibodies which adhere to the surface of micro-organisms which are thus said to be 'opsonized.' This reaction is said to make the bacteria 'ready for the table.'

Complement fixation. Antigen-antibody complexes have the ability to 'fix' or bind complement and this may be used as a test for antibodies using a known antigen or vice versa. A haemolytic system must be used to make the effect visual. Thus, serum (inactivated by heating to destroy its own complement) which is suspected of containing a particular antibody is mixed with the corresponding parasite antigen and added to fresh guinea-pig serum (as a complement source). 'Sensitized' sheep red blood cells (i.e. those mixed with (heated) antiserum prepared against them in a rabbit) are added, and if haemolysis occurs, the complement level has remained high and the serum is therefore negative for parasite antibody. If no haemolysis occurs, the complement has been bound by the parasite antigen antibody complex and the serum under test is positive for antibodies to the parasite.

Immunodiffusion: in vitro detection of antibodies and antigens

Double diffusion. Detection of antibody and antigen has been greatly facilitated by the fact that the precipitation reaction between them can be visulaized in gels, usually agar. In immunodiffusion techniques, antigen and anti dy are allowed to diffuse towards each other in agar and - according to the number of types of antibodies and antigens present - one or more precipitation bands are formed.

In the simplest techniques, the antibody and antigen are mixed with agar, placed in tubes separated by a further layer of pure agar, and allowed to diffuse towards each other through the pure agar. Precipitation band (s) form at the interface. This method is sensitive but does not give much separation of different bands.

Much better 'resolution' is obtained in plates, in which antigen and antibody are allowed to diffuse towards each other from troughs or wells cut in agar. Such plates are known as Ouchterlony plates, after the inventor.

Immunoelectrophoresis. In immunoelectrophoresis, the antigen, placed in a circular hole in agar, is subjected to electrophoresis. The components of the antigen separate at rates depending on the net charge in their molecules. At the alkaline pH's used, the proteins, which carry a negative charge, might be expected to move towards the anode. However, agar itself carries a negative charge and since it

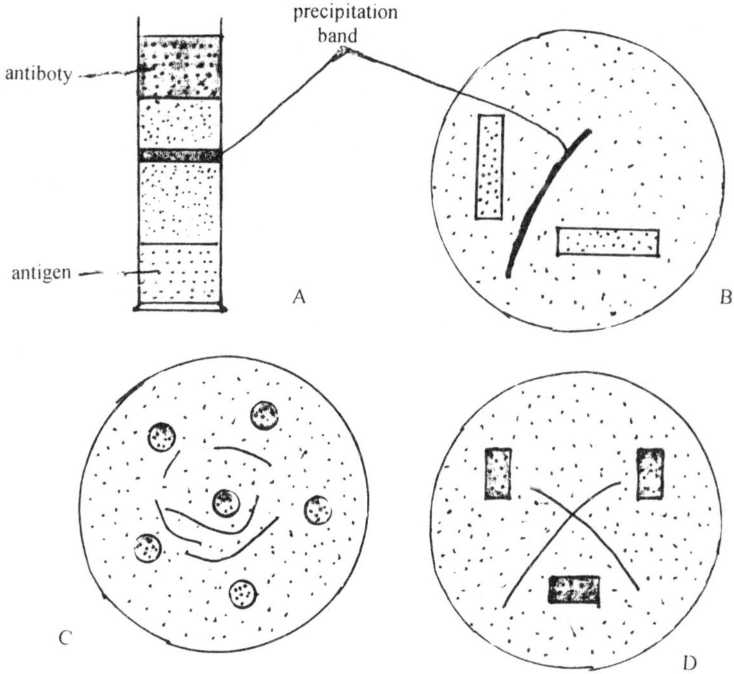

Fig 18.7 : Various forms of double diffusion-in-gel techniques. A, Tube technique of Oudin. B, C, D, Variations of the plate technique of Ouchterlony.

is not able to move towards the anode, pressure (endosmosis) on the liquid in the gel makes the protein fraction move in the other direction towards the cathode (endosmosis).

Antibody is then placed in the trough (Fig. 18.8) and allowed to diffuse into the agar towards the separated antigen components. Where antigen and antibody meet, precipitation arcs develop.

Hypersensitivity

In addition to immunity, an antigen-antibody reaction may give rise to a state of *hypersensitivity*; this term is widely used as being synonymous with *allergy*. Hypersensitivity is essentially a heightened response to an antigen, in an animal which has already been immunologically primed, or *sensitized* to that antigen. Thus, a guinea-pig injected with a small quantity of ovalbumin becomes 'sensitized' to that substance. If a further challenging dose of ovalbumin is given, 2—3 weeks later, the animal suffers a violent general systemic reaction (*anaphylaxis*) which may result in death within a few minutes. This would be an example of *immediate hypersensitivity*.

Table 18.3. Comparison between humoral and cell-mediated immunity.

	Humoral	*Cellular*
Sensitizing material	proteins, polysaccharides, lipids	proteins or protein-hapten
Reaction time'	immediate': minutes to hours	'delayed': 24-48 hours
Initiating event	union of antibodv with antigen	Reaction of 'sensitized' lymphocytes with antigen
Transfer	circulating antibody in serum	cells and cell derived 'transfer' factor
Effector mediators	complement; vasoactive amines (e.g. histamine)	secreted soluble factors (lymphokines)

By restricting the antigen to a small area of skin, local hypersensitivity may be produced. Intadermal reactions based on this procedure are of value in the diagnosis of certain diseases of helminth origin (see below). Other hypersensitivity reactions in man include asthma, hayfever, urticaria, eczema or (more rarely) anaphylaxis. 'Delayed' hypersensitivity (= cell mediated immunity) has already been dealt and its main characteristics are mentioned briefly below.

One fundamental difference between immediate and delayed hypersensitivity is that whereas immediate hypersensitivity (a humoral reaction) can be transferred with serum antibodies (which can be detected by the common serological techniques of precipitin tests, complement fixation etc.), delayed hypersensitivity (since it is a cell mediated reaction) is transferable *only with lymphoid cells* or their derivatives.

Hypersensitivity is a complex subject, and the principles on which it operates can only be summarized briefly here. For detailed consideration see texts such as Roitt (1971), Bellanti (1971) and Gold & Peacock (1970).

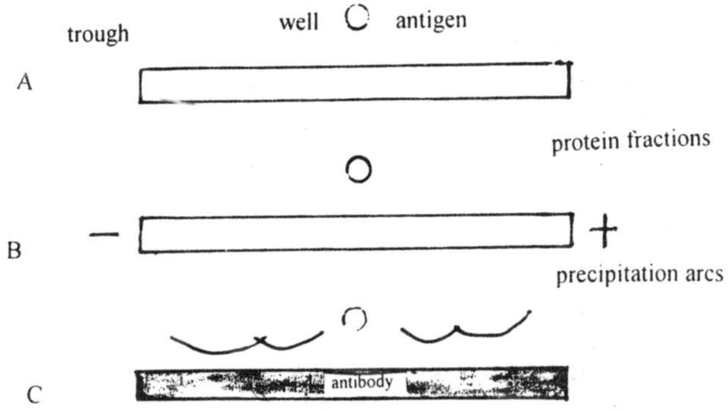

Fig. 18.8 : The principle of immuno-electrophoresis.

A. Antigen added to well in agar.

B. Electric field applied, antigen fractions separate.

C. Antiserum containing antibody added to trough. Diffusion of antibody results in precipitation of antigen-antibody complexes forming arcs.

The following five types of hypersensitivity have been recognized:

Type *I Anaphylactic-type hypersensitivity (anaphylaxis)*. Antigen reacts with a special type of antibody known as *reaginic (homocytotropic)* antibody bound to mast cells or circulating basophils. This results in degranulation of the mast cells and release of vasoactive amines (e.g. histamine). SRS-A (slow reacting substance of Anaphylaxis) is also released during anaphylaxis, but its origin is unknown.

Reaginic antibodies are characterized by their high affinity and permanent attachment to skin. They are best tested for by utilizing their high affinity for cells in *passive cutaneous anaphylaxis* (PCA) with several days' interval between intradermal injection of antibody and intravenous injection of antigen. Serum from an infected animal is injected into the skin of a normal, parasite-free animal of the same species. Several days later, a mixture of antigen and Evans blue is injected intravenously. Within 20-30 minutes a discrete blue area 10-20 mm in diameter appears at the initial site of injection in the case of a positive serum. In this way, reaginic antibodies are distinguished from other antibodies which do not have an affinity for cells and which diffuse away from the site of injection. Reaginic antibodies appear to be characteristic of certain helminth infections, e.g. *Nippostronylus brasiliensis*.

Type II *Cytotoxic-type hypersensitivity*. In this type, either phagocytosis or lysis of a cell takes place. This is due to the action of antibodies (IgG or IgM) against antigens on the surface of target cells (e.g. blood cells). Lysis takes place when the operation of the full complement system can occur.

Type III *Complex-mediated hypersensitivity*. In this type, complexes or microprecipitates formed between antigen and humoral antibodies can lead to (*a*) cell damage, (*b*) activation of the complement system, (*c*) platelet aggregation and (*d*) activation of the Hageman factor. A series of complex reactions follows including vasoactive amine release, polymorph attraction and activation of the *kinin* system. The latter involves the release of small peptides known as kinins with potent pharmacological properties. Kinins have been known to be released in various *Trypanosoma* infections. Examples of Type III hypersensitivity are Arthus' reaction, serum sickness and malarial nephrosis.

Type IV *Cell-mediated (delayed type) hypersensitivity*. As already discussed in this type, T-lymphocytes (thymus 'processed'), with specific antibody on their surface, on contact with antigen, are stimulated to release factors which induce delayed-type hypersensitivity. A specific inflammatory reaction develops in the area where the antigen is localized. The reaction takes 24-48 hours-hence the term 'delayed.'

Type V *Stimulatory hypersensitivity*. Antibodies directed against certain cell surface components may stimulate cellular activity.

Hypersensitivity reactions in parasite diagnosis. Intradermal reactions may be useful in the diagnosis of certain diseases of helminth origin. The test is carried out by injecting an extract of parasite or placing some powdered parasite on a small area of scarified skin. A positive result is indicated if a weal develops in the test region within 10—20 minutes.

Immunity in lower vertebrates

The immune response in lower vertebrates has been relatively little investigated; Rowlands (1969) has reviewed the literature in this field. In general, antibodies take longer to appear after the antigen invasion, but persist longer, than in mammals. Their development is also very dependent on temperature. Only a very brief account is given here. The phylogeny of immunity has been reviewed by Smith *et al.* (1969).

Fish. Lampreys possess immunoglobulins and are capable of antibody response. This antibody activity appears to be confined to immunoglobulins with sedimentation coefficients between 17S and 8S. Higher fish have more complex immune systems and immunoglobulins have been detected in elasmobranchs and teleosts. Antibody has also been detected in fish mucus. In general, however, the immune response is weak and poorly protective. Snieszko (1970) and Hoar & Randall (1970) have reviewed the problems of fish immunology.

Amphibia. The immune response and immunoglobulins in amphibia more nearly approach those of mammals than fish. Two immunoglobulins (18S and 6.7S) have been identified.

Reptilia. Humoral antibodies have been little studied 18S, 11S, 7S and 4S components of serum have been identified.

Immune reactions in invertebrates

Immune reactions have been very little studied in invertebrates, the relevant literature begin reviewed by Tripp (1969) and Huff (1940). Most work in this field has been carried out on arthropods (especially insects).

Invertebrate body fluids do not appear to contain substances comparable to antibodies of vertebrates. Nevertheless, substances which agglutinate or haemolyse cells have been reported from the body fluids of some invertebrates. The main reaction of invertebrates to foreign material is to remove it by phagocytosis, although large particles may be surrounded and encapsulated. Little information appears to be available on the factors necessary for phagocytosis in invertebrates and this is a field which would undoubtedly reward further study.

Chapter—19

The Host: Immunity to Specific Parasite Infections

Few fields in parasitology have expanded so rapidly or have aroused such widespread interest as that dealing with the immunology of parasitic organisms. Much of this interest has been concerned with some of the world's worst diseases of parasitic origin, such as malaria, schistosomiasis, or hydatidiasis and has centred on attempts to develop vaccines or to improve diagnostic methods of the causative organisms. At a more basic level, the evolution of antigenic patterns in host and parasite has received some attention. It has been found that parasites possess antigens which may have similar molecular configurations to those of (*a*) closely related species, (*b*) species from other classes of parasites and (*c*) their hosts. This last phenomenon appears to provide some protection for the parasite since the host is unable to recognize the parasite surface as 'non-self.' This *host unresponsiveness* has aroused great interest among parasitologists.

Immunity to parasitic organisms, although showing many features in common with microbiological infections, is, in many cases, made immensely complicated by the occurrence of several hosts in the life cycle. Thus, immunity is rarely a simple reaction to a single stage in the life cycle of a parasite, but frequently one involving responses to stages of the parasite in perhaps several invertebrate and/or vertebrate hosts - stages which may show a continually changing antigenic pattern. Moreover, in some infections - particularly protozoan infections - completely 'sterile' immunity is not achieved. Instead relatively small numbers of parasites persist at a 'latent' or 'subclinical' level - a phenomenon referred to as *premunition*.

Parasite Antigens and Functional Immunity

General Account

It has been pointed out earlier that invasion of a vertebrate generally results in local tissue reactions and the appearance of antibody in the host serum. It must be emphasized that the appearance of antibody in the serum does not necessarily imply that thereby the resistance of the host to the parasite has been increased. A degree of immunity will only have been developed if the antibody acts in such a way as to inhibit the development of the parasite. This inhibition (or death) of a parasite can result from: (*a*) stimulation of host tissue reactions (especially cell mediated reactions) resulting in phagocytosis or lysis of the parasite or encapsulation of it; (*b*) production of precipitates which either interfere with metabolic processes (perhaps by interacting with enzymes), especially those taking place at surfaces, or which block functional openings such as the mouth or excretory pores and prevent their normal operation; (*c*) reactions interfering with glandular secretions from the parasite.

Antibodies which confer some measure of protection on a host are known as 'protective' antibodies and the antigens which stimulate their production are sometimes referred to as 'functional' or 'essential' antigens. Thus these antigens are essential to the parasite and if the processes in which they are involved are modified in some way by antibody, the parasite cannot grow and develop normally.

A complex terminology has grown up around parasite antigens although all authors do not necessarily use these terms in quite the same way. Thus antigens of the tissues are sometimes referred to as *somatic* antigens — also known as *endogenous, structural, bound* or *internal* antigens. The terms *surface* or *external* antigens are sometimes used for those presumed to be present on the surface. Antigens released during growth, development or tissue penetration as metabolic products (especially secretory or excretory materials) are referred to as *metabolic* antigens - also known as *exogenous antigens* or *exoantigens*. The distinction between somatic and metabolic antigens, however, may not always be clear.

Cell mediated immunity and reaginic (homocytotropic) antibodies in parasitic infections.

Much of the early work on immunity to parasitic organisms was concerned with the effects of humoral antibodies, and cellular phenomena were generally neglected. It has now become increasingly apparent that cell cediated immunity and reaginic (homocytotropic) antibodies may play an important role in the immune processes in parasitic infections. In considering immunity to a limited extent e.g. *Schistosoma mansoni*. Knowledge of these topics, which is expanding rapidly, has been comprehensively reviewed by Soulsby (1972b) and Sadun (1972).

Protozoa

Haemoflagellates

Immunity to trypanosomes in general has been reviewed by Neal, Garnham & Cohen (1969), Gray (1967), Goble (1970), Desowitz (1970), D' Alesandro (1970) and Lumsden (1972a).

Trypanosoma lewisi

This species presents one of the classical cases of acquired immunity and one of the earliest established for protozoa. The life cycle has already been described. Typically, in a new infection, the trypanosome multiplies in the peripheral blood of rats for five or six days, after which time retardation followed by complete cessation of reproduction occurs. The organisms agglutinate in rosettes and a 'crisis' occurs on or about the tenth day when the majority are destroyed. A second 'crisis' occurs about the 35th day characterized by complete disappearance of all organisms from the bloodstream. It is thought that three antibodies might be involved in the development of immunity (D'Alesandro, 1970); (*a*) a reproduction inhibiting antibody, termed *ablastin,* (*b*) a trypanocidal antibody (IgG) specific for division forms and (*c*) a trypanolytic antibody (IgM) responsible for termination of infection by killing the adult forms.

> All these effects, however, could be explained also on the assumption that only one antibody, ablastin, is produced. According to this view, reproduction is inhibited by ablastin which inhibits nucleic acid synthesis, and the organisms become sticky and agglutinate. The agglutinated organisms are filtered out by the reticulo-endothelial system, especially by the liver and spleen, with the result that the majority suddenly disappear. This accounts for the first 'crisis.' As the agglutinated organisms disappear, it becomes increasingly difficult for those remaining non-agglutinated to meet, and the *complete* removal of all trypanosomes is accomplished by normal phagocytic mechanisms.

An increase in the number of granules in *T. lewisi* just before the onset of the 'crisis' has been noted but the association between their appearance and the onset of immunity is no understood; these granules appear to consist of a phospholipid. That the '10-day crisis' and the subsequent elimination of parasies is due to an immune reaction has been dramatically demonstrated by splenectomizing rats before infection — an operation which removes the most active site of antibody formation. In these splenectomized rats a very long parasitemia without death developed. Although the peak infection (294 000 per cm^3 $\pm$ 38 700) developed on the tenth day - as in the non-splenectomized controls - by the 25th day there was still a very high parasitemia (125 690 per $cm^3 \pm 20\ 000$) and, even after 60 days, 6 per cent of the experimental rats showed some parasitemia (1250 per cm^3).

Human trypanosomiasis. Immunity to *T. rhodesiense* and *T. gambiense* which cause sleeping sickness in man or related forms in cattle such as *T. brucei* has been studied extensively. Such species have an exceedingly complex antigenic structure made up of somatic, surface and metabolic antigens. Moreover, each species contains numerous 'strains' and each strain gives rise to numerous 'variants,' up to 20 being described in one instance. Occurrence of such a high degree of antigenic variation is responsible for the chronic relapsing course of trypanosomiasis and the appearance of distinct variant specific antibodies in the host serum. Moreover, there is a tendency for variants to appear in a characteristic order. The mechanism host. In some species (e.g. *E. tenella*) in chicks, a single infection is sufficient to provide complete protection against reinfection. The mechanism of immunity is not understood. It appears that penetration of cells can occur in some hosts but further development is prevented. In other host species, penetration may be prevented by the secretion of an antibody against an unknown proteolytic enzyme utilized by the penetrating sporozoite. This view could receive support from the experiments of Leathem & Burns (1971) who found that stages of *E. tenella* recovered from the epithelial cells of immune hosts at 12-24 hours were still infective when transferred to susceptible chicks, whereas those from a 48-hour infection were not. This has been interpreted as showing that within 24-48 hours after invasion, development had been irreversibly inhibited by the immune mechanism. Alternatively, of course, the 48-hour sporozoites may have developed beyond the infective stage.

Humoral antibodies have been shown in many host species infected with coccidea, and may, in some cases, be protective. Passive immunity can be induced by the inoculation of serum from immune hosts, but only if the intervals between serum and parasite inoculations are short.

Helminths

General considerations

Helminths offer a number of advantages over protozoans, bacteria or other micro-organisms for basic investigations in immunology, e.g.

(*a*) with a few exceptions (e.g. *T. crassiceps,*) they do not multiply in the body (in contrast to bacteria or protozoa) so that precise control of an effective dose is possible

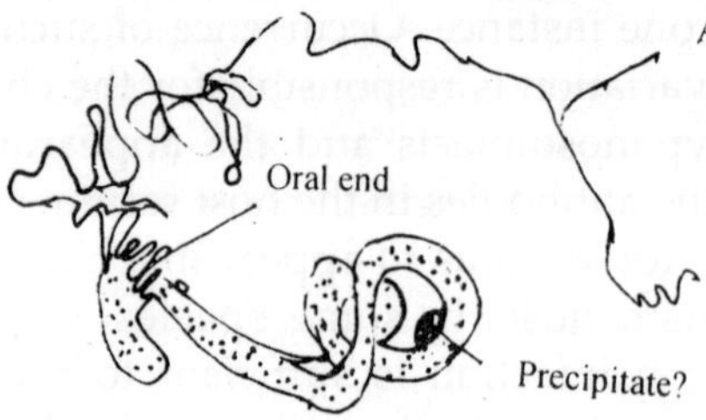

Fig. 19.1 : Third stage larva of *Ascaris lumbricoides* var. *suum* removed from the lungs of an experimentally infected mouse and incubated for 25 days in immune guinea-pig serum; a strand of precipitate originates from the oral opening. some precipitation also took place around the anal region.

(*b*) they tend to mature in a specific site within the body, or, if they undergo a tissue migration, the route is well defined

(*c*) they are relatively enormous in size, compared with a single bacterium or protozoon, so that individual tissues can be isolated and antigens prepared from them

(*d*) immunological effects, such as the formation of an envelope around a cercaria in immune serum, can sometimes be directly observed and followed under the microscope.

Although complete sterile immunity against helminths may be developed, an immune response may be evidenced in other ways such as stunting, retardation and inhibition of reproduction.

Since many helminths are large, immune phenomena such as precipitation or phagocytosis may have no effect on them. Much, however, depends on the type of worm. Thus, an antibody reacting with the antigenic cuticle of a nematode may virtually be without effect, as this is a non-absorbing surface. On the other hand, the tegument of a cestode is highly absorptive and transport activities could be greatly affected by the presence of antibody precipitated on the surface.

In nematodes, however, precipitations may occur in immune serum at the oral or excretory pores and these may be inhibitory with some protozoa, many antibodies to helminths appear to be directed against metabolic antigens; in the case of *Nippostrongylus brasiliensis* released lipases and esterases may serve as metabolic antigens.

The realization that metabolic products of helminths can serve as antigens has given new importance to the elaboration of techniques for the *in vitro* cultivation of helminths.

Irradiated larvae as antigens. A number of attempts have been made to utilize extracts of worms as antigens, but although these invoke the production of antibodies the degree of immunity developed is generally of a low order. Immunity develops much more strongly in the presence of living whole worms, presumably because of the production of metabolic antigens. Means have, therefore, been sought to treat worms in such a way that they would be attenuated and yet still retain their ability to invoke somatic and metabolic antibodies. It was found that when treated with X-rays or radiation from cobalt 60, some species of larval worms became sterilized and yet retained their ability to induce immunity. Much of the early work was carried out on larval nematodes as many of these undergo extensive migration in the definitive host, and although sexual maturation was usually prevented by irradiation, the migratory phase (during which the immune reaction is evoked) was comparatively little affected. This has resulted

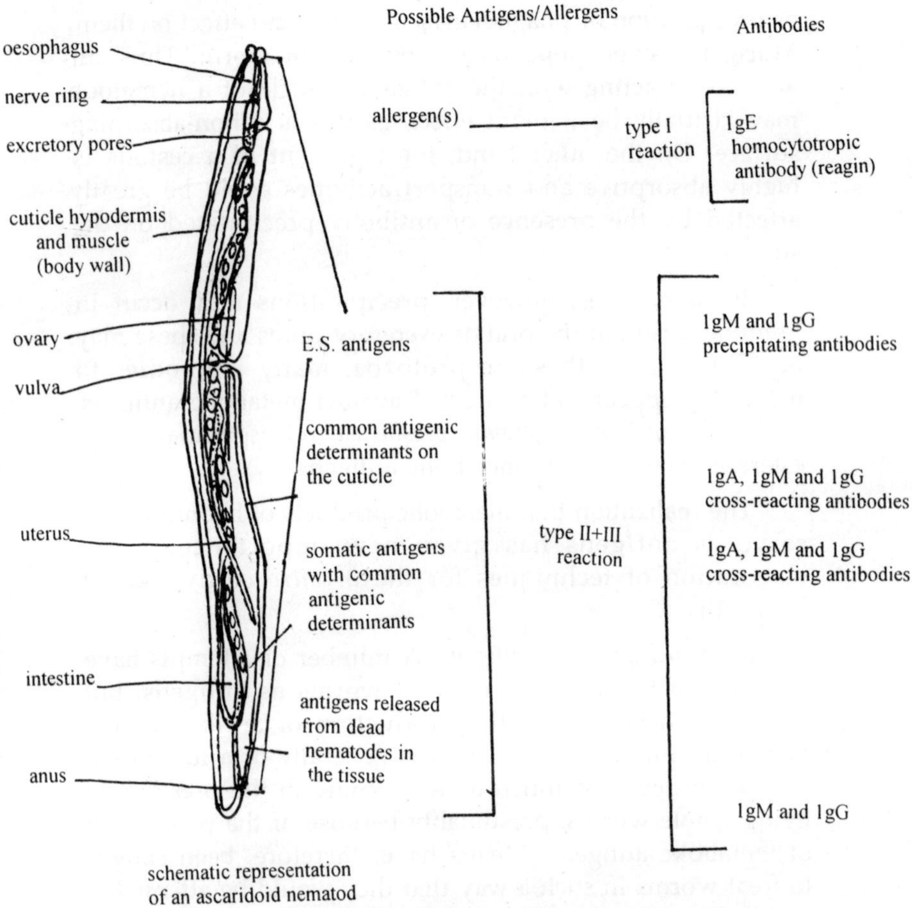

Fig. 19.2 : The possible site of some nematode antigens/allergens and their antibody responses.

in the development of vaccines which can produce some degree of immunity against several nematodes of economic importance.

This approach has now been attempted with a number of other helminths of economic importance, e.g. *Schistosoma mansoni,* generally with only limited success; this work is discussed further below.

Trematodes

Immunity to trematodes in general has been reviewed by Kagan (1966), Platzer (1970) and Smyth (1966).

Fasciola hepatica

There appears to be some conflicting evidence as to whether acquired resistance to this species ever occurs. Thus, based on experiments with normal and X-radiated metacercariae, Boray (1967) concluded that acquired resistance to this species does not develop in sheep or cattle. Nor can passive immunity be demonresistance has been claimed in rabbits and mice although only relatively small numbers of experimental animals were used. On the basis of lymphocyte infiltration into the liver and the histopathology of the latter it was concluded that delayed hypersensitivity may play a role in the degree of immunity developed. This view is supported by the fact that the transfer of lymphoid cells from rats infected with *F. hepatica* has been claimed to confer some immunity on isogenic hosts subjected to primary infections.

Schistosomes

Immunity in schistosomiasis has been extensively investigated, various aspects of the field being reviewed by Brujining (1967), Jordan & Webbe (1969), Smithers & Terry (1969a, b) Smithers (1968, 1972) Kagan (1966) and Lewert (1970).

Antigen-antibody interactions. Antibodies against both adult and larval schistosomes have been detected in serum and a number of diagnostic tests based on the usual antibody manifestations have been developed. These include precipation, flocculation, haemagglutination, complement fixation and skin tests. In addition, three unusual reactions have been developed. These are (*a*) the *Cercarienhullenreaktion* = CHR), of Vogel & Minning (1949), (*b*) the *miracidial immobilization* test and (*c*) the *circumoval precipitin* test (= COP).

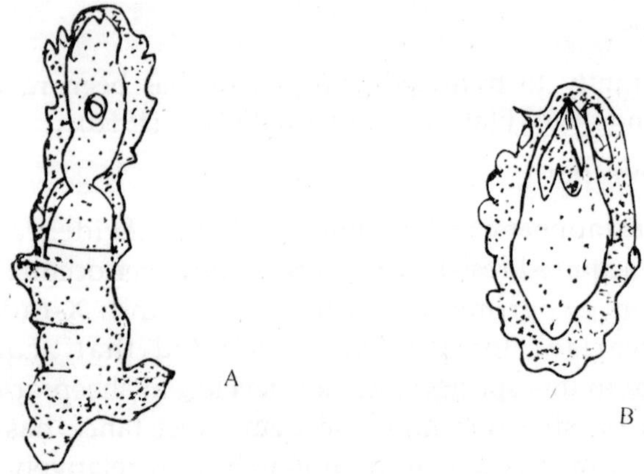

Fig. 19.3 : Immunological reactions of *Schistosoma mansoni* in antiserum. A. *Cercarienhüllenreaktion* (CHR). B. Circumoval precipitin test (COP).

(*a*) *The CHR test.* is of special biological interest. It is manifested by the appearance of a transparent membrane around the tail of a schistosome cercaria when placed in immune serum. This reaction is destroyed by heating the serum to 56°C for 30 minutes. Electrophoretic studies have revealed that the CHR membrane develops only in the gamma and alpha fraction of the serum. The CHR reaction has been applied to both *S. mansoni* and *S. douthitti* with rather similar results. As a diagnostic tool the method has obvious limitations. It is not as sensitive as some of the other reactions and also necessitates maintaining living cercariae in the laboratory.

(*b*) *The miracidial immobilization test* is based on the immobilization of miracidia in immune serum, and it has been found to be an exceedingly sensitive one. The method has not yet been widely used. All stages of schistosomes have been found to produce immobilization antibodies.

(*c*) *The COP test* is based on the fact that when a live schistosome egg is placed in antiserum, a globular precipitate, contiguous with the edge of the egg-shell, appears. The precipitate is best seen after 24 hours in serum at 37°C but can be detected as early as two hours. Host serum becomes positive about the 40th day after infection at the time the eggs are produced. It should be noted, however, that even serum from *unisexual* infections, whether males or females gives a weak COP reaction. The source of the antigen in eggs is not known, but it is probably of metabolic origin, for if eggs are allowed to hatch in a small volume of water, the metabolic products are strongly antigenic and will produce a weal if injected intradermally into an infected host or will cause a precipitate in a tube of antiserum.

Practically every stage of the life cycle has been used for antigen - adult worms, cercariae, infected snail livers, miracidia and eggs. The method of extraction has varied considerably, saline, alcoholic or ether extracts being favoured by different workers.

Of all the tests mentioned above perhaps complement fixation has been the most widely used and it is still probably the most sensitive test for the diagnosis of early infections. Unfortunately, the skin test at present developed is not specific and lacks standardization, but on account of its relative simplicity would be a diagnostic tool of value. Kagan (1958) gives an excellent review of these tests.

Acquired immunity to schistosomes. Although precise experimental evidence is lacking, it is generally accepted that man and animals develop resistance to schistosomiasis. This conclusion is based on work describing a reduction in the number of cercariae developing into mature worms in subsequent infections compared with the numbers developing in the primary infection. In man, a very high degree of resistance is developed in adults over about the age of thirty-five. It is generally held that this resistance has developed as a result of previous exposures, although the evidence on which this conclusion is based is largely circumstantial in man. The mechanism of such resistance in man is not understood, but from experiments with monkeys there is evidence that it is the living *adult* worm which provides the major stimulus to acquired resistance although apparently this resistance is not effective against the adult worms already established (see concomitant immunity, below). It was formerly thought that humoral factors did not play a part in schistosome immunity, for no 'protective' action of the numerous antibodies reported could be identified. *In vitro* experiments have shown that hyperimmune sera from rhesus monkeys (i.e. animals exposed to two to four infections of *S. mansoni* cercariae) contain an antibody (IgG) lethal to schistosomula. Moreover, this antibody is dependent on a labile factor in serum—probably a component of the complement system. It is thus likely that, in hyperimmune monkeys, invading schistosomula would be rapidly killed.

If, however, schistosomula are cultured *in vitro* in normal serum or are maintained for the same period in a mouse they are almost completely protected against the action of the lethal antibody *in vitro.* A possible explanation of this is that a worm could 'acquire' a layer of host antigens and hence in its adult condition be 'protected' against host antibody. Newly penetrated schistosomula would not, of course, have such a protection (see below).

These results clearly need confirmation in other host species, and more supporting evidence is needed before we can conclude, unequivocally, that protective antibodies operate in schistosome immunity. It is significant to note, however, that a specific schistosomal antigen has been detected in hamsters infected with *S. mansoni.*

Vaccination against schistosomiasis. Numerous attempts have been made to develop vaccines against schistosomes using either dead worms, extracts from them or irradiated cercariae. Dead materials are useless as antigens but an initial infection with normal cercariae induces immunity in experimental hosts. Results have been very variable, especially if the initial exposure is less than 100 cercariae or of the challenge occurs less than 16 weeks after the initial infection. In some cases, a high degree of resistance has been induced with as few as 25 cercariae. In contrast to nematodes irradiated cercariae even in doses of 25 000 may induce no immunity.

Concomitant immunity to Schistosomes. The existence of an infection of an adult worm in a host induces a condition known as *concomitant immunity* in which such a host is immune to a second infection. The term arose first intumour transplantation to describe a condition in which an animal bearing one tumour was sometimes resistant to a second graft. The question must be asked: why it is that the established worms are not destroyed by the immune response? It appears, as suggested above, that, in some way not understood, adult schistosomes accumulate host antigens on their surface which serve to 'protect' them from antibody attacks. This question is discussed further under *molecular mimicry*.

Cell mediated immunity in schistosomiasis. Cell mediated immunity appears to operate in relation to the immunopathological events associated with the presence of eggs in the tissues. Thus, in these tissues, the typical granulomas which form around the eggs appear to be a form of delayed cell mediated hypersensitivity. This reaction has been shown to be not only species-specific but also stage-specific. Thus, there is no cross sensitization between *Ascaris suum* and *S. mansoni* eggs and little between the latter and eggs of *S. haematobium* and *S. japonicum*. Furthermore, sensitization is transferable by cells but not by serum.

Cestodes

Various aspects of immunity to cestodes have been reviewed by Gemmell & Soulsby (1968), Smyth (1969), Weinmann (1966, 1970) and Gemmell & MacNamara (1972).

Since the body surface of adult and larval cestodes presents a virtually 'naked' cytoplasmic surface, it may be expected that in host tissue sites, at least, cestodes would be highly immunogenic. In general this appears to be the case and, in mammalian hosts, even a few larvae may induce a high and durable level of resistance. Moreover, acquired immunity may develop very rapidly as early as one to two days in *Hymenolepis nana* or *Hydatigera taeniaeformis* in rodents. It is important to stress that in cestodes, in particular, the strain of both host and parasite appear to be of special importance in relation to the degree of immunity developed. Thus, cysts of *Echinococcus multilocularis* when injected intraperitoneally into different strains of mice will develop into hydatid cysts in some but not in others.

In the best known cestodes of man and other mammals (i.e. *Taenia* spp.) the host/parasite contacts established during the life cycle are (*a*) the oncosphere, during the initial stages of penetration of the intestinal mucosa of the intermediate host, (*b*) the larva in its final tissue site (often the liver) and (*c*) the scolex of an adult worm attached to the intestinal mucosa. Most of the work on cestode immunity has centred on species of economic importance (such as *Taenia* spp.) or those readily maintained in laboratory animals (e.g. *Hymenolepis* spp).

Taenia (= Hydatigera) taeniaeformis

Much of the early work on cestode immunity centred around this organism and it was shown that even a few cysts in the rat liver were sufficient to prevent the establishment of further infections when challenged with this organism. This protection persists for at least two months, even after encapsulated larvae are removed from the liver. In common with many cyclophyllidean cestodes, oncospheres must be eliminated either at the site of penetration of the intestine or during migration, a process known as *early* or *pre-encystment* immunity or at the final visceral site, referred to as *late* or *postencystment* immunity.

Taenia spp.

Many other larval taeniid species present a similar immunological pattern, e.g. *T. pisiformis T. saginata, T. ovis* and *T. hydatigena*. It might be expected that such a high level of immunity could be passively transferred and the protection of rabbits against *C. pisiformis,* by injection of antibody containing serum from infected rabbits, has been demonstrated; immunity has similarly been transferred by lymph node cells. Activated oncospheres have been used with some success as vaccines against both homologous and heterologous species. Immunity has also been established against. *T. ovis* and *H. taeniaeformis* using larvae grown *in vivo* in filtration membrane diffusion chambers.

Echinococcus granulosus

Most of the work on this species has been carried out on the hydatid stage with relatively little work being done on the adult stage in the carnivore host. Complete immunity to larval infections is probably never developed but in subsequent infections fewer cysts become established. There is great individual resistance and even in the same liver, for example, both dead and viable cysts may occur. Some degree of 'post-encystment' immunity to *Echinococcus* can develop, however, and can be induced artificially by injection of oncospheres. Host immunity to different strains of *Echinococcus* may prove to be very specific as exemplified by the fact that the 'horse strain' is common in Ireland and yet sheep in that country are not infected. This suggests strong immunological and/or physiological differences between the horse and sheep strains. This hypothesis is supported by the fact that the *in vitro* system in which the sheep strain will grow to maturity fails to support development of the horse strain.

Table 19.1. The percentage efficiency of homologous and heterologous eggs and activated embryos as vaccines in protecting intermediate hosts against the establishment and subsequent survival of the cystic stages of various *Taenia* species.

		Challenge infection with:							
		T. hydatigena		*T. pisiformis*		*T. ovis*		*E. granulosus*	
Vaccine		*Immunity*		*Immunity*		*Immunity*		*Immunity*	
Parasite	*Antigen*	*Early**	*Late†*	*Early*	*Late*	*Early*	*Late*	*Early*	*Late*
T. hydatigena	Emb.	99.9	100	28.0	48.6	48.0	10.7	31.3	89.0
	Eggs	71.3	99.5	19.2	0.0	54.3	51.8	0.0	6.9
T. ovis	Emb.	51.0	100	45.1	25.1	85.5	94.6	38.4	97.4
	Eggs	39.7	8.7	0.0	8.7	55.6	94.7	19.8	0.0
T. pisiformis	Emb.	23.0	35.0	92.4	88.8	0.0	0.0	0.0	19.3
	Eggs	22.5	0.0	97.9	58.5	21.6	0.0	0.0	0.0
E. granulosus	Eggs	—	—	—	—	—	—	91.2	96.0
	Eggs	—	—	—	—	—	—	91.9	100

*At the intestinal level. †At the site of election.

On account of their medical importance, serological tests for the diagnosis of hydatid cysts have been much studied; these include complement-fixation, haemagglutination, bentonite-flocculation and latex-agglutination, immunofluorescence and intradermal tests. The efficiency of these has been reviewed by Kagan (1968).

The immunology of the adult stages in dogs has been poorly examined. In general there appears to be considerable variation between individual dogs. Thus, in experiments involving 20 experimental dogs and 20 controls, Williams & Esandi (1971) found that a thermolabile homocytotropic skin-sensitizing antibody was detectable in five of the infected dogs but not in the remainder or in the controls. Circulating reagins were not detectable until the seventh week of infection, i.e. just after the worms had become gravid.

Hymenolepis spp.

H. diminuta, H. nana and *H. microstoma* are commonly maintained in laboratory rodents and have been most widely studied.

Hymenolepis diminuta

This species exhibits 'premunition,' that is, protection against reinfection as a result of an existing infection (in contrast to 'immunity' which is related to a *previous* one), a phenomenon common amongst cestode species. It has been shown that the size of worms in a secondary infection is inversely proportional to the number of primary worms harboured. When a primary infection is eliminated by an operation, the rate of growth of the worms in the secondary infection was normal. This suggests that 'premunition' in cestode infections may not be due to the action of an immune mechanism but to a crowding effect, which probably has mainly a nutritional basis.

Hymenolepis nana

This species has been widely used as a model for studying immunity to cestodes. It is unusual among cestodes in that cysticercoid may develop directly within the villi of the rodent host. Since the host tissues are thus invaded it can be expected that a high level of immunity will develop and indeed it has been shown experimentally that immunity is detectable as early as nine hours after the initial infection and becomes almost absolute after about 48 hours. In 'immune' mice a few challenge eggs may hatch, penetrate and develop to cysticercoids within the villi, bu none develop to adults. Antibodies may be detected in infected mice serum. The immune pattern described above refers to an egg challenging dose. If a cysticercoid challenge dose is used instead, some adults do become established although only about one-fifth of a normal infection. Hence, a preceding tissue (cysticercoid) infection provides some degree of lumen protection possibly through the action of antibody in intestinal mucus.

Table 19.2. Development of resistance to *Hymenolepis nana* in rats. An immunizing dose of 500 eggs was followed by a challenging dose of 5000 after the time stated : 6-12 rats were used in each experiment.

Time before challenge (hr)	*No. of cysts at 100 hr*		*Reduction of cyst count*
	Immunized	*Controls*	
12	226	251	10%
24	103	225	54%
48	36	218	84%

Passive immunity has also been demonstrated for *H. nana,* the antiparasitic activity being associated with the IgG (7S) immunoglobulin fraction. That immunity in *H. nana* involves a thymus dependent mechanism has been shown by the suppression of immunity in thymectomized mice, and also by injection, into normal mice, of rabbit antimouse thymocyte serum.

Nematodes

Immunity in nematodes has been extensively studied, most attention having been paid to gastrointestinal nematodes of economic importance (e.g. *Haemonchus contortus*), species pathogenic to man (especially *Tricinella spiralis*) and species easily maintained in laboratory animals (especially *Nippostrongylus brasiliensis* in rats). Various aspects of immunity to nematodes in general have been reviewed by Mulligan (1968); Michel (1968); Soulsby (1966, 1970, 1971); Thorson (1970); Otto (1970); Dobson (1972) and Olson & Izzat (1972).

Morphologically, nematodes differ markedly from cestodes and trematodes in possessing a tough, relatively inactive outside surface, in contrast of the 'naked' (and hence strongly immunogenic) cytoplasmic surface of trematodes and cestodes. Thus, nematodes would be relatively unaffected by antibody-antigen precipitations or leucocyte attachment at the surface. On the other hand, precipitations at the oral or excretory pores would affect metabolism and this appears to be borne out in practice. Only a very limited number of examples of immunity to nematodes are discussed below. A characteristic of many nematode infections is the phenomenon known as 'self-cure' - 'the sudden elimination of a burden of parasites as a result of an immune response on the part of the host'. The term 'self-cure' was first introduced in relation to *Haemonchus contortus* but has been extended to many other species such as *Nippostrongylus brasiliensis*.

Nippostrongylus brasiliensis

The immune reactions against this species have probably been more extensively studied than any other helminth species; the extensive literature has been reviewed by Ogilvie (1969) and Ogilvie & Jones (1971). The value of this species as a model is that a strong, active immunity is generated and this immunity may easily be passively transferred by antiserum so that the study of protective antibodies is feasible. Moreover, high levels of reaginic antibodies are stimulated in rats.

The pattern of development of immunity is as follows.

In heavy infections, eggs appear first in the faeces on the sixth day post-infection and reach a peak of production on the tenth day, falling rapidly during the next week or so to zero. Fall in egg production is accompanied by a massive expulsion of worms (i.e. a 'self-cure' reaction) from the intestine but a small number of worms survive the expulsion phase and remain in the intestine for long periods. A curious feature of these residual worms is that the majority of these are male, the females remaining being largely sterile. With small initial infections (approximately 50 larvae) worm expulsion is more gradual and may extend over 30 days.

Rats which have thrown off a primary infection exhibit a marked resistance to a challenge infection, as exhibited by inhibited development of the worms or by complete failure of the secondary infection to become established.

Kassai & Aitken (1967) point out that in using *N. brasiliensis* as a model for immunological studies, two measurements have special significance. These are the actual size of the established worm population, i.e. the 'ten-day' take (measured on the eighth and tenth days) and the rate of loss of worms which determines the size of the '20-day' take (measured on days 18-20). By formulating a 'take ratio' calculated as:

$$\frac{\text{20-day take}}{\text{10-day take}} \times 100$$

these two values can be related. This should not be confused with the 'percentage take' which is the take expressed as a percentage of the total infecting dose. Kassai & Aitken demonstrated that the adult ability to eliminate worms from a primary infection did not develop in rats until they were about five weeks old; infections originating in four-week-old rats survived for up to ten weeks. The percentage take was also much higher in three-to four-week-old rats (12-16 per cent) tan in rats aged five weeks or more

(1.4-4 per cent). The sex ratio also differs markedly with the age of the rats, being about 50: 50 male: female in young rats but rising to 97: 3 in eight-week-old rats.

The mechanism of 'self-cure' in *Nippostrongylus* has been much investigated; there is evidence that it is a two-step process involving (*a*) action of antibodies and (*b*) expulsion. The antibodies are mostly associated with the $7S_{y1}$-immunoglobulin and it has been suggested that this may interfere with the acetylcholinesterase activity of the worm, although this has not been unequivocally established. The gut cells of the worms are also apparently damaged by antibody so that normal feeding is disturbed; there is also resorption of sperm and cessation of egg production.

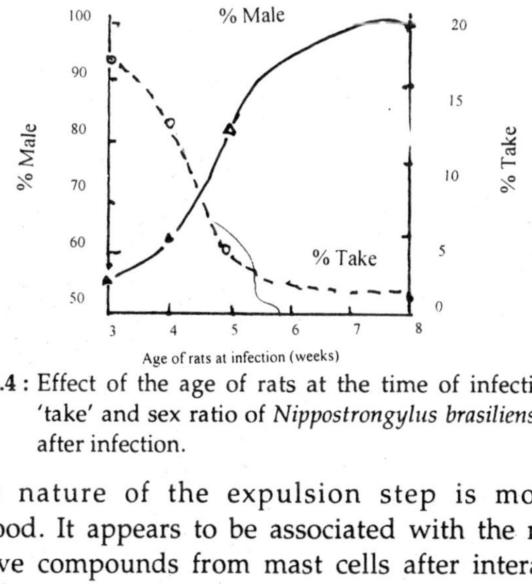

Fig. 19.4 : Effect of the age of rats at the time of infection on the 'take' and sex ratio of *Nippostrongylus brasiliensis* 20 days after infection.

The nature of the expulsion step is more fully understood. It appears to be associated with the release of vasoactive compounds from mast cells after interaction on the cell surface of bound reaginic antibodies (which reach their maximum titre at 20 days post-infection) with worm allergen.

It has been suggested that *Nippostrongylus* stimulates the synchronous development of new populations of mast cells and plasma cells which produce immunoglobulins of at least three classes: IgE, IgA and IgG. IgE is associated with the allergen-reaginic antibody system as indicated above. The pharmacological mediators released by the mast cells are then believed to interact with (*a*) plasma cells containing antibodies of other classes, to release antibody and (*b*) the mucous membrane, to create an intercellular pathway across the mucosa. Such a mechanism could deliver large amounts of locally produced (IgA) and/or systemically produced (IgG) anti-*Nippostrongylus* antibody to sites specifically occupied by the worm, thus effecting a maximal protective response.

Most of the work on *N. brasiliensis* has been concerned with single large infections of larvae which terminate abruptly as outlined above. When, however, 'trickle' infections are given (i.e. small daily infections) no such dramatic expulsion occurs and the worms develop beyond 14 days and, although stunted, are able to remain patent for more than three months. Even rats previously immunized with a high primary infection of *N. brasiliensis* could be infected by low-level 'trickle' infections. Hence it appears that under certain conditions worms can become 'adapted' to the immune response. When these 'adapted' worms were transferred to immune rats—in contrast with normal worms—they established themselves in the intestine and produced viable eggs. The nature of this adaptation is not understood but presents an intriguing problem.

Haemonchus contortus

In this much-studied nematode of sheep 'self-cure' followed by protection was first observed by Stoll (1929). Since then the phenomenon has been extensively studied, as have similar phenomena with other sheep nematodes such as *Trichostrongylus* spp. and *Ostertagia* spp. The relevant literature has been reviewed by Gordon (1967), Mulligan (1968), Michel (1968) and Soulsby (1966).

In this species, about two weeks after infecting sheep, egg production begins and continues increasing for about ten weeks; thereafter there is a fall to low egg counts within two to three weeks. Unlike *Nippostrongylus*, the expulsion reaction appears to be precipitated by the intake of a new batch of larvae. This immune response does not occur in lambs under about four months of age, which is comparable to the situation in young rats. Sheep have been vaccinated against *H. contortus* by using third stage larvae partly inactivated by radiation doses of 40 000—60 000 roentgens. These irradiated larvae are still able to undergo a limited migration and retain enough metabolic activity to stimulate a high degree of immunity to reinfection. Double vaccination is even more effective. Unfortunately, differences in responses of various breeds of sheep have rather invalidated the use of this irradiated vaccine. There is some evidence that the antigen initiating the self-cure reaction is the exsheathing fluid produced at the third ecdysis. Metabolites from third and late fourth-stage larvae cultured *in vitro* and fractionated, have been used with some success as vaccines, and immunodiffusion precipitation reactions have been detected against immune serum and metabolites.

Table 19.3. Immunity of lambs against *Haemonchus contortus*. Groups of 5-7 lambs were infected with larvae irradiated with various doses as indicated. They were challenged with 8000 normal infective larvae 117 days later, and autopsied after a further 4-6 days.

Group	*Roentgen dose*	*Mean no. of worms on challenge*
1 (B)	10 000	5
2 (B)	20 000	0
3 (B)	40 000	200
4 (B)	60 000	0
5 (B)	100 000	442
6	Double infection	620
7	Control	2042

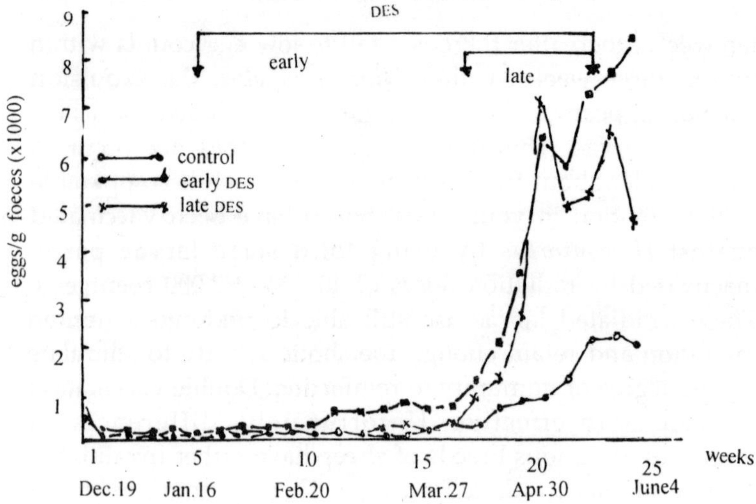

Fig. 19.5 : Effect of administration of the hormone diethylstibestrol (DES) to unbred ewes infected with *Haemonchus contortus*; the onset of the 'spring rise' is not apparently affected. lactation and DES may temporarily suppress the immune capacity of the host.

The 'spring rise' phenomenon. As already mentioned That *H. contortus* together with some 30 other species of gastro-intestinal nematodes exhibits a phenomenon referred to as the 'spring rise' or 'postparturient rise' in lactating ewes during early spring. An extensive literature exists on this phenomenon. It is generally accepted that most of this increased egg production is due to maturation of larvae picked up in autumn and arrested in their development throughout the winter, at the early fourth stage. The mechanism of this phenomenon remains obscure but there appear to be two schools of thought. The first believes that the immune status of the host is chiefly responsible and that fluctuation in this leads to the maturation of the worms.

The alternative view is that arrested development of the fourth-stage larvae is induced by environmental conditions acting on third-stage larvae on the pasture during autumn and that during winter arrested development is maintained by a process analogous to diapause in arthropods; development recommences spontaneously when these processes are completed. This view further postulates that these developing larvae trigger a 'self-cure'-like reaction in the sheep with expulsion of worms; it is speculated that this immune response is temporarily suppressed in lactating ewes and is associated with hormone levels. The latter hypothesis is supported by the fact that injections of the hormone diethylstiblesterol raises the egg production but do not advance the date of onset of spring rise.

Dictyocaulus viviparus

The pattern of immunology in this species is somewhat similar to that of *H. contortus*. The use of irradiated larvae as vaccines against the parasitic bronchitis caused by this worm has been successfully produced. Third-stage larvae partially inactivated by exposure to 40 000 roentgens form a suitable vaccine, and calves doubly vaccinated with this vaccine are almost completely resistant to infection when challenged with 10 000 normal larvae. Vaccination is presumably effective against this species because in the natural life cycle the worm population is normally regulated by the protective immunity of the host.

Other trichostrongylids

Immunity to numerous other species of trichostrongylids (e.g. *Ostertagia* spp., *Cooperia* spp., *Trichostrongylus* spp.) has been extensively studied. The relevant literature has been reviewed by Michel (1969) and Fitzsimmons (1969).

Instead of using X-radiation, larvae have been attenuated with the cytotoxic agent triethylene melamine (TEM) and these, too, act as satisfactory vaccines.

Trichinella spiralis

This species has been intensively studied as a model of helminth immunity. It has been clearly demonstrated in a number of hosts (e.g. rats, guinea-pigs) that animals which recover from an infection of *T. spiralis* resist further infection. Single infections appear to be rather erratic in inducing immunity but multiple infections appear to be very effective.

Trichinella antibodies appear in serum within 24 hours post infection and precipitation has been reported to occur as early as the fifth day in the intestine, mouth, anus and excretory pore when worms are placed in the serum recovered. A degree of resistance is also transmitted from immune mothers to their young. Acquired immunity manifests itself by (*a*) stunting of growth, (*b*) retardation or complete inhibition of development, (*c*) inhibition of reproduction, (*d*) elimination of previously established worms and (*e*) refractoriness to infection. This acquired immunity manifests itself in about 18-22 days after an initial infection of about 1000 larvae. Immunization must be considered in two phases (*a*) parenteral (stimulated by the tissue migration of the larvae), and (*b*) intestinal or enteral (stimulated by older larvae or adults in the intestine). Rather surprisingly, it has been shown that the immunity acquired by mice subjected to an enteral-only infection was at least as strong as that acquired by mice in which the full, biphasic enteralparenteral infection had run its course. The origin of the functional antigens inducing protection is not known; since newborn larvae appear to be incapable of producing frunctional antigens it has been argued that these only occur in trichinellids which possess a stichosome.

There is no agreement regarding the mechanism of immunity or how the enteral phase is able to induce it. Previously it was thought that both humoral and cellular factors operated in immune animals. However, even in the presence of high serum antibody titres hosts were still found to be susceptible. This has led to the hypothesis that elimination of the adult worms is based on delayed hypersensitivity. This would fit in with the experimental evidence that it is the metabolic (excretory and secretory) products which act as functional antigens. The antigens of *T. spiralis* have been extensively analysed and various serologic tests developed.

Larvae irradiated with X-rays or cobalt 60 at doses sufficient to produce sexual sterilization, but not sufficient to prevent growth to adult size serve as reasonably effective vaccines. These do not induce complete immunity, but the number of adults maturing in the intestine, and the number of larvae which reach the muscles, can be measured in tens rather than in thousands.

Table 19.4. Effects of re-infecting rats, forty days after an initial infection of *Trichinella spiralis* using: A, untreated larvae; B, larvae irradiated with 10 000 R from Co^{60}.

	Group A (No Co^{60})		*Group B (Treated Co^{60})*	
Days after re-infection	*Adult trichinae in intestine*	*Larvae recovered from muscles*	*Adult trichinae in intestine*	*Larvae recovered from muscles*
2	17	181 000	1912	15
	165	32 200		
4	0	166 000	848	11
	8	25 800		
6	0	164 000	42	0
	4	47 240		
8	52	101 000	9	0
	0	640 000		
10	0	156 000	0	22
15	0	190 000	0	0
	0	-		
20	0	96 000	0	496

Other nematodes

The immune reactions to numerous other species of nematodes have been extensively examined; in some cases effective antigens have been examined or vaccines developed. Some of the most actively investigated include: *Ancylostoma caninum, Uncinaria stenocephala Litomosoides carinii Ascaris suum Toxocara canis.*

Immunity in cold-blooded vertebrates and invertebrates

Most of the immune reactions dealt with in this chapter have been concerned with warm-blooded hosts. Immune mechanisms do occur, however, in cold-blooded vertebrates and invertebrates although these difficult areas of immunology have been rather neglected.

It is beyond the scope of this text to discuss these fields in detail; some relevant reviews are as follows: Cold-blooded vertebrates: Rowlands (1969); fish: Hoar & Randall (1970), Snieszko (1970); invertebrates (in general): Tripp (1969); molluscs: Brooks (1969), Feng (1967), Tripp (1963); insects: Salt (1963), Shapiro (1969), Stephens (1964), Weiser (1969); arthropods (in general): Poinar (1969).

Immune unresponsivencess: molecular mimicry

Perhaps one of the most remarkable aspects of the host-parasite relationship is that parasites are able to survive so long in an 'immunologically hostile' environment such as that provided by (say) the bloodstream of the host. How this is achieved is a matter of great controversy, but - at least in some cases - parasites appear to be able to do this by avoiding stimulation of the host's immune response. This is achieved by presenting to the host a surface (and/or released metabolic products) which, antigenically, so closely resembles that of the hos, that the latter, regards it as 'self' and does not respond to its presence. This concept of antigen sharing by host and parasite has been termed 'molecular mimicry' and there is much dispute as to how this phenomenon develops; the various views are discussed further below. Other mechanisms of avoiding the host's response may also operate and a phenomenon referred to as 'immunological enhancement' in tumour biology may also, in some cases, be involved.

Antigen sharing. The idea that hosts and their parasites might share some antigenic determinants was first put forward by Sprent (1962). Subsequent investigations, especially by Capron and his co-workers have shown that the phenomenon is widespread and that helminths do not only share antigens with their hosts but also with other species within their class, and even with helminths from a different class. Thus, for example, in the trematode *Dicrocoelium dendriticum,* 19 antigens have been identified by immunoelectrophoresis, of which six are shared with *Fasciola hepatica;* the latter shares five antigens with *S. mansoni. F. hepatica* also shares four antigens with its intermediate snail host (*Lymnaea truncatula*) and a further six with its definitive cattle host. The occurrence of molluscan antigens in adult trematodes tnabled Capron *et. al.* (1968) to detect the presence of the latter (*Schistosoma mansoni*) in the definitive mammalian host by testing for antimollusc antibodies! A similar pattern of antigen sharing occurs in cestodes and nematodes.

Table 19.5. Antigens of trematodes, analysed by immunoelectrophoresis, showing the numbers of antigens shared between heterologous species.

	Fasciola hepatica	*Dicro- coelium dendri- ticum*	*Parago- nimus wester- mani*	*Clono- rchis sinesis*	*Schisto- soma mansoni*	*Schisto- soma haema- tobium*	*Schisto- soma japo- nicum*
Fasciola hepatica	25	6	5	5	5	5	4
Dicrocoelium dendriticum	6	19	5	5	3	3	2
Paragonimus westermani	5	5	20	5	4	4	2
Clonorchis sinensis	5	5	5	21	5	5	4
Opisthorchis felineus	4	3	4	11	2	2	1
Schistosoma mansoni	5	3	4	5	21	19	11
Schistosoma haematobium	5	3	4	5	19	21	11
Schistosoma japonicum	4	2	2	4	11	11	23